Applied Measurement Engineering

*How to Design Effective
Mechanical Measurement Systems*

Charles P. Wright

Prentice Hall P T R
Englewood Cliffs, New Jersey 07632

Library of Congress Cataloging-in-Publication Data

```
Wright, Charles (Charles P.)
  Applied measurement engineering : how to design effective
mechanical measurement systems / Charles Wright.
    p. cm.
  Includes bibliographical references and index.
  ISBN 0-13-253477-0
  1. Engineering instruments. 2. Measuring instruments.
3. Electronic instruments. I. Title.
TA165.W75 1995
681'.2—dc20                                    94-1636
                                                  CIP
```

Editorial/production supervision: *Harriet Tellem*
Manufacturing manager: *Alexis Heydt*
Acquisitions editor: *Michael Hays*
Editorial assistant: *Kim Intindola*

ISBN 0-13-253477-0

Prentice-Hall International (UK) Limited,London
Prentice-Hall of Australia Pty. Limited, Sydney
Prentice-Hall Canada Inc., Toronto
Prentice-Hall Hispanoamericana, S.A., Mexico
Prentice-Hall of India Private Limited, New Delhi
Prentice-Hall of Japan, Inc., Tokyo
Pearson Education Asia Pte. Ltd., Singapore
Editora Prentice-Hall do Brasil, Ltda., Rio de Janeiro

Dedication

To my wife, Malinda, who loves books and made room in our lives for me to write this one.

To Peter Stein, who throughout his 40 year career has educated two generations of engineers in making valid measurements.

Contents

Contents

Chapter 6   Self-generating Transducers   152

Thermocouple Theory and Practice   176

Chapter 7   The Transducer Model and the Problem of Noise   196

Preface

THE REAL MEASUREMENTS PROBLEM WE FACE

During my 25 years as a professional measurements engineer worrying about the design and use of measurement systems, I've consistently noticed something about the overwhelming majority of engineers working in the field. Ninety-nine percent of them are *retrofitted* after arrival in the field. They migrated into the test measurements field, sometimes inaccurately called instrumentation or data acquisition, from another field. The most popular migration routes are from mechanical and electrical engineering. Other routes include civil engineering, physics, or, worse, computer science—but they migrated. They did so without the benefit of a formal university educational experience in the engineering of measurement systems. They got into this business through its large back door and had to learn an entirely new set of integrated skills and knowledge *after* doing so.

The less than 1% who came in the evidently small front door entered via an undergraduate and graduate mechanical engineering curriculum in Measurements Engineering. That program, at a single university, educated over 900 engineers at the undergraduate level, and produced 58 Masters and Ph.D.s in Measurements Engineering. It is my privilege to be one of those 58. That program ended in the late 1970s. Since then, not a single engineer has graduated with an advanced degree in Measurements Engineering, or anything remotely associated with it, from any university in the United States. That is a shame.

My experience is that it takes a retrofitted engineer, or brand new graduate, 5–10 years to earn his or her professional spurs in this technical specialty. That time is required to turn a smart, interested mechanical or electrical engineer into a productive, expert, creative measurements engineer who can work at the systems level in the design of effective systems for experimental mechanics test measurements.

This large number of retrofitted engineers is working at a reduced level of effectiveness and productivity until that 5–10 year apprenticeship is complete. This inefficiency costs your organizations, companies, and at an integrated level this country, countless billions of dollars per year in lost opportunities and product development time. It significantly harms our national ability to compete in ever more competitive world markets.

WHO THIS BOOK IS FOR

Since I can't, with a wave of my hand, create the needed university graduate programs in the engineering of measurement systems, this book is my next best bet. This book is for those thousands of retrofitted engineers working in test measurements. My intention is to help you shorten your apprenticeship in measurement systems design and operations, and reach your highest level of expertise more quickly. This book is about what it consistently takes to design and operate measurement systems that are effective on the test laboratory floor in supporting your customer's real-time, test time decision-making processes. This is an underappreciated and undertaught engineering field that desperately needs more professionals and fewer amateurs.

It is estimated[1] that at any given recent time in the United States there are more than 100,000 engineers working in fields directly allied to mechanical engineering testing. These test fields include the closely associated ones of mechanical, dynamics, and thermal/vacuum/solar simulation testing (the latter two sometimes called environmental testing as if mechanical inputs were not test environments too!), aero- and hydrodynamics, and field and flight testing on all sorts of products from toothbrushes, washing machines, and Fords, to space shuttles and high-power chemical lasers. The test disciplines include test conception and optimization, planning and conduct, and the necessary measurements engineering functions to get the experimental data. If this analysis is in any way valid, there are a large number of engineers in the United States working directly on the creation of experimental mechanical engineering data. Perhaps the number is 5,000 of the 100,000+. Perhaps it is 20,000. It is, in any case, a large number.

Having spent my engineering career worrying about the effective gathering and assessment of valid experimental test data, my view is that this country produces lousy, inaccurate, and worse, invalid, low-utility test data *by the ton* on a daily basis. This is experimental data of assessable accuracy, but very dubious validity. Everybody else also does this, by the way, since Americans have not cornered the market. It certainly isn't all junk, but too much of it is. My sense is that the situation in the tests measurements field is growing steadily worse—not better. Why? There are at least five reasons.

1. There have been no professional level graduate university programs in the engineering of measurements systems in almost 20 years.
2. American university engineering faculty do not, in general, realize that measurements engineering is a unique, unified, mathematically supported, systematic, legitimate, teachable academic specialty within a mechanical engineering curriculum. As customers for that system, we are ultimately responsible for this condition.

3. Our universities, with our collusion, have succeeded in educating almost a generation of mechanical engineers, steeped in analysis, who act like computer operators without the vision to see past the keyboard. They tend to initially have little engineering common sense and wouldn't know a perceptive test if it kicked them in the pants. We are also responsible for this condition.

4. The amazing advances of the last 10 years in the desktop computer or workstation-based data acquisition "system," with ever more complex, commercial data acquisition software, allow the amateur in the field to generate reams of invalid, unassessed, squiggly-lines-on-paper test data faster and cheaper than ever before.

5. It is hard to produce *demonstrably valid experimental test data.* It is easy to *say* that you do it, but difficult to *prove* that you do it.

WHAT THIS BOOK IS ABOUT

This book is about thinking about experimental mechanics measurement systems at the systems level, so you know that your data is valid and why it's valid. The effort is about asking the right questions before the test runs to assure that the data is valid when the test is over, and about getting the data the first time unambiguously and on purpose.

My technical career has been founded on and steeped in measurements engineering. The unified approach[2] defining the theoretical background and details of that engineering field are not what this book is about. I will include the necessary references so you can pursue it if you desire. This book is about what the application of those unified engineering principles looks like on a daily basis to a designer and user of effective measurement systems. In that sense, this book's subject matter could be called *applied* measurement engineering.

Information is included here because we continually worry about these issues in our daily system design and operational work. Our long-term track record in the design of effective measurement systems and the presentation of large amounts of valid test data is very good. At least we've been told so when we discuss them publicly with other peer engineers and customers who should know. We are professionals at what we do and publish 40,000–50,000 pages of valid test data per year. There is a causal relationship between the measurements engineering subject matter we consistently worry about and that success and customer satisfaction.

I want to convey information to you that allows solving measurement system design problems on the back of an envelope. That seems to be where most good design work occurs anyway. If you ask an engineer "What do you think will happen if I push over here on your structure?" and he runs away to his office to consult his computer model, it is an admission that the computer, not the engineer, understands the problem. The rule works in the other way too. Proficient measurement engineers can design and analyze systems and their problems on the backs of envelopes. If you have to go to a computer to design or analyze a measurement system (with certain restrictive exceptions like digital filter synthesis), you are still serving your apprenticeship. Experienced measurement engineers design on write-on, wipe-off boards on their office walls. Think of them as large, reusable envelopes!

HOW THIS BOOK WORKS

Each of this book's 17 chapters builds on those before it, and fills in more of the mosaic surrounding the design of measurement systems that are really effective on the test floor.

Chapter 1, *Basic Concepts,* presents several subjects that are foundations for applied measurement engineering. These set the context from which later chapters flow. Subjects include measurements versus instrumentation and data validity versus accuracy (they are different), the often neglected subject of energy flow in measurement systems, response syndromes, and Fourier analysis.

Chapter 2, *Measurement System Transfer Functions and Linearity,* presents the concept of the complete system transfer function (also called the frequency response) and output/input linearity for both monotonic and resonant systems.

Chapter 3, *Frequency Content . . . Or . . . Waveshape Reproduction,* defines the rules for the reproduction of the waveshape or the frequency content of the information at the system's input in terms of transfer function and output/input linearity criteria. Chapter 3 also includes a section on what happens if your design does not follow these rules. Using these design rules for information reproduction, the chapter concludes with a discussion of filtering in measurement systems and examples of how much trouble you can get into if this process occurs without understanding.

Chapter 4, *Non-self-Generating Transducers and How They Really Work,* introduces a six-terminal model of the most general transducer type in use today—the non-self-generating transducer. This class includes all transducers for which information about the process is carried on a change in internal impedance. The chapter includes discussions on the criticality of boundary condition control at all transducer interfaces and a structure for interpreting necessarily complex and generic transducer specifications. The last section introduces the subject of information conversion. Here, non-self-generated information can be moved to advantage around the frequency spectrum under design control. This useful ability for noise level separation and stability augmentation is unique to this transducer class.

Now that you've got non-self-generating transducers, what do you do with them? Chapter 5, *Everything You've Ever Wanted to Know About the Wheatstone Bridge,* introduces the oldest (1833) and most general circuit component for supplying electrical boundary condition control for non-self-generating transducers, the Wheatstone bridge. Discussions include the bridge's history, using the bridge as a computer, bridge equations so outputs can be calculated, shunt calibration methodology, wiring standards, and proper setup and operational methodology.

Chapter 6, *Self-generating Transducers,* presents the characteristics for the other transducer class that includes all piezoelectric transducers and thermocouples. I've included here a complete article by Dr. Robert Moffat of Stanford University explaining thermocouple temperature measurement much better than I could. Why reinvent the wheel?

Now that non-self-generating and self-generating transducers have been presented, an integrated transducer model including both sets of properties is included in Chapter 7, *The General Transducer Model and the Problem of Noise.* Using this overall model, the deterministic mechanisms of noise generation in transducers and other components can be explained and attacked. The concept of public noise level documentation is explained.

Chapter 8, *Noise Level Reduction Techniques,* discusses the three methods of noise level control and minimization. Differential measurement systems are discussed with the concepts of normal and common mode measurements.

Chapter 9, *Noise Level Control by Information Conversion Techniques,* provides information on using waveforms other than DC for the excitation of non-self-generating transducers to the designer's advantage. There are classes of measurements problems that can only be solved using this methodology. Sine wave and pulse train carrier systems are discussed and their definite design advantages explained. Examples are shown using typical laboratory instrumentation components.

Chapter 10, *Frequency Analysis,* discusses the most frequently used analysis technique for experimental data—the analysis in the frequency domain. Fourier spectra, power spectral density, octave and one-third octave, shock response spectrum analysis methods, and spectral averaging are discussed.

In Chapter 11, *Sampled Measurement Systems,* digital data acquisition systems make their first appearance. The focus is on methods to assure that valid data comes into your digital system by proper antialias filtering and sampling methodology. Design methodologies are given for answering the 12 questions you need answered before your sampling method can be set. Two opposing sampling methodologies are defined and explained.

Chapter 12, *Measurement System Operational Methods,* gives a set of guidelines that can be used to determine how a measurement system is operated once it's been deployed.

Chapter 13, *Data Validation Methodologies,* presents methods for validating experimental data based on frequency response, linearity, rise time, and undershoot/overshoot considerations. Examples of using the actual physics of the problem for data validation are given from vibration and structural testing.

Knowledge-Based Measurement Systems Design Principles are introduced in Chapter 14. Knowledge-based systems operate at the highest level of measurement system capability by integrating theory with experiment in real time. The concepts of built-in automated tools for data validity checking and utility enhancement are presented with examples from dynamics and thermal testing.

Chapter 15 finally arrives at *The Subject of Software* in automated measurement system design. In-house versus commercial, or third-party software development is discussed. I'll present what we've learned about software development over the last 15 years and what I'd change about it all!

Chapter 16 presents *Leasership and Management Issues* unique to large measurement system design and implementation. With the advent of Total Quality Management in the engineering workplace, it is mandatory for the measurement system designer to accurately identify his or her measurement soul so it is retained in the TQM process. Successful tips are given on educating your senior management and getting your customer to do your internal marketing for you. The chapter concludes with the measurements contract that defines the mutual responsibilities between the measurement system designer/operator and the customer for the data. This contract answers the question "Who owes what to whom in a successful relationship between you and your customer?"

Chapter 17 presents *The Words They Never Talked About in College: Skill, Craft, Excellence, Responsibility, Vision, and Professional.* This is the stuff they never teach engineers in school The subjects are crucial to an engineer wishing to be considered successful and a professional expert in the test business.

A lot of knowledge has been accrued in my years in this fascinating profession. I've also accrued my share of burnt tail feathers that were fascinating as well! It's time to pass that experience on to those coming next.

Charles Wright
Rolling Hills Estates, CA

NOTES

1. P. K. Stein and C. P. Wright, *Our Engineering Education—The Not-So Scientific Method;* Proc. Thirteenth Aerospace Testing Seminar, Institute of Environmental Sciences/The Aerospace Corporation/USAF Space Division, Manhattan Beach, CA, October 1991.

2. P. K. Stein, *The Unified Approach to the Engineering of Measurement Systems, Part I—Basic Concepts;* Stein Engineering Services, Phoenix, AZ, 1992.

Chapter 1

Basic Concepts

1.1 THE PROFESSIONAL CONTEXT FOR MEASUREMENTS ENGINEERING

Context, n.: the interrelated conditions in which something exists or occurs

Since the field of measurements engineering is no longer taught in universities, I can take some small liberties with its contextual definition. The definition we use is one that has evolved over the last 20 years or so. What has been added to the context over that period is an extension of the field to include digital data acquisition and its supporting software, and the customer for the data.

In some circles these added digital and software subjects seem to overshadow the original focus on those principles supporting valid analog measurements, *as if they were somehow dated, different, or no longer necessary.* On the contrary, these additions to the recipe underline the importance of the original subject matter on the engineering of measurement systems defined by Peter Stein's 40-year body of work and research.[1] The reason that the original body of work is so important is that present digital data acquisition capabilities are so easy to use that fundamental principles can be violated very easily in the absence of fundamental understanding. "The answer must be right. The computer said so. And it was so *easy* to get! All I had to do was hook up a couple of those transducer things and punch a couple of keys. Wow."

The professional context for measurements engineering as practiced in our organization is shown in Figure 1.1. The areas of necessary technical expertise show as defined areas in the figure. The expertise areas are topologically related to the other areas of expertise they touch. In this way, the technical subject matter of energy and information flow in

SYSTEM LEVEL THINKING

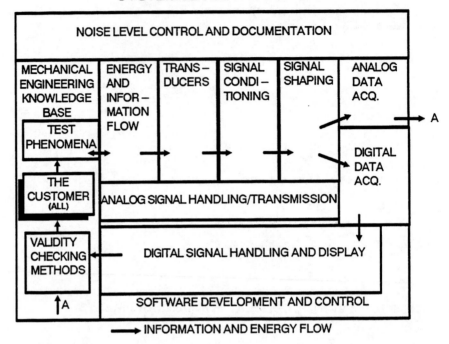

Figure 1.1 The professional context for measurements engineering

measurement systems is the natural and necessary boundary between the test phenomenon and the transducer set used to measure its properties and behavior. The arrows define the flow of experimental data through the various interrelated expertise areas.

The fundamental knowledge base for the successful practice of measurements engineering is mechanical engineering knowledge (Note: Civil engineering is included in this base—no offense to the civil engineers!). I will repeatedly assert in this book that such knowledge is necessary for proper measurement system design, use, and data validation. In the absence of this knowledge base, a measurement system designer or user is flying absolutely blind. I include the customer in the mechanical engineering knowledge base. This inclusion is unique in the literature. The customer's knowledge base is crucial to the measurement system designer. The customer must be able to give the designer an idea about what is going to happen in the phenomenon or design is impossible. Design for what? Is it going to occur at 1,000,000 psi or 2 psi? Will the frequency content end at 50 Hz or 50,000 Hz? The designer must have an idea of the answer from the customer before the measurement system design can even begin. We also feel that an abiding concern for the customer's understanding and welfare is crucial to the system designer's success.

The test phenomenon interacts with the sensing transducers through a complex set of relationships governing the transfer of information and energy between the two. The information is the data we want. The energy transfer is the unavoidable problem that prevents perfect information flow. These relationships are strong and eventually nonlinear functions of amplitude and frequency and must be understood.

Transducers are used to "sense" the information available from the phenomenon. Their problematical complexity is underlined by the formal definition of a transducer: an energy conversion device. The transducer is even defined in terms of the problem it creates by just being there—energy transfer—and not by the solution we seek—information transfer. Transducers must be understood as both energy *and* information transfer devices if valid data is to be produced. Later chapters will discuss the only two classes of transducers of interest to the experimentalist: self-generating and non-self-generating.

Assuming the information transfer into and through the transducer is valid (not a trivial assumption at all), some sort of signal conditioning is necessary to extract and service the generally electrical waveform. (Measurement engineering is a general and unified part of engineering. As such, its theoretical basis is not limited to measurements by electrical means. Its treatment of self-generating brittle coatings and non-self-generating photoelastic coatings for stress analysis are examples of this.) These services might include excitation, isolation, voltage/current transformations, impedance matching, electrical ground control, zero control, span (sensitivity) control, linearization, calibration, spectral shaping, and information conversion. They are used to prepare the voltage or current waveform for eventual scaling, reading, display, and storage.

Signal shaping includes the technical expertise associated with all types of filtering for all purposes. These purposes include spectral shaping for frequency suppression or enhancement and antialias filtering prior to the sampling process.

Analog data acquisition covers the expertise necessary to scale, record, and archive the information in the analog domain. Here, the expertise associated with oscillographic recording, magnetic tape record/reproduction, and FM/FM multiplexing exist.

Touching the mechanical engineering knowledge base, and the expertise areas for information and energy transfer, transducers, signal conditioning and shaping, and analog data acquisition is the overarching expertise associated with the physical control and documentation of noise levels in the data. Noise levels include all the stuff you don't want to measure—today. These noise coupling mechanisms are just as important and meaningful as the data coupling mechanisms. They must be understood, brought under control by design, and documented if the data are to pass the later validation tests.

Digital data acquisition includes the expertise needed to control and optimize the digital sampling function that changes the analog information into numbers for eventual display, assessment, conversion, and archiving.

Digital signal handling and display includes that body of knowledge necessary to handle the signal's information in the digital domain including the performance of all frequency analysis functions and algebraic manipulations. Although the entire frequency analysis function can be performed in the analog domain (and was for most of the past 75 years!), these analyses are today done almost exclusively in the digital domain.

The now necessary software skills that support the digital data acquisition and signal handling functions show at the bottom of the figure. Note that this expertise area touches the mechanical engineering knowledge base. This is because effective software in the measurements business must embody and institutionalize mechanical engineering knowledge. This is where the term "knowledge-based" measurement system comes from.

The information flow now completes the loop back into the mechanical engineering knowledge base via validity checking methods. These methods can be either analog or digital, or both. Validity checking is that crucial process in which the measurement engineer assesses the worth of the now transduced, noise contaminated, signal conditioned, shaped, modulated, demodulated, acquired, sampled, and converted data product. Here, the engineer decides if the resulting data has anything whatever to do with the measureand needed in the first place. Only when data has passed these validity checks does it go to the paying customer, which then increases the mechanical engineering knowledge base about the phenomenon.

My staff and I spent a lot of time developing and discussing this powerful figure. The result gives a fairly thorough picture of the interrelatedness of the various bodies of necessary knowledge. We've found that most customers are amazed when shown this figure. They usually have little prior appreciation for the breadth and depth of knowledge required to be successful in the measurements business. Smart customers, however, learn fast.

Experienced measurements engineers seem unique to me in their incredible breadth of knowledge—from the details of anisotropic stresses in composites, to the electromotive theory governing thermocouple behavior, to the microelasticity of metals, to pulsed excitation signal conditioning, to sampling theory, to digital signal processing for FFT analyses, to computer architectures and displays, all in the space of 15 minutes of conversation. Perhaps that's why the field has kept me interested for so many years.

When I got out of graduate school with a full dose of naivete in hand, I thought I'd learn all the answers in 3 years on the job, and then become a consultant. I discovered very quickly that there were more questions than answers. If you look at Figure 1.1 for a while you'll understand why. I've been in the business 25 years and there are still more questions than answers! The field has remained eternally fascinating for me. I certainly intend that this book hold some fascination for the reader as well.

1.2 WHAT IS A MEASUREMENT SYSTEM?

In this book the term measurement system refers to the entire measurement chain. What do I mean by *entire*? To answer that question I first have to define *system*. A system in this book includes both the hardware and the software—the stuff—AND the people who operate the systems in performing their work AND the customers for that work. In short, a SYSTEM = STUFF + PEOPLE. My working system definition is sociotechnical since it includes the involved people.

Now, what do I mean by *entire*? My working definition of a measurement system includes:

- The total phenomenon under investigation including the test environment
- The boundary between the phenomenon and the transducer

- The transducer itself
- All that other (usually) electrical stuff (cabling, signal conditioners, amplifiers, modulators, demodulators, recorders, multiplexers, samplers, filters, analyzers, etc.)
- Any software and computer devices
- The staff operating the system
- The customer for the data
- The customer's customer

That is an all-inclusive definition of measurement system that you will find nowhere else. Figure 1.2 shows the components and their relationships. Why include all this stuff? The subtitle of the book is *How to Design Effective Measurement Systems*. The operative word in the subtitle is *effective*. In order for measurement systems to be truly effective on the test floor—where it counts—all the components above must be included in the design. If you fail to do this, you will forfeit effectiveness.

Why include the total phenomenon under investigation? If you do not understand solid mechanics, statics, elasticity/plasticity and materials, how are you going to provide valid measurements of the needed measureands during a static test? How will you *know* whether the data are valid? How will you know how much the measurement system affects the process? How will you know what validity checks to apply to the outputs of the system?

How can you hope to make valid dynamics measurements on a vibration, shock, acoustics, transportation, or deployment test or a modal survey if you don't understand dynamics and kinematics? How do you *know* whether the data are valid? What checking tools do you use?

How can you hope to make valid thermal measurements if you don't understand heat transfer mechanisms and thermodynamics? How do you judge the validity of your data? Again, what checking tools do you use?

In my experience, mechanical engineers and physicists make the best measurement engineers because they come into the game with an education in these technical areas. They also can speak on an engineering level with the customer for the test who is usually from a mechanical engineering background. They speak *customerese*. Electrical engineers have a little more trouble since they have little educational grounding in the front-end technical areas shared with the customer. They, initially at least, do not speak *customerese*. This must be a learned skill. EEs come into the game with other technical strengths—circuit design and synthesis, understanding of transfer functions, possibly digital signal processing or software.

The transducer to phenomenon boundary is the most technically sensitive in any measurement system. In other publications about "instrumentation," the system starts inside the transducer. A fundamental and covert assumption is made that the transducer is already in the right state and that no energy has transferred between the phenomenon and the transducer. This is a very dangerous assumption to make. This definition can only answer the question "What did the meter read?" The designing measurement engineer must understand and control the complex physics at this boundary if the data product is to be demonstrably valid.

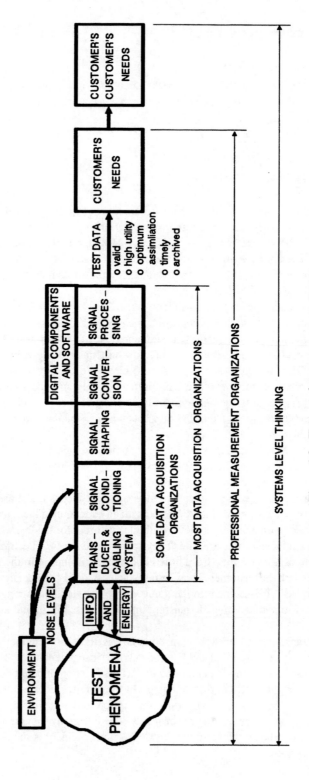

Figure 1.2 The complete measurement process

6

There should be no controversy about the inclusion of the other electrical stuff in the measurement system, nor about software and its impact.

What about the people who are operating the system and the customers for the data? Why are they included? You must include the engineers and/or technicians who will operate the system for a couple of reasons. They are going to operate your system, so you need to design for them so that those operations assure the efficient production of valid, noise-free, highly useful data on the test floor. There is an entire chapter devoted to just this subject.

Why include the customers? Well, for one thing they are *paying the bill!* More important, you should be including validity checking and data utility enhancement features in your design that support these customers on the floor in *their decision-making process.* You want to craft your system design so that you deliver to them demonstrably valid data in a form that they can easily assimilate. Data which are valid, but not understood by the customer have little use and may even be detrimental to the test process. This subject also will be discussed at length in a later chapter. The last reason to include the customers in your design process is that without them there may be no design process. Let your customers sell your design for you. If your customers go to your leadership and say "We think you need to help Joe develop the XYZ measurement system for our upcoming test. He's got some good ideas we want to see implemented," that will have much more weight than if you go to your leadership with the same message.

An allied question which needs to be asked is "What is included in the test process?" My definition of the test process includes: (1) pretest setup, noise level verifications, calibrations; (2) the actual time the test is in progress (loads, or inputs, or excitation applied to the test article); and (3) the time associated with data validity checking, assimilation, and real-time decision making on the test floor. Effective measurement system designs support this entire test process, particularly the customer's decision making.

1.3 WHAT MEASUREMENT SYSTEMS ARE WE TALKING ABOUT?

The three main uses for measurement systems are shown in Figure 1.3.[2] The added emphasis in the figure is mine. The first use shows in the upper right and is as a part of a control application. Here, the output of the measuring system is used in an overall system control algorithm. The control systems engineer's question is *"What is the output of the controlled phenomenon or process?"* The focus in this use is not on the validity or accuracy of the measured data, but on whether gasoline comes out when you put crude oil in—the overall control process OUTPUT that doesn't even show in the figure. If the data validity and accuracy are off, the control system will be tweaked to compensate.

Moving to the center, the second use is in metrology for calibration purposes. The article under test here is the measuring system or component itself. The metrologist's question is *"What are the detailed characteristics of this system under specific boundary conditions?"* The answer is the measurement component's characteristics resulting in the calibration data and certificate. The overriding focus is on accuracy of the INPUT and

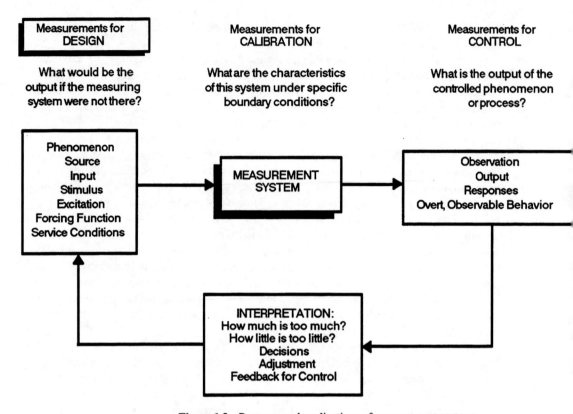

Figure 1.3 Purposes and applications of measurement systems

OUTPUT and absolute control of the boundary conditions—fixed end, free end, cantilevered, infinite pressure supply volume, isothermal conditions, zero source impedance, infinite load impedance, etc.

The last case on the left is measurement for design purposes. Here, the investigator uses the known characteristics of the measuring system (known because he or she designed the system) and views the output of that system with the process undergoing whatever stimulus the test provides. Using the measuring system and its output, the investigator makes interpretations of what happened at the system INPUT from what he or she sees at the output. What happened at the system INPUT tells about the stimulus from the process, operating conditions in the process, service conditions, and loads. That is the process-generated experimental data he or she seeks. The measurement engineer's question is *"What would the measurement system's output have been if it had not been there exchanging energy with the process causing it to change?"* This is a fundamentally different question than the previous two and requires a differerent skill and knowledge set to answer.

My entire effort in this book is devoted to the last of these cases, *measurements for DESIGN purposes*. The purpose of these measurements is the design verification of the article or process undergoing test. I invite metrologists and controls specialists to read this book knowing my focus is elsewhere.

1.4 INSTRUMENTATION . . . OR . . . MEASUREMENTS?

It took me 10 years to change the name of the organization I run from "Instrumentation" to the Measurements Engineering Department. Why did I fight that battle so long? Why was it so important? It was important because there is a fundamental difference between instrumentation and measurements. I also think a continuing focus on the difference is useful to the designer.

1.4.1 System Level Focus

If you look around at a lot of instrumentation systems and then at measurement systems, you find that the design contexts have different qualities. The people who design them tend to think differently and have different values.

Designers of *instrumentation systems* tend to focus on the arrangement of preselected individual links (instruments) of the measuring chain into an operating unit. Their emphasis tends to be on the individual links and their accuracy. You hear things like "Oops. We forget this 0.09% error over here. We've got to fix that." The focus tends to be on the electrical, statistical and, these days, the bits and bytes aspects of the system.

Designers of *measurement systems* tend to think about the application of unified scientific and engineering principles to the design of the entire measuring system, including the phenomenon and the folks who interact with it. Their emphasis is on the entire system and on data validity. They define valid data as that which faithfully represents the phenomenon as if the measuring system was not there. They just come at the problem from a different perspective. My education and 25 years of experience show me that this latter context, measurements engineering, is a much more powerful one than that of instrumentation. My intention is to convince you that assertion is valid.

1.4.2 Hierarchy of Terms

I'll define some terms for you that will be used throughout this book. This will eliminate a little of the mystery for you. *Instrumentation* refers to the individual bits and pieces of measuring systems. Digital voltmeters, amplifiers, meters, multiplexers, and tape record reproducers are instrumentation. Instrumentation engineers are the people who design these items.

Data acquisition systems are the next level up the hierarchy. They are arrangements of pieces of instrumentation and software into operating systems. Data acquisition engineers design and operate these systems. My experience tells me that the context for data acquisition systems tends to exclude much understanding of the phenomenon under investigation. People operating in this context say things like, "Oh, I don't care what's on the other end of the lines from the test cell. They're just volts coming into my data acquisition system to me. Anyway, the guy who works in the test cell is in a different union." I actually had a data acquisition engineer in a reputable company say that to me verbatim.

On top of the pile is the *measurement system*. In this book measurement systems have a very broad definition that I'll get to in the next section. In short, the definition promotes thinking at the systems level that includes the total phenomenon, all the electromechanical and computer stuff, and the involved people. You hear measurement engineers say things like, "I think we've got to change the strain gage installation. This design will reinforce the structure too much causing the readings to be low. Besides that, the strain gradients in the part will drive us to shorter gage lengths to lessen strain averaging. We may even have to go to pulsed excitation to overcome self-heating. And since this is an anisotropic composite, the standard principal stress reduction equations won't work either." Do you hear the understanding of the phenomenon coming out clearly in that conversation? That is a major part of what measurements engineering is about.

So, my hierarchy is in descending order: measurement systems, then data acquisition systems, then instrumentation.

1.4.3 What's in a Name?

Individual	Job Function	Tool Used
football player	play football	football
carpenter	build things	hammer
baker	bake cookies	flour
measurements person	make measurements	instrumentaion

Note that football players are not called footballs. Carpenters are not called hammers. Bakers are not called flour. *Measurements* people should not be called *instrumentation* anything. People's jobs are named by the function they perform—not by the tools they use.

We *make* measurements. We *use* instrumentation. That is why the specialty involved with the proper design and use of measuring systems is, in this book, measurement engineering. The clear focus of this book will be measurements—not instruments!

1.5 DATA VALIDITY . . . OR . . . DATA ACCURACY?

Almost all publications I read on this subject speak of accuracy of data rather than validity. Well, I've got a problem with that and, I think, a better way to design. I'll give you an example.

A person has the job of measuring the gas temperature in the exhaust of a gas turbine. He or she goes into the instrument laboratory, looks around and finds on a table a thermocouple with a green, 1st level calibration tag on it. "Hmmmm, this must be accurate. It has an in-date calibration sticker on it that says it's been calibrated to 0.1%." Looking around some more, the person finds a thermocouple indicator and cable. The indicator also has a green sticker on it saying it has been calibrated to 0.1%. It even has an input for the type of thermocouple in the other hand. He or she walks into the test cell, installs the thermocouple in the turbine exhaust, and connects the cable to the indicator. Everything checks out fine. Upon running up the turbine, the indicator reads 1050.13°F. The person dutifully writes down 1050.13°F and reports to the boss.

What has happened here? The person put two highly accurate and traceable devices together. They must have been. They had calibration stickers on them. The calibration accuracies at the system level look like 0.2% worst case, or 0.14% RSS. That is surely highly accurate.

Except the real temperature of the gas was more like 1270°F and the person made a *220°F error!* Errors of this magnitude are easy to make in the high temperature rotating machinery business. This was due to the inappropriate design of the thermocouple probe— inappropriate design of the measurement system. The person made an accurate measurement of a temperature 220°F in error! The data were highly accurate but totally invalid. A much better solution would have been to make a valid measurement of the 1270°F temperature with a ±10°F uncertainty.

1.5.1 Accuracy

Systems where only accuracy is the goal have several characteristics. The system is, hopefully, designed to reflect accurately the *sensed measureand within the transducer.* In other words, measure the thermocouple temperature accurately. The system answers the question "What did the meter read?"

Changes in the process under investigation caused by the mere presence of the measurement system are not accounted for and may be appreciable. Systems designed from this standpoint show an overweaning attention to the minutia of the calibration laboratory. "But wait a minute, we haven't accounted for this 0.024% error over here!"

The boundary conditions at the transducer/phenomenon interface are not necessarily accounted for or controlled.

The entire use process requires little knowledge of the phenomenon under investigation.

1.5.2 Validity

Measurement systems designed with data validity as the goal look differently. Here, the measurement system goal is the *faithful reproduction of the measureand as if the measurement system was not there.* Validity, therefore, is a higher level requirement than accuracy—although it can include accuracy. Accuracy does not necessarily include validity. Ask the guy with the inappropriate thermocouple.

In a validity-based system, changes in the phenomenon caused by the measurement system's presence are negligible in an engineering sense. Proper design and operation make them negligible.

Boundary conditions at the transducer/phenomenon interface must be understood, accounted for, and controlled. This requires understanding of the fundamental knowledge of mechanics, statics, dynamics, materials, kinematics, heat transfer, and thermodynamics. To make valid strain measurements you had better understand solid mechanics and materials. To make valid thermal measurements you had better understand heat transfer and thermodynamics.

1.5.3 Philosophy

This book comes from a context where the validity of the experimental data is more important than the accuracy. In the case of the exhaust gas temperature measurement example, I would much rather be 10°F from the right temperature than 0.14% from a temperature 220°F in error.

There are often cases where the requirement is validity first and accuracy second. These cases can be the most difficult. My philosophy has always been that measurements provided by my department will be demonstrably valid by definition. They need only be as accurate as meets the test requirements. Why this last caveat? Because accuracy costs money—sometimes lots of money. The cost of accuracy goes up as about the third power of the requirements!

As an example, we have three channel load measurement systems for the determination of spacecraft weight and center of gravity. These systems are accurate to ±0.1% of full scale and guaranteed to be so. They do it time after time. The real uncertainties are more like ±0.06% of full scale. The weight and CG data are valid *and* accurate and we can to prove it.

What is the price paid for this very real accuracy? They are calibrated as systems and remain configured as such. They are never disconnected. They are used for no purpose other than mass properties. They sit in velvet lined boxes when not in use. Only specially trained engineers and technicians may use them. And, the calibration and use cycle costs are about *50 times* that of any other load measuring system in our laboratories.

I remember the speed shop owner when I was working on my first hot rod and needed some parts. He said, "Son, speed costs money. How fast do you want to go?"

Accuracy costs money. How accurate do you want to be?

1.6 ENERGY AND INFORMATION FLOW
IN MEASUREMENT SYSTEMS

If there is a most basic concept in measurement this is it:

It is a physical law of the universe that information (data) cannot be transferred from a process to a measurement system without the simultaneous transfer of energy between the two. The uncorrupted information transfer is the goal. Energy transfer is the problem.

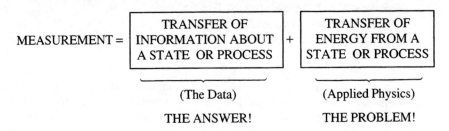

Why is the energy transfer between the process and the measurement system a problem? The process of measurement will, by definition, draw energy out of the phenomenon causing that phenomenon to change. In some cases it may transfer energy into the phenomenon with the same result—a changed phenomenon. If you change the phenomenon you want to observe, you are observing (measuring) the wrong phenomenon. So, don't the laws of physics doom you to failure? What is the point?

In an absolute mathematical sense, failure is inevitable before you start. That is why mathematicians make lousy measurement engineers. In an engineering sense, however, by exercising careful control over the energy flow—BY DESIGN—you can get close enough. Remember, engineering was once defined as the *art of successful approximation!*

The fundamental question of measurement engineering is not: What did the meter read? This question is for metrologists. The fundamental question of measurement engineering is: *What would the measurement system have read if it had not been there transferring energy with the process and changing it?* That is the question you have to answer for your design every time you use it.

Here are some examples of this from our work here at TRW where a major product line is spacecraft and space-based systems.

- The strain measured on a graphite epoxy composite structure under load is not the strain that would have been there absent the gage installation.
- The surface temperature of a satellite black box measured during a thermal vacuum test is not the temperature that would have been there absent the temperature transducer installation.
- The measured dynamic acceleration history of a solar panel during deployment is not that which would have been there absent the accelerometers cabling.

Ad infinitum.

1.6.1 Components of Energy

What goes on at the boundary between the phenomenon under investigation in your test and the transducer installation? What goes on *here?*

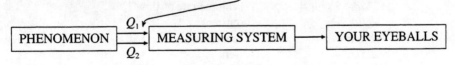

By definition, energy and information are transferred. You do not get a vote. What kind of energy? Energy is composed of pairs of physical quantities (Q_1 and Q_2) that necessarily coexist due to the laws of physics. The multiplicative product ($Q_1 \times Q_2$) of these two quantities always has the units of energy, or is energy related. The terms[3] for these quantities in measurements engineering are generalized force and generalized displacement. Here are examples of these generalized, dual system inputs from various fields.

14   Basic Concepts   Chap. 1

ENERGY FORM = (Units)	GENERALIZED × FORCE	GENERALIZED DISPLACEMENT
Electrical = (joule)	Voltage ($L^2MQ^{-1}T^2$)	Charge (Q)
Magnetic =	Magnetic potential (QT^{-1})	Magnetic flux ($L^2MQ^{-1}T^{-1}$)
Mechanical linear = (Ft Lbs)	Force (LMT^{-2})	Displacement (L)
Mechanical rotational = (Ft Lbs)	Moment (L^2MT^{-2})	Rotation (no units)
Acoustic/fluidic = (Inch Lbs)	Pressure ($L^{-1}MT^{-2}$)	Volume (L^3)
Thermal = (calories)	Temperature (Θ)	Entropy ($L^2MT^{-2}\Theta^{-1}$)

Where L = length, M = mass, T = time, Θ = degrees

Your measureands will be one of these pairs of quantities, or time, space, or time and space derivatives of these quantities. The product of your pairs must be energy, or be energy related. Here are some examples of these derivative quantities.

SPACE DERIVATIVES
dX/dx	strain
dE/dx	electric field strength
$d\Theta/dx$	temperature gradient
$d^2F/dy.dz$	stress
$d^2Q/dy.dz$	charge density
$d^3(energy)/dx.dy.dz$	energy/unit volume

TIME DERIVATIVES
dX/dt	velocity
dV/dt	volume rate of flow
dQ/dt	current
d^2X/dt^2	acceleration
d^3X/dt^3	jerk
$d(energy)/dt$	power

TIME AND SPACE DERIVATIVES
$d^2X/dx.dt$	strain rate
$d^3Q/dy.dz.dt$	current density
$d^3(energy)/dy.dz.dt$	intensity
$d^3(calories)/dy.dz.dt$	heat flux

There is a much more thorough and erudite discussion of this elegant concept in the reference.

1.6.2 You Can't Have One Without the Other

To measure a load you must allow a deflection to occur in the load cell flexure or in the test article itself. That force acting through that deflection requires energy (force × displacement = work = energy) from the phenomenon. You have just changed the phenomenon.

To measure a deflection with a gage head LVDT (linear variable differential transformer—a very popular deflection transducer) you have to allow a force to deflect the spring loaded shaft. That force acting through that displacement requires energy from the phenomenon. You have just changed the phenomenon.

To measure a pressure you must allow the change in pressure to change the internal volume of the pressure transducer causing the diaphragm to deflect. That pressure acting through that volume change requires energy from the phenomenon. You have just changed the phenomenon.

To measure the surface temperature of a widget you must allow heat to flow into, or out of, the temperature transducer so that it takes on the surface temperature. That heat flow transfers energy into, or out of, the phenomenon changing its temperature. You have just changed the phenomenon.

To measure the dynamic acceleration of a vibrating component, or of a solar array boom during deployment, the accelerometer and cabling system must be accelerated drawing energy from the phenomenon. You have just changed the phenomenon.

To measure a strain change on a test article you must allow the strain gage to change its dimensions. That requires force applied from the test article. The force times the gage's dimensional change is energy from the phenomenon. In addition, the strain gage installation itself reinforces the structure and changes the stress trajectories locally. You have just changed the phenomenon. Shall I go on?

Remember the fundamental question: *What would the measurement system have read if it had not been there transferring energy with the phenomenon?*

1.6.3 The Challenge of Measurement Engineering

So, if it is inevitable that by measuring you will change the phenomenon, what is to be done? Your job as the designer of a system for making valid measurements is to allow the transfer of valid information (your only product) from the phenomenon and to minimize the transfer of energy. Minimize means reduce the energy transfer until it is so small as to be insignificant in an engineering sense. You perform both tasks with your design.

There are large numbers of people in this business who do not appreciate this basic and most crucial measurements concept. These are the people who stick inappropriate thermocouples calibrated to 0.01% into gas flows and read temperatures in error by 200°F. These are the people who install large strain gages in areas of high strain gradient and make order of magnitude errors and don't know it. These are the people who run vibration and shock tests with 100 gram accelerometers and wonder why the answers don't make sense.

These are the people who use 0.1% pressure transducers at the end of 8 feet of stainless steel tubing to measure 40KHz pressure variations in a rocket engine. These are the people who measure *negative* pressures during depth charge explosive tests. Do they get answers? Yes they do. They get accurate, totally invalid answers. I am interested in valid answers that are only as accurate as required. That is what this book is about.

These examples may seem bizarre to you. These errors, and worse, are made every day in the measurements business. These are documented cases. They arose because the crucial relationship between energy and information flow in the measurement system was not understood.

By the way, no amount of fancy computerized posttest data analysis is going to get you out of most of these problems. These are true cases of garbage in-garbage out. The worst of these cases were made by people whose entry into the measurements business was via the computer keyboard. That is a very rocky way to enter the field.

That you are going to screw up these measurements to a certain extent is written in the laws of physics. We, as system designers and users, do not get a vote in the matter. No amount of management pressure is going to change that. What we can do is be very aware that this is going to occur and use that awareness to minimize the errors by design before they occur.

1.7 MEASUREMENTS SYSTEM RESPONSE SYNDROMES

Or, why you want less than 1/16th of the information available.

> *Syndrome* n.; a group of signs or symptoms that occur together and characterize a situation

One of the most difficult tasks for the measurement system designer is to understand all the complex interactions that occur when the test environment, including the parameter of interest, impinges on the measurement system at test time. The designer's job is to anticipate these interactions and provide mechanisms in the design that capture only those desired interactions—the data—and suppress the undesired interactions—the noise.

An equally daunting task occurs after the test if the task noted above does not occur in the design phase. This is the task of figuring out what the system's response to the environment means. Is this squiggle on this plot data, or noise, or both? How do I know? Where did it come from? Does this make sense? How did this happen?

Presented in this section is a method[3] for unscrambling this complex omelet of noise and data interactions. This is an analysis tool called measurement system response syndromes. It should be used in the design phase to clarify and focus your thinking about how your system will interact with the *entire test environment.* As Webster's definition states, a syndrome is a group of symptoms that occur together and characterize a situation. The operative words are "occur together." These interactions occur together and had best be understood before the fact—at design time. It is very expensive, and potentially embarrassing, to generate that understanding after the test is in progress—or worse, after it is over. The response syndromes tool can help you, however, in both cases.

1.7.1 Response Syndromes

Whenever your data acquisition system is making measurements of a test phenomena, there are several levels of concern regarding how the system is performing as the test progresses. As a measurement system user, you must know how your system is performing during the test if you are going to validate your data. Unvalidated data might as well be garbage—and expensive garbage at that.

      Level 1—Environment effects.   Earlier in this book I made the statement that *all components of a measurement system respond to all parts of the environment all the time, every time.* As a measurement system designer your job is to, by design, assure that the system is responding optimally to that part of the total environment that you want to measure, you call that data—and responding negligibly to all other parts of the environment—you call that noise. If you think your measurement system is not responding to all parts of the environment, my experience tells me that you are not looking at your "data" closely enough. These effects are certainly there. The question is, "How much is too much?"

      Here is an exotic example of what I'm talking about. Presume that your measurement job today is to measure dynamic strains in a hot gas turbine engine case. What other parts of the measurement environment might be around while your test is running?

- Time. Time is *always* there. I invite you to run a test that does not occur in time. Drift and aging effects occur because of this.
- Temperature. You are working on a very hot turbine case.
- Pressure. The strain gages are inside the turbine case. The gages and leadwires are subject to large static and dynamic pressures.
- Motion. The turbine is vibrating itself to death, which is why they want the strain measurements in the first place.
- Radiation. The turbine is part of a space-qualified nuclear power reactor.
- Corrosion. The hot gas passing at high velocity over the gage installations is highly corrosive.
- Magnetic fields. There is a superconducting magnet 8" away in one direction and a large, rotating electrical generator 8" away in the other direction.
- Et cetera #1. Even worse.
- Et cetera #2. You don't even want to know about this one.

The *only* thing you want to measure is $\Delta R/R$ resistance changes in the strain gages due to strain. All the other noted parts of the measurement environment are now noise levels to the measurement system. You don't want your system to respond to radiation-induced zero shift, magnetic fields induced changes in gage resistance and gage factor, temperature-induced effects on gage resistance and gage factor, time or corrosion induced zero shifts, or any other nasty combinations of environment and performance parameters.

These types of complex environment/system interactions, *only one of which you want as data,* occur every time you measure anything. The only difference is today's definition of what is the desired data and what is undesired noise. Tomorrow everything will be rearranged because you will want to measure temperature with a platinum resistance thermometer and strain will be a noise level! But, the design rules are generic and will not change.

Remember, you are trying to play this game with a *real* measurement system that responds in some real fashion to everything. You do not have the privilege of running this job with your *ideal* measurement system—the one the vendor sold you that responds only to what you want to measure today. That system does not exist. If it did, it would be in use in the next laboratory and not available for your job.

Level 2—Response type.   The transducer and its cabling system are the components of the measurement system most highly coupled to the test phenomenon and environment. There are only two types of transducers: self-generating and non-self-generating. All transducers fall in one category or the other. This will be discussed in some detail in later chapters.

Non-self-generating transducers carry the information about what you want to measure on a latent internal change in impedance. Strain gages, for example, carry strain information on the ratio of change in resistance to original resistance ($\Delta R/R$). An external application of energy is required to carry the latent impedance change information out of the transducer and make it available for use in later system components. This you call the excitation, or the secondary or interrogating input. Examples of non-self-generating transducers include strain gages and all strain gage based transducers, piezoresistive transducers, linear variable differential transformers (LVDTs), servo transducers, angular rate gyros, resistance temperature transducers, rotary potentiometers, variable reluctance transducers, and photoelastic coatings for stress analysis.

Self-generating transducers are those that give an input-related response directly with no requirement for an additional secondary energy input. Examples include thermocouples, piezoelectric transducers, magnetic velocity transducers, acoustic emission transducers, and brittle coatings for stress analysis.

Now, there is a nefarious situation here. *All measurements systems exhibit both responses, self- and non-self-generating, simultaneously to all the environments in a predictable fashion all the time, every time.* Do you sense a trend toward bad news here?

That is bad news to the system designer. How are you ever expected to sort out these responses and identify the one that you want as valid data today? This also will be discussed in a later chapter.

Level 3—Response evidence.   As if the preceding news was not bad enough, *each time a system responds in either fashion, self- or non-self-generating, to any part of the test environment, there are two kinds of evidence for that response. The evidence can be either temporary or permanent.* Motion-induced generation of charge in a piezoelectric accelerometer is an example of temporary evidence. When the motion ceases, the generation of charge ought to cease, or you no longer have an accelerometer—you have an acceleration switch. The strain-induced resistance change in a strain gage is another example of a temporary response. When the strain returns to its original pretest zero value, the resistance change ought to return to zero with it. The change in voltage generated by a temperature gradient in a thermocouple circuit ought to vanish when the gradient is removed.

The other evidence type is permanent. Here, the evidence remains when the stimulus is taken away. Long-term zero shift with time and temperature in a transducer or signal conditioner is an example. If you significantly exceed the Curie temperature of a piezoelectric transducer it will cease to be a transducer and become a paperweight—permanently. We use small, deformable cones for interference measurements on some vibration tests to see if parts are impacting. Post yield strain gages are used to measure permanent plastic strains in test articles. Temperature-sensitive paints are used in thermal testing to record maximum temperature excursions. The evidence is carried on the permanent change of color.

Again the measurement universe is perverse. It gets worse.

Level 4—Response effect. *Either evidence type, temporary or permanent, can have an effect on the measurement system's output that is either additive or multiplicative.* In other words, the response can affect the system's zero or its gain, usually both.

Strain-induced resistance change in a strain gage is an example of a desired additive response—the strain causes a system output to be added to whatever is already there. All piezoelectric transducers exhibit pyroelectric effects. A change in the temperature generates charge that adds to that generated by what you want to measure. In most measurement scenarios, the desired system response is this additive response effect on the system's zero.

A strain gage's internal measure of sensitivity, its gage factor, changes with temperature and magnetic fields. This is an example of a multiplicative effect. Another occurs if you operate piezoelectric transducers at temperatures above their operating ranges. The basic sensitivity in picocoulombs per engineering unit will decrease permanently. If you significantly exceed this temperature, the sensitivity will go to essentially zero. This is the case noted in Level 3. Thermocouples can change their calibration, their thermoelectric coefficient or gain, when the thermocouple metallurgy changes, perhaps by use in a corrosive atmosphere. These are all examples of multiplicative effects—changes in system gain. In most measurement scenarios, this multiplicative response is a noise level.

1.7.2 So What Does It All Mean to You?

What it means shows in Figure 1.4. Here you can see the response levels of concern listed down the left side: environment, response type, response evidence, and response effect. Use the example of the measurement of strain inside the turbine engine case again.

Here, the desired environment is the strain change on the turbine case—that is what you want to measure today. The undesired environments include everything else—temperature, time, vibration, corrosion, radiation, etc. In Figure 1.4, only temperature shows as the undesired environment because of the width restriction on this page. All the rest of the undesired environments are really there, in your design process, stretching out about 6 feet from the right edge of the figure.

Each response to a part of the environment will cause both a self-generating and

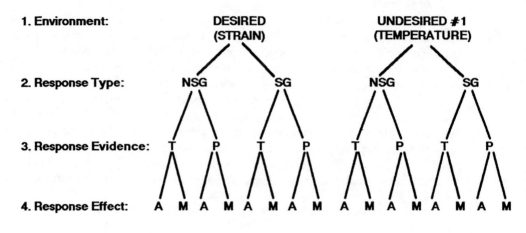

1. Environment:

2. Response Type:

3. Response Evidence:

4. Response Effect:

NSG = nonself generating   SG = self generating
T = temporary   P = permanent
A = additive   M = multiplicative

Figure 1.4   Measurement system response syndromes

non-self-generating response in the measurement system. In the case of the desired response, strain in our example, you want only one of these to occur—the non-self-generating response. A self-generated response to the desired environment, strain, is a noise level in this example. In fact, this response would be an example of the parameter you want to measure causing its own noise level! All responses from the undesired environments are noise levels by definition. Your job is to design a system which documents that it was not subject to outputs from any other path.

Marching right along. Each of the above response types will cause temporary and permanent evidence. You want the temporary evidence of the non-self-generating response. All other responses are noise levels.

Each temporary and permanent response will cause outputs that are multiplicative and additive. In our case, we want the additive output cause by the strain change. In summary, for this example, the only system output you want is the (4) additive, (3) temporary, (2) non-self-generating response from the (1) desired environment, strain. Outputs from all other combinations or paths are noise levels and are, to an engineering approximation, not allowed or your data is invalid and corrupted by noise levels.

As you can see in Figure 1.5, we have traced a tenuous path in getting where you wanted to go. The only data path you want to document is the single one noted above from the chart—the chart that is 6 feet wide in your design! Because this is the only path to valid data, to the degree that you stay on this path you will make valid strain measurements on your turbine engine case.

What about some other data paths? Figure 1.6 shows the path for system zero shift (noise level) due to time at high temperature causing corrosion of the gage foil alloy result-

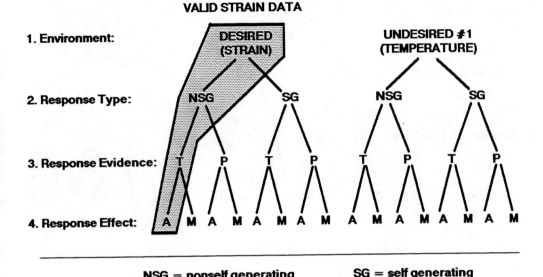

VALID STRAIN DATA

1. Environment:

2. Response Type:

3. Response Evidence:

4. Response Effect:

NSG = nonself generating   SG = self generating
T = temporary   P = permanent
A = additive   M = multiplicative

Figure 1.5   Measurement system response syndrome path for valid strain data

ing in increased resistance—an additive, permanent, non-self-generating response to the undesired temperature environment. Figure 1.7 shows the path for temperature-induced changes in the strain gage's sensitivity (noise level), its gage factor—a multiplicative, temporary (generally), non-self-generating response to an undesired temperature environment. Figure 1.8 shows the complex data paths for system responses driven by high-level vibratory motions coupled with strain—as its own noise level! The research is incomplete in this area. As of this writing it is not known whether these effects are caused by rigid body motion or vibratory strains in the test article—or both. Both are shown. These responses are very complex, very difficult to separate, and look as though they were "piezoelectric" in nature. There are at least three separate mechanisms active on this path and will be discussed in a later chapter. These responses are additive, temporary self-generating responses to the undesired motion environment.

Working one's way through what is data and what is noise (let alone the source of the noise) can be a daunting task for the measurement engineer. These treelike measurement system response syndrome charts offer a very powerful method for generalizing and organizing your design and diagnostic thinking about data and noise level sources. They can help to take the arm-waving out of the process and allow information-based decision making and diagnosis. They also can be done on the back of an envelope. Every piece of data, noise level, or effect you have ever produced in a measurement system, or seen or heard about can be located and pinpointed on this chart by labeling the environments across the top appropriately for your test.

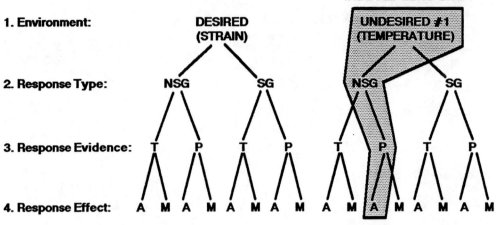

Figure 1.6 Measurement system response syndrome path for temperature induced zero shift

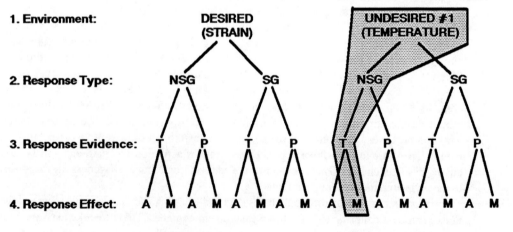

Figure 1.7 Measurement system response syndrome path for temperature induced gage factor change

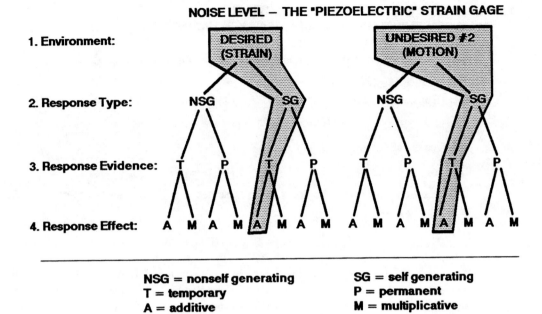

NOISE LEVEL – THE "PIEZOELECTRIC" STRAIN GAGE

1. Environment:

2. Response Type:

3. Response Evidence:

4. Response Effect:

NSG = nonself generating   SG = self generating
T = temporary   P = permanent
A = additive   M = multiplicative

Figure 1.8   Measurement system response syndrome path for "piezoelectric" strain gage

1.7.3 You Actually Want Less Than One-Sixteenth of the Information

You can see that there are 16 paths to get to the bottom of the chart if you consider the desired environment, strain in our example, and only one of the undesired environments, say temperature. There are always other undesired environments—this is inevitable. At least time is always there. Some of the most difficult noise level problems imaginable are caused by long-term effects of time on a measurement system. The other undesired environments are always there even if they fall off the right hand side of the page through inattention. You must not allow them to fall off your design page. Your measurements system will respond to these response paths in some fashion. You can write them down to make sure you consider them when you should—during the design process.

At best, if there were only a single undesired environment, you want only one-sixteenth of the information available to you at your system's output. Your customer is only paying for that unique one-sixteenth. In reality, there are always other environments acting on your system whether you account for them in your design or not. So in truth, you always want less than one-sixteenth of the information available—sometimes significantly less!

The proper and effective design and use of measurements systems is, in its simplest form, nothing more than the ability to navigate consciously through this maze of possibilities to arrive at the desired destination—the valid data and only that data.

1.8 FOURIER ANALYSIS: THE CONNECTION
BETWEEN THE TIME AND FREQUENCY DOMAINS

Let me quote to you from Bendat[4] and Piersol on the subject of Fourier series.

Fourier Series
Consider any periodic record $x(t)$ of period T. Then for any value of t

$$x(t) = x(t \pm kT) \text{ where } k = 1, 2, 3, \ldots .$$

The fundamental frequency f_1 satisfies

$$f_1 = 1/T$$

With few exceptions periodic data can be expanded in a Fourier series according to the following formula:

$$\sum_{k=1}^{\alpha} x(t) = (a_0/2) + \Sigma(a_k Cos2\Pi f_k t + b_a Sin2\Pi f_k t)$$

What does this mean to you as a measurement system designer? It means that any periodic waveform can be constructed from a series of sine and cosine waves of the appropriate amplitude and phase. This spectrum will have discrete frequencies. Conversely, any non-periodic waveform also can be constructed from a series of sine and cosine waves of the correct amplitude and phase. This spectrum will, however, be continuous. Figure 1.9 shows one period of a 10-unit peak sine wave of frequency $1/f$ in the time domain. You can see its

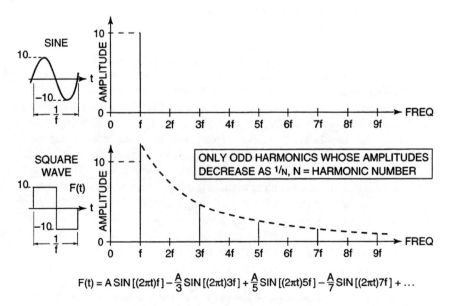

$$F(t) = A \, SIN \, [(2\pi t)f] - \frac{A}{3} SIN \, [(2\pi t)3f] + \frac{A}{5} SIN \, [(2\pi t)5f] - \frac{A}{7} SIN \, [(2\pi t)7f] + \ldots$$

Figure 1.9 Spectra of sine and square waves

Fourier spectrum to the right. There is a single line at frequency f, 10 units tall. There is no surprise here.

Below you can see a square wave of 10 units peak at the same repetition frequency f. Since it can be devolved into a series of sine and cosine waves, you can see the Fourier series just below the figure. The spectral representation shows spectral lines at f and its odd harmonics, whose amplitude decreases as one over the harmonic number. This representation shows clearly in the series equation. Note that: (1) there are only odd harmonics ($3f$, $5f$, $7f$, etc.); (2) the first harmonic amplitude is larger than 10 units; (3) the amplitude goes down as one over the harmonic mumber; and (4) the phase alternates ($+$, $-$, $+$, $-$). These are the distinctive characteristics of the square wave. They make it very useful in system design and checkout. A measurement engineer should have these relationships memorized.

Figure 1.10 shows the same square wave Fourier series plotted three-dimensionally—amplitude versus frequency versus time. This figure shows how the devolved square wave frequency components can come together to give back the time history of the square wave.

Why bring this up now?   I bring the Fourier series subject up now because it is extremely useful to the measurement system designer. A successful designer and system user will be highly facile in transferring his or her thinking from the time to the frequency domain and back. If you show an experienced measurements engineer a time history, he or

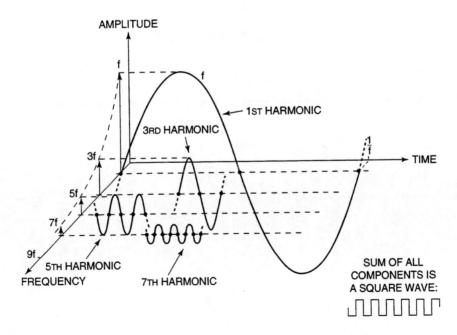

Figure 1.10 Relationships among time, amplitude, and frequency for a square wave of repetition rate f

she is immediately thinking about its description in the frequency domain, and vice versa. The engineer can discuss the issue in either domain with ease.

Second, the square wave represents one of the most useful diagnostic tools for the system designer. You will hear of it and see it repeatedly in this book. Measurement engineers should always walk around with a mini-toolkit in their pockets. That toolkit should include, at least, a shorting bar, a GR to BNC adapter, and the mental idea of the Fourier transform of a square wave!

NOTES

1. Publications detailing this body of work are available from Stein Engineering Services, 5602 E. Monte Rosa Street, Phoenix AZ, 85018.
2. Peter K. Stein, *A New Conceptual and Mathematical Transducer Model Application to Impedance-Based Transducers Such as Strain Gages*; Proc., Second Discussion Meeting Technical Committee on Measurement of Force and Mass, International Measurement Confederation (IMEKO), The Hague, Netherlands, September 1971.
3. Peter K. Stein, *"A Unified Approach to Handling Noise in Measuring Systems,"* AGARD LS-50, Flight Test Instrumentation, NATO, Neuilly sur Seine, France; September 1972, pp. 5–1 to 5–11.
4. Julius Bendat and Allan Piersol, *Engineering Applications of Correlation and Spectral Analysis;* John Wiley and Sons, New York, 1980.

Chapter 2

Measurement System
Transfer Functions
and Linearity

2.1 WHAT THE HELL IS A dB ANYWAY?

I get asked that question more times than I would like. The question is generally asked by those familiar only with "static" measurement systems. These are usually used in structural testing and thermal testing of large objects. In my business, spacecraft temperatures do not change very quickly. Zero frequency structural tests are "static" tests after all. Measurement systems that support them can be thought of as "static." But the thought is dangerous.

Facility in the use of the decibel in measurement system design is fundamental. I would not want to go into a system design process with a staff not having a back-of-the-envelope understanding of the use and meaning of decibels in system design. If some engineers do not have that understanding, they get it very quickly. You may think it's painful or boring—but you need it anyway. It's a little like taking foul-tasting medicine, so we'll get it over quickly and move on to the more interesting stuff.

The definition of the decibel began a hundred or so years ago with two power measurements—a system input power P_1, and an output power, P_2. The ratio is the system power gain and was defined as a Bel.

$$\text{Bel} = \text{Log}_{10}(P_2/P_1)$$

For systems with lots of power gain they defined ten times the Bel as the *deci*bel.

$$\text{Decibel} = 10\,\text{Log}_{10}(P_2/P_1)$$

Since power is a squared quantity, $P_2 = Q_2^2$ and $P_1 = Q_1^2$, where Q_2 and Q_1 are the base measureands (like volts for instance). Now

$$\text{Decibel} = 10 \ \text{Log}_{10}(Q_2/Q_1)^2$$

Or

$$\text{Decibel} = 20 \ \text{Log}_{10}(Q_2/Q_1)$$

This is the basic definition of the decibel as used in measurement engineering. It is the definition you can assume throughout this entire book.

There is only one exception to this known to me in the test measurement business. In random vibration work, where the answers are in terms of power or auto spectral density, the $10 \ \text{Log}_{10}(P_2/P_1)$ definition is used. This always results in some confusion. The vibration engineer is thinking about PSD levels in 10 dB decades and the measurement engineer is working a problem on the same test in 20 dB decades! One must pay attention to keep one's sanity in this environment.

Examples of the 20 dB definition are:

$$\text{if } Q_2/Q_1 = .1, \ \text{then Decibels} = -20$$
$$\text{if } Q_2/Q_1 = 1., \ \text{then Decibels} = 0$$
$$\text{if } Q_2/Q_1 = 10, \text{then Decibels} = 20$$

Table 2.1 shows, on the right, the decibel ranges of +120 dB (a factor of 1,000,000) down to −120 dB (a factor of 1/1,000,000). This 240 dB range is more than you will ever have to worry about in your measurement system design. On the left Table 2.1 is expanded and shows the relationship between the 20 decibels that divide the interval in Q_2/Q_1 from 1 to 10.

There are several convenient values of decibels between −20 and +20 that should be memorized. These favored values will occur repeatedly in system designs and the discussions that surround them. Table 2.2 shows them for reference.

2.2 THE AMPLITUDE PORTION OF A MEASUREMENT SYSTEM'S TRANSFER FUNCTION

Note: A component or system's transfer function is comprised of descriptions of amplitude (or gain) and phase between output and input, each as a function of frequency. These are more usually called the frequency and phase responses in the literature. I am going to use the term transfer function more often in order to underline the fact that the amplitude and phase portions are inseparable and are the two sides of the same mathematical coin. In that sense, the term transfer function is a more complete system level performance definition.

All input waveforms for any measurement system have frequency content. From the previous discussion of the Fourier transform, you learned that any waveshape can be devolved into a series of sine and/or cosine waves of the appropriate amplitude and phase, and a constant. The constant is the DC, or zero frequency, component of the waveform. The mathematics of the expression hold for zero frequency. The DC component is just the sine wave component at zero frequency. The DC component is, therefore, no more or less important than any other frequency component.

Table 2.1 dB to ratio scale

Table 2.2 Useful dB to ratio conversions

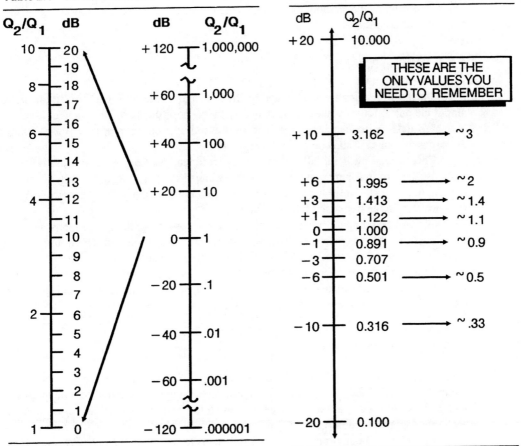

The manner in which the measurement system handles all of these frequency compo-nents, in terms of their individual amplitudes, is a crucial determinant of measurement sys-tem performance. If the measurement system discriminates across the frequency range of interest by amplifying or suppressing some frequencies more than others, errors in the out-put waveshape will occur, causing the waveshape to be invalid. Such distortion of the out-put waveshape is called *frequency distortion*—which you can now see is a very poor de-scription of the problem. One would want, therefore, a measurement system which has constant gain over a frequency range of interest—a system which has "flat" frequency re-sponse. A system which exhibits no frequency distortion will output an exact replica of its input (assuming no errors caused by linearity or phase considerations). Such a system does not exist. All real measurement systems exhibit some discrimination of frequencies. The best you can do is, by design, place this discrimination where it causes your data no harm.

The important thing to remember as you read this section is that you control the amount of frequency distortion by your design. As you read about slopes and frequency limits, roll-offs and undershoot, remember that you place them there via your design. They are under your control.

2.2.1 General Remarks

All of the discussion in this section, and the figures, are in terms of logarithmic plots of the amplitude and frequency axes. This is necessary because this particular subject involves both high and low frequencies and high and low amplitudes on the same plot. A logarithmic display is just an easy and convenient way to do this.

This discussion is initially limited to nonresonant systems. We'll discuss resonant systems later. Nonresonant systems are those systems, usually second order systems, where the amplitude portion of the transfer function does not rise above the flat portion of the response within the frequency range of interest.

We will be using a principle in this discussion which is called the superposition of linear systems. All this means is that we are going to break the system down into smaller parts and discuss the parts first. We will discuss the problems at the low end of the frequency range as if there are no problems at the high end—and vice versa. Then we will put the entire system performance together by superposing the low end and high end characteristics and see how the entire system performs in a later section.

2.2.2 A Word on Slopes

A complete discussion of the amplitude portion of the system transfer function can occur within the context of slopes on log-log graphs. These slopes are the lines on the graphs to which the system's actual performance is asymptotic. Two lines which are asymptotic to each other approach but theoretically never meet. They do meet in an engineering sense— we can get close enough for design purposes. This is fairly important stuff. Practiced measurement engineers can operate with these slopes on the back of an envelope. They are, in a real sense, a shorthand, effective method of determining the overall system performance. You don't need a computer model to work with this subject—you need a pencil and an envelope.

These characteristic slopes can be seen in Figure 2.1. A slope of +1 on a log-log plot rises at a rate of one decade of amplitude, or 20 dB, for every decade increase in frequency—a slope of 20 dB/decade. It also doubles, approximately 6 dB, in amplitude for every doubling of frequency (an octave)—a slope of approximately +6 dB/octave. Positive slopes are shown at the left of the figure for the low frequency end of the spectrum. Negative slopes are shown at the right, or high frequency, end of the spectrum.

For all intents and purposes, these slopes occur in measurement systems in even multiples of approximately a slope of one (that is 1, 2, 3, 4, etc.). A slope of 2.63 is a difficult, although possible, beast to design. Further, these slopes increase in multiples of 6 dB per octave (doubling of amplitude for a doubling of frequency) or 20 dB per decade (increase

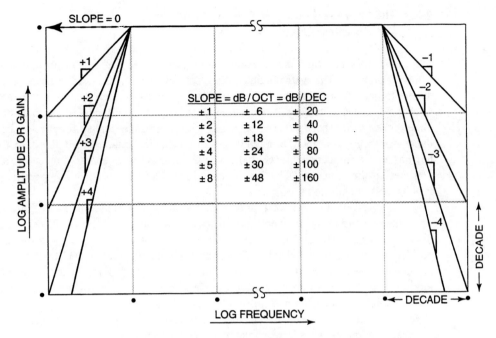

Figure 2.1   Characteristic slopes of frequency response curves

of a factor of ten in amplitude for an increase of a factor of ten in frequency). This can be seen in Figure 2.1 and below.

The following statements about slopes on log-log curves are equivalent and inter-changeable:

$$SLOPE = \pm 1 = \pm 6 \text{ dB/octave} = \pm 20 \text{ dB/decade}$$
$$\pm 2 = \pm 12 \text{ dB/octave} = \pm 40 \text{ dB/decade}$$
$$\pm 3 = \pm 18 \text{ dB/octave} = \pm 60 \text{ dB/decade}$$
$$\pm 4 = \pm 24 \text{ dB/octave} = \pm 80 \text{ dB/decade}$$

And so on.

You will see these these slopes stated in the literature in all three forms. In order to be an effective designer you need to be able to understand and shift between the three forms easily. As you can see, it is easy to do this. I am going to use the simplest form in this book and talk about slopes in integer values (slopes of +/− 1 or +/− 4).

What causes these characteristic slopes to occur in measurement systems? They are functions of the structure of the measurement system—a function of your design. They are caused by the physics occurring in the phenomenon/measurement system chain and by the internal circuitry in the system (its tendency to filter or amplify waveforms). Examples of this will be found later in this book.

2.2.3 The Low Frequency End of the Spectrum—
The Simplest Case

In the absolute sense, the simplest form of low frequency roll-off in a measurement system is no roll-off at all. The system gain continues at its constant, "flat" value all the way down to zero frequency or DC. A typical example of this type of system would be the thousands of channels of DC excitation, Wheatstone bridge signal conditioners in use in laboratories all over the world. These are used primarily with strain gages and strain-gage-based transducers for the measurement of a whole raft of mechanical quantities. In this case, the system has no low frequency effects—what you put in at low frequencies you get out. If this is your design situation, say "Thank you" and proceed to the discussion of upper frequency limit issues which follows.

However, let me describe first the simplest non-DC system. Such a system will roll-off at the low end of the frequency spectrum with a slope of +1 (or +6 dB/octave, or +20 dB/decade—same slope). This slope is characteristic of a single energy storing element governing low frequency behavior. This system will not respond to a DC level at all. There are numerous examples of systems with this characteristic. Some of them are probably lurking in your test laboratory. An oscilloscope with an "AC" coupling position on a trace's vertical amplifier is such a system. In this case, the single energy storing element which governs low frequency behavior is the blocking capacitor in the signal high side path that the manufacturer switches in when you push the "AC" button.

The microphone of choice in the acoustics test business is usually the Bruel & Kjaer condensor microphone—really nothing more than a capacitive, low level, dynamic pressure transducer. These microphones show this low frequency characteristic due to a mechanical capacitor. There is a tiny hole behind the diaphragm used to bleed off static pressure so that the diaphragm is not damaged. This hole acts as a capacitor in the transducer's internal mechanical structure.

The amplitude portion of the transfer function of such a system is shown in Figure 2.2. Here I have plotted the system's gain versus frequency with both parameters non-dimensionalized for simplicity. Remember, this is a generic, second order system response. I will not bore you with the system diferential equations because you don't need them after having seen them once in the university. Further, it does not matter whether the gain is in picocoulombs per G, picocoulombs per PSI, millivolts per microstrain, volts per pound, or anything else. The system gain is general. Where it is flat—call that a gain of one. The same thing holds true for the frequency axis.

This is the output plot you would get if you swept the system with input sine waves of constant amplitude and varying frequency, and noted the output level as a percentage of the input and as a function of the frequency. The ratio of the system's output level to its input level is the system's gain.

Note the slope on the graph. Starting at high frequency toward the right hand side of the graph, and moving down the frequency axis you can see the response is, in its "flat" region, at a gain of one. The slope here is zero. At some frequency, f_1, you can see a second slope begin where the response is no longer constant, but is decreasing with decreasing

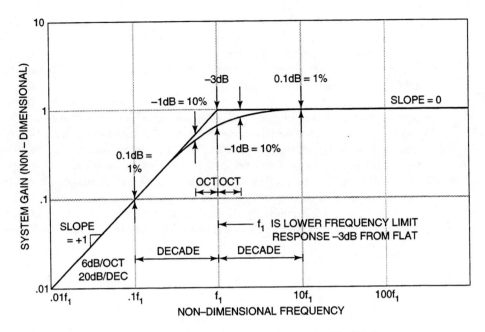

Figure 2.2 Simplest non-DC low frequency roll-off (+1)

frequency. This slope is +1. The actual system response is the curved line which is asymptotic to the two slope lines.

Let's inspect this characteristic in some detail, because the details will pay off later in your designs. There is a frequency, here called f_1, where the output response is -3 dB down from the gain in the "flat" region. This frequency is called the lower frequency limit. Here the response is 3 dB down from the flat portion, or an amplitude error, if uncorrected, of about 30%! This is a general definition no matter what the characteristic looks like.

The lower frequency limit of a measurement system is that frequency where the amplitude response is down 3 dB from the value in the flat portion of the amplitude portion of the transfer function.

Looking at frequencies higher than the lower frequency limit—what do you see? An octave above f_1 $(2 \times f_1)$—the response is still 1 dB down. This is a *mere* 10% error. You have to be a decade in frequency above the lower frequency limit before you are within 1% of its final, "flat" value. An octave down from the lower frequency limit, at $.5f_1$, you are still 1 dB below where you are supposed to be on the asymptote—again a 10% error. A decade in frequency below the lower frequency limit you are within 1% of the asymptotic value. Here your system actually has a gain of 0.099 when it should be 0.1.

The point of this explanation is this. For typical measurement systems with this low frequency characteristic you have to be "far" above the lower frequency limit (more than a decade, more than a factor of ten) before your data's amplitude is affected less than 1%. The effects of a lower frequency limit of this type (slope = +1) extend much farther above the lower frequency limit than you think they do. You must take this into consideration in your system designs. Just to put this into perspective, if you were to make that inadvertent 1% error you would have just thrown bits 9–16 in your 16-bit A/D converter away as error. A 1% A/D error occurs at about bit 8 in a binary A/D converter. So tell me all about your need for 16-bit A/Ds in your system design!

All of the above discussion is about how a generic measurement system with a low frequency slope of +1 responds to steady state sine wave input. But what about other waveforms? At the other end of the waveshape continuum from sine waves are pulses, which are the most difficult waveform to reproduce. For this very reason they are just dandy for checking out measurement system performance. We use them on a daily basis for just that reason. Pulses, or pulse trains, tend to push measurement systems to their performance limits and show you very clearly and quickly their capabilities.

Look at Figure 2.3. At the top you will see the same characteristic as in the previous figure—flat frequency response above the lower frequency limit and a roll-off of +1 below that frequency.

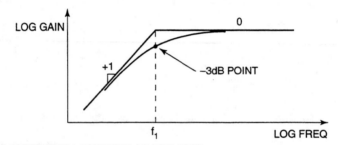

THIS SYSTEM WILL RESPOND AS FOLLOWS:

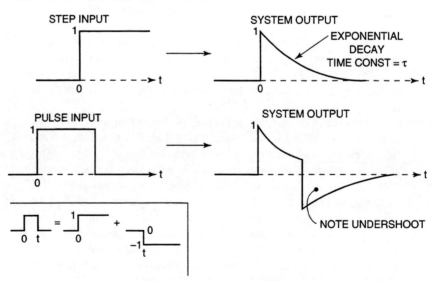

Figure 2.3   System with low frequency roll-off, slope = +1

At the lower left hand corner of this figure there is a small graphic which shows that a pulse is merely a positive step followed later in time by a negative step of the same magnitude. Since this is so, we can construct the system's response to a pulse from its response to successive steps. Below the frequency response curve is the step input waveform. The first part of a step is a very fast rise from zero to its final value of one. If your system is to follow this rise, it must have lots of frequency response or bandwidth. Does this system have that? Yes it does. We are assuming the response is flat out to a very high, as yet unspecified, frequency. So, the system should follow that fast rise, and it does—it jumps up like a shot. The second part of a step is an infinite amount of time at some DC level, a level of one unit in this generic case. In order for our system to follow this level it must have response to zero frequency, to DC. Does it? Nope—it is rolling off at +1 on the low end. As a matter of fact, it has zero response at DC.

In the case of a system with a low frequency roll-off of +1, that drop-off will be exponential as shown in the output waveform. The time constant of this exponential drop-off is directly related to the value of f_1. The lower the value of f_1, the longer the time constant. In the limit, f_1 is infinitely low (namely zero or DC), and the time constant is infinitely long.

So, how does the system respond to a pulse? You can see that the system's response to a pulse, which is the sum of a positive and a negative step, is the sum of the system's responses to those two steps.

Note the undershoot here. There is no negative data in the input. Yet our system shows negative data at its output. Is this really negative pressure in our blast pulse or has this been created in the measurement system? Is this really negative load being read from out piezoelectric load cell during a transient loading test, or has it been created in the measurement system? For the answers to these questions, read the section which follows later in the book on data validation techniques.

These responses to sine waves, steps, and pulses from a system with a low frequency roll-off of +1 are generic. They must occur. Knowing that, you can use these characteristics to tell how the system behaves. If you want to know about a system's low frequency characteristics, don't bother with complicated sine waves—that is a pain. Simply input a step and see how the system responds. If you see an exponential decay to a step, you have all the information you need to define the entire system's low frequency behavior. The shape has told you the system rolls off at +1 and the time constant will tell you the value of the lower frequency limit.

2.2.4 Systems with More Complex Low Frequency Behavior

What about more complex systems? Do you have any? Yes, you probably do. Every channel which uses a piezoelectric transducer and charge amplifier is a more complicated system with a roll-off on the low end with a slope equal to, or greater than, +2. These transducer/signal conditioner combinations exist by the thousands of channels in the dynamics test areas which deal with vibration, acoustics, shock, kinematics, and transportation testing.

In each case mentioned here, there are at least *two* mechanisms (not one!) controlling low frequency behavior. In the case of a piezoelectric accelerometer or pressure transducer, the capacitance of the transducer coupled with the huge, but not infinite, impedance of the charge amplifier make a simple RC high-pass filter with a slope of +1 at the input of the charge amplifier itself. However, there is usually a later non-DC coupled amplifier stage within the charge amplifier with a slope of at least +1. In this log-log arena, terminal slopes add algebraically so the measurement system has an overall slope of +2 or more. The manufacturer usually doesn't even tell you about this second roll-off! The condensor microphone mentioned earlier has a mechanical capacitor built-in via the bleed hole behind the diaphragm. This accounts for this first slope of +1. The following electronics are not DC coupled and have their own slope of +1 or more. So, the overall measurement system displays a slope of +1+1 = +2, or more at the low frequency end.

Figure 2.4 shows a comparison of a simple system with a low frequency roll-off of +1 and a system with a more severe roll-off slope of +2 or more. The simple system's slope lines and actual response curve are shown as solid. Those for a more complex system are shown as dashed. Both systems have the same low frequency limit, f_1. As the frequency is decreased from a high frequency toward the low frequency limit f_1, note that the simple system begins to drop off first. Both systems arrive at the same point at the low frequency limit—they both are -3 dB down from the "flat" portion of the curve. They get there, however, by entirely different paths.

There is a much more severe "knee" in the curve from the more complex system. Approaching f_1 from above, the more complex system shows "better" frequency response—it hangs in there longer. Furthermore, at frequencies below the low frequency limit, the complex system suppresses much more amplitude than the simple system. The complex system acts as a much better filter down here.

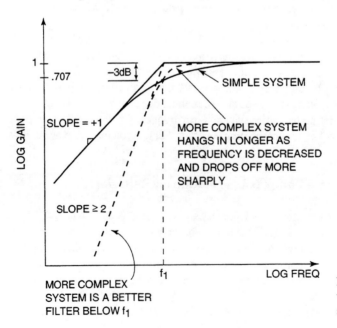

Figure 2.4 Comparison of systems with various low frequency roll-off slopes

That all sounds great, right? We should always use more complex systems because they have "better" frequency response—right? Maybe. The answer to that question is not as simple as the question. Wait for the answer until the discussion on phase response which will follow.

Figure 2.5 shows the response of these more complex systems to steps and pulses. You can see that these more complex systems even undershoot to a step. The performance is totally different than that of simple systems. The responses shown are for a system with a low frequency roll-off of exactly +2. Note that the step response shows undershoot, but after that the response does not cross the zero line again. Systems with low frequency slopes of more than +2 will cross the zero line and will seem to actually ring. The number of times the zero line is crossed is a measure of the slope at the low end.

There is a very important message here. We have just seen two systems with the *same* low frequency limit behave in a totally different manner. Why? Because they have different

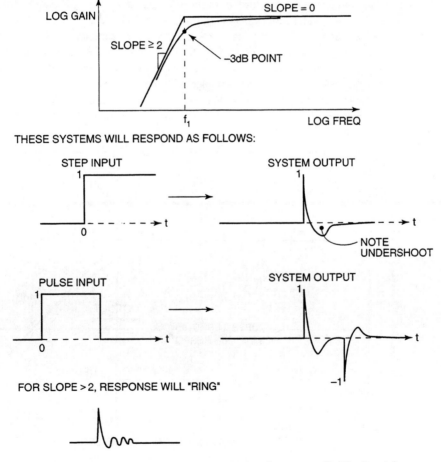

Figure 2.5   Systems with more complex low frequency roll-offs, slope ≥ 2

slopes on the low end of the frequency spectrum. These slope values, and the cutoff frequency f_1, totally determine the low frequency behavior of the system.

Both the low frequency limit and the terminal slope outside the passband must be specified before system performance can be characterized. Either without the other is useless for predicting or verifying system performance or data validity.

2.2.5 The High Frequency End of the Spectrum—
The Simplest Case

On the low frequency end of things we had an option. If you are not comfortable with all that bizarre behavior, you could simply use a DC coupled system. This might give you some other operational problems—but you could do it. At the other end of the frequency spectrum, the high frequency end, you have no such convenient choice. All systems show roll-offs at the high frequency end of the spectrum. For an XY plotter, this might be 2 Hz. For a system designed to record valid acceleration data from a detonating nuclear weapon or high speed rotating machinery, high might be 500,000 Hz! But both systems roll-off somewhere and can be analyzed in the same, generic manner. As I said earlier, a good measurements engineer can do this stuff on the back of an envelope—generically. In the long run, it is a good thing our systems do roll off on the high end. If they didn't, it would be very difficult to optimize performance with regard to noise levels.

Figure 2.6 shows the amplitude portion of the transfer function for a system with a high frequency roll-off at a slope of -1, the simplest system you can have since a slope of

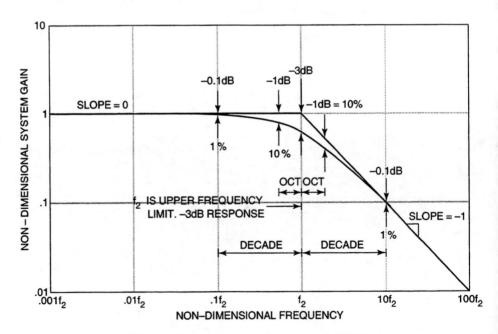

Figure 2.6 Simplest upper frequency roll-off (−1)

zero is not available to you. An example of this type of system would be the vertical amplifier of an oscilloscope with a cute little "filter" button on it. Here, the filter is usually a capacitor in parallel with the signal inputs to "short out" high frequency signals. That is just what it does too, looking like a short circuit to high frequencies at the device input. Thus, the term "filter."

This characteristic looks like the mirror image of the low frequency behavior of a system with a slope of +1. The upper frequency limit is defined as f_2. The upper frequency limit of a measurement system is that frequency where the amplitude response is down 3 dB from the value in the flat portion of the amplitude portion of the transfer function.

The numbers are the same as for the simple system with slope = +1 at the other end of the spectrum. An octave below and above the upper frequency limit (at $.5f_2$ and $2f_2$) the response is within 1 dB (about 10%) of the asymptotes. One full decade below and above the upper frequency limit, the response is within .1 dB (about 1%) of the asymptotes. The point again is that the system upper frequency limit must be far above the highest frequency of interest in the signal (at least one full decade) before the amplitude errors are less than 1%.

If the highest frequency of interest to you is 2KHz, and you are interested in holding frequency distortion errors to 1% or less, you must place the upper frequency limit at least 20KHz for this simple system!

Figure 2.7 shows this system's response to steps and pulses. As you can see, the response to an input step is a rising exponential, as the response at the other end was a falling exponential. This system does not have an infinite upper frequency limit so it simply cannot respond to the fast rise of the step. It can pass DC, however (since we have not yet precluded it), so it eventually must get to the DC level of the step. This system will display an exponential rise as shown. Again, the response to a pulse is just the sum of the system's responses to two succeeding steps, one plus and one minus as before. Here, the time constant of the exponential rise is directly related to the upper frequency limit f_2. Again, these features can be used to characterize such a system. All you have to see is that exponential rise and note the time constant of that rise and you have all the information needed to determine system performance.

2.2.6 Systems with More Complex High Frequency Behavior

A more complex system is shown in Figure 2.8. Here, the terminal slope is −2 or more. This system will respond to a step at its input with much more complex behavior. This system will show a characteristic "rise time" as shown in the graphic. The rise time is defined as the time between the systems's reaching 10% and 90% of its final value. Further, the response has a clear point of inflection (where the first derivative changes sign). This means that the slope of the time history begins to increase as the response level comes up, then begins to decrease as the response approaches the DC level.

Note that the response slope in the simple system in Figure 2.7 is always decreasing—and has no point of inflection. This is the way to tell if a system has an upper roll-off slope of two or more. Put in a step—if you see a point of inflection you know the slope is two or more. Unfortunately, that is all you can tell. There is no way to go from a rise time

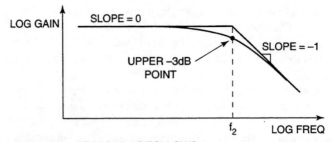

THIS SYSTEM WILL RESPOND AS FOLLOWS:

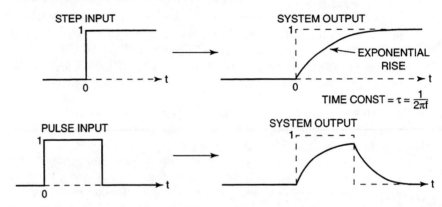

Figure 2.7   System with most simple high frequency roll-off, slope = −1

measurement to the slope of the upper frequency roll-off. That must be determined by other methods, such as the injection of sine or random signals.

Again, we have two systems with the same upper frequency limit giving radically different responses to the same input. What gives? The same explanation holds here. Both the upper frequency limit and the terminal slope outside the passband must be specified before system performance can be characterized. Either without the other is useless for predicting performance or verifying data validity.

The crucial points when thinking about a measurement system's amplitude portion of the transfer function are:

- Errors caused by high and low frequency limits in measurement systems extend farther into the system's passband than you think. The bottom line is that you generally need a lot more frequency response than you think to assure that amplitude errors are less than a specified amount. If you are not sure—find out.
- Inattention to the first item can cause serious errors. Undershoot, rise time, and frequency distortion are examples.
- In order to specify a system's performance, both the frequency limits (the −3 dB points) and the terminal slopes outside the passband must be known.

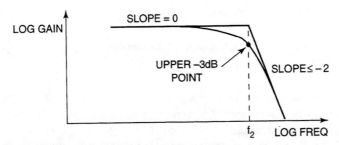

THESE SYSTEMS WILL RESPOND AS FOLLOWS:

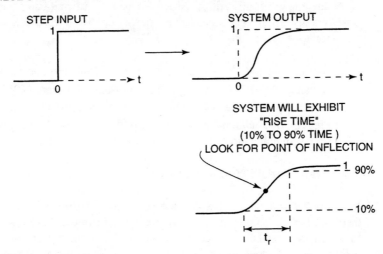

Figure 2.8   Systems with more complex high frequency roll-offs, slope ≤−2

- High and low frequency behavior of measurement systems can, for the most part, be determined by noting the system's responses to steps and pulses.
- All measurement systems display this behavior.

2.3 WHAT ARE THE SYSTEM'S FREQUENCY LIMITS?—OR—WHO SHOT THE −3 dB POINT?

System frequency limits should be specified, in the amplitude domain at least, by the frequencies at which the response is down 3 dB (−3 dB point) from the flat portion of the response curve and by the terminal slopes outside those frequency limits. With these two pieces of information at the high and low frequency limits you can, to a very close approximation, predict the system's performance. Without these two parameters you can't predict anything. Two systems with the same −3 dB points can have wildly different performance due to the effects of the terminal slopes.

So, given these clear and simple requirements:

- −3 dB points at each end of the frequency spectrum
- terminal slopes at each end

How does the vendor world view this issue? These are the people from whom you are going to buy the bits and pieces of your measurement systems or the entire systems themselves. You had best understand how they think.

Look and see how three competent vendors, who supply instruments you might use every day, approach this critical issue. We have significant and positive experience with these vendors. These vendors make high quality hardware which is in continuous use in our test laboratories. They are:

- Neff Instruments, Inc.—manufacturers of multiplexers and signal conditioning.
- Dynamics Division of Waugh Controls—manufacturers of Wheatstone bridge and piezoelectric signal conditioning, amplifiers, and filters.
- Metrum Information Storage, Inc. (used to be Honeywell Test Instruments Division)—manufacturers of analog and digital tape recorders.

2.3.1 Neff Model 620/400 Multiplexer

In Figure 2.9 you can see that on each of this multiplexer A/D converter's 512 channels there is a low-pass filter. It has a terminal slope identified (−12 dB/octave or −40 dB/decade), but the frequency is specified by saying that at the breakpoint frequency, 10 Hz, the response is 6 dB down—not 3 dB! They tell you nothing at all about the system's high frequency response. I give them a C on the low end and an F on the high end specification.

2.3.2 Dynamics 7600/7860 Bridge Signal Conditioner/Amplifier

Figure 2.10 is taken directly from the 7600 manual. You can see in paragraph 2–20 that the −3 dB frequency limits of the filter section are specified nicely. No mention is made at all of the terminal slope outside the passband. They got the other half of the specification from Neff. Dynamics gets a C on their high end specification. They make no statement about the low frequency limit since this system is DC coupled.

2.2.3 Metrum (Ex-Honeywell) Model 90 Tape Record/Reproducers

The Model 90 tape record/reproducer is a 7-, 14- or 28-track laboratory grade instrumentation tape recorder. Figure 2.11 is from Metrum's technical specifications brochure for the Model 90. Here you can see the absolutely meaningless marketeer's nonspecifications which I've outlined in the box for you. But all is not lost. Figure 2.12 shows the actual

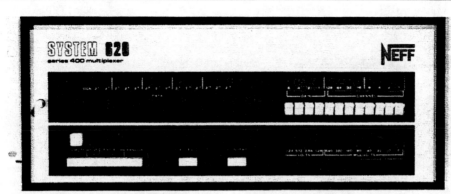

Differential Multiplexer Card (620450)

As shown in the block diagram, the Differential
Multiplexer card includes data filters and differential
channel switches for 16 channels.

Low-Pass Filter

To reject superimposed noise and unwanted signal
frequencies, the bandwidth of each channel is
controlled by a two-pole passive R-C filter providing a
terminal rolloff of 12dB/octave. The standard cutoff
frequency (−6dB) is 10Hz which provides greater
than 25dB attenuation to input noise components at
the 60Hz power line frequency.

A Differential Multiplexer card with direct input (no
filter) is also available (620451). It is equipped with
bifurcated terminals on which the user can install
filter components of his choice.

Reduced Aliasing Errors

Input channel bandwidth limiting is important when
complex signals or signals with a higher frequency
noise component are sampled at rapid sampling
rates. Signal frequencies in excess of half the channel
sampling rate, if not considerably attenuated, can
cause aliasing errors that prevent accurate
measurement of the desired signal. The input filter
greatly attenuates signal components and noise likely
to produce aliasing errors.

Integrated CMOS Switches

Input signals are coupled through the low-pass filters
to differential input CMOS/FET switches which select
the desired input channel under program control. The
integrated circuit CMOS/FET switches in the series
multiplexing circuit provide fast switching, low offset
voltage and good thermal stability. They maintain a
high degree of isolation between the analog input
circuit and digital switch drive — a feature which
minimizes the leakage current known as "pump-out".
Since pump-out current is directly proportional to
channel sampling rate, CMOS/FET switches provide a
faster, more accurate sampling of each channel.

A shunt-series switch set at the output of each
four-channel input group increases the basic isolation
of the CMOS/FET channel switches. The effect is
increased rejection of cross-talk from non-selected
input channels. Crosstalk rejection of 120dB is
obtained for input signals from DC to 100Hz.

The CMOS/FET input switches are capable of
accurately switching a wide range of input voltage
levels. Maximum full-scale input is ±10.24 volts;
however, up to ±30 volts may be applied without
damage. Overload protection of ±100 volts is
available as an option (620452).

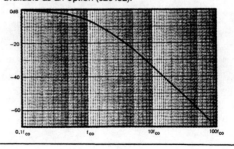

Figure 2.9   Specifications for Neff 620450 multiplexer card (Printed through the courtesy
of Neff Instruments, Inc.)

⬤ DYNAMICS

7600

2-17.   <u>GALVO LEV CONTROL</u>

2-18.   Option T provides a control, labeled GALVO LEV, for adjusting the galvo
          (main) output from 0 to 100% of full scale (±10 V). This control is
          recessed and screwdriver-operated. When present, the GALVO LEV control
          is located above the ZERO RTI control.

2-19.   <u>FILTER SWITCH</u>

2-20.   Option M provides a five-position single-pole output filter with -3-dB
          frequencies of 10, 100, 1K, and 1K Hz selected by the FILTER switch,
          which is located above the TAPE and/or GALVO controls. The fifth
          position is W.B. (wideband), which places the -3-dB frequency outside
          the range of interest and thus effectively removes the filter from the
          system.

2-21.   Option N provides an output filter similar to that of option M but with
          a seven-position FILTER switch to allow selection of two customer-
          specified cutoff frequencies. The customer-specified frequencies may
          be in the range from 0.1 Hz to 30 kHz.

2-22.   When the filter is supplied (option M or N), internal switches select
          either the tape or the galvo, or both, outputs to be filtered. These
          switches are located on the printed circuit board and are clearly
          labeled. By removing the right side cover, the switches can be operated.

2-23.   <u>OUTPUT MONITOR POINTS</u>

2-24.   Monitor points, labeled OUTPUT, are provided at the top of the front
          panel so an instrument can be connected to the amplifier for monitoring
          the output locally. The test points are connected to the galvo (or main)
          output.

Figure 2.10   Page from Dynamics 7600 Wheatstone bridge conditioner specification
(Printed through the courtesy of Dynamics Division of Waugh Controls)

Figure 2.11   Metrum Model 90 Tape Record/Reproducer specifications (Printed through the courtesy of Metrum)

Dynamic Characteristics, 600 kHz System:

Tape Speed (ips)	Bandwidth ±3 dB (kHz)	Signal rms/Noise rms (dB) ①	
		14TK	28TK
120	0.3 -600	40	37
60	0.3 -300	40	37
30	0.15-150	39	36
15	0.1 - 75	37	34
7.5	0.1 - 37.5	37	34
3.75	0.1 - 18.7	37	34
1.875	0.1 - 9.3	37	34
0.9375	0.1 - 4.7	36	33

Dynamic Characteristics, 1.5 MHz System:

Tape Speed (ips)	Bandwidth ±3 dB (kHz)	Signal rms/Noise rms (dB) ①	
		14TK	28TK
120	0.4-1500	31	28
60	0.4- 750	32	29
30	0.4- 375	32	29
15	0.4- 187	32	29
7.5	0.4- 93	32	29
3.75	0.4- 46	31	28
1.875	0.4- 23	30	27
0.9375	0.4- 11.5	29	26

Dynamic Characteristics, 2.0/4.0 MHz Systems:

Tape Speed (ips)	Bandwidth ±3 dB (kHz)	Signal rms/Noise rms (dB)① 2.0 MHz		4.0 MHz	
		14TK	28TK	14TK	28TK
240	0.8-4000	N/A	N/A	26	23
120	0.4-2000	28	25	26	24
60	0.4-1000	29	26	29	26
30	0.4- 500	30	27	30	27
15	0.4- 250	30	27	30	27
7.5	0.4- 125	30	27	29	26
3.75	0.4- 62.5	29	26	27	24
1.875	0.4- 31.25	28	25	25	22
0.9375	0.4- 15.6	25	22	22	19

① Measured at the output of a bandpass filter having 18 dB/octave attenuation beyond bandwidth limits and using recommended tapes.

FM RECORD/REPRODUCE

FM record/reproduce amplifiers may be operated in any of the IRIG ±30% or ±40% modes. All filters may be operated with either flat-amplitude or transient-response characteristics. The specifications listed are for operation in the flat-amplitude mode. In the transient mode, the amplitude at bandedge will be down 6 dB.

Total Harmonic Distortion .... With a single record-level setting, the harmonic distortion of a 0.1 bandedge frequency will be no more than 0.4% for ±40% systems; and no more than 2% for ±30% systems having a modulation index greater than 3 and 3% for modulation indexes less than 3 for all frequencies to bandedge.

Linearity .... ±0.5% of full deviation from best straight line thru zero.

DC Drift .... ±0.5% of full deviation over 8 hr and 20° F after 10 minute warmup.

Input Level .... 0.5 to 10V pk

Input Impedance......................... 20 kΩ, 600Ω, or 75Ω selectable by pin jumper.

Output Impedance 50Ω

Output Level .... 1V rms into 50Ω

Record Speed Switching .... 8 electrically selected center frequencies plus manual mode selection for up to 10 center frequencies.

Reproduce Speed Switching .... 8 solid state selected plug-in filter assemblies plus manual mode selection for up to 10 filter assemblies.

Dynamic Characteristics, ±40% FM IRIG Wideband Group I:

Tape Speed (ips)	Center Freq (kHz)	Data Bandwidth (Hz within 1.3 dB)	SNR rms (dB)	(dB)②
120	432	0-80k	50	47
60	216	0-40k	50	47
30	108	0-20k	50	47
15	54	0-10k	48	47
7.5	27	0- 5k	48	47
3.75	13.5	0- 2.5k	48	47
1.875	6.75	0-1250	47	46
0.9375	3.38	0- 625	45	43

Dynamic Characteristics, ±40% FM IRIG Intermediate Band:

Tape Speed (ips)	Center Freq (kHz)	Data Bandwidth (Hz within 1 dB)	SNR rms (dB)	(dB)②
120	216	0-40k	51	49
60	108	0-20k	51	49
30	54	0-10k	50	49
15	27	0- 5k	50	49
7.5	13.5	0- 2.5k	50	48
3.75	6.75	0-1250	49	48
1.875	3.38	0- 625	49	47
0.9375	1.68	0- 312	47	45

Dynamic Characteristics, ±40% FM IRIG Lowband:

Tape Speed (ips)	Center Freq (kHz)	Data Bandwidth (Hz within 1 dB)	SNR rms (dB)	(dB)②
120	108	0-20k	55	52
60	54	0-10k	55	52
30	27	0- 5k	54	51
15	13.5	0- 2.5k	53	51
7.5	6.75	0-1250	53	50
3.75	3.38	0- 625	52	49
1.875	1.68	0- 312	51	48
0.9375	0.84	0- 156	50	48

Dynamic Characteristics, ±30% FM Wideband Group II:

Tape Speed (ips)	Center Freq (kHz)	Data Bandwidth③ (kHz + 1,-2 dB)	Data Bandwidth③ (kHz + 1,-3 dB)	SNR rms (dB)④
120	900	400	500	36
60	450	200	250	35
30	225	100	125	34
15	112.5	50	62.5	33
7.5	56.25	25	31.2	33
3.75	28.125	12.5	15.6	32
1.875	14.062	6.25	7.8	31
0.9375	7.031	3.125	3.9	28

② SNR rms when ±40% FM is used with 28 track heads.

③ IRIG reference frequency 1 kHz

④ Applies to 50 mil track widths, degrade 2 dB for 25 mil track heads.

MAGNETIC HEADS

Configurations ....¼-inch 4 tracks, ½-inch, 7 tracks and 1-inch 14 tracks per IRIG 106-75 with annotation tracks.1-inch, 28 track IRIG format, 25 mil track width. Other formats available upon request

Gap Scatter .... 100 μinch band

Interstack Spacing .... 1.500 ±0.001 inch

Gap Azimuth .... Record heads perpendicular to mounting plate ±1 min arc. Reproduce head azimuth adjustable ±12 min arc.

Head Warranty ....5 years or 200,000,000 feet of tape irrespective of mode or speed on standard density IB or WB heads.

ORDERING INFORMATION

The Model 90 Magnetic Tape Recording System is available with many combinations of data electronics and accessories. Contact your nearest METRUM Sales Engineer to assist you in selecting the configuration suited to your particular requirements.

Figure 2.12 Metrum Model 90 Tape Record/Reproducer specifications (Printed through the courtesy of Metrum)

technical specifications on the device for several types of FM electronics. Four configurations are quoted:

1. 40% Deviation Wide Band Group I
   "Data bandwidth . . . within 1.3 dB (author's note: 15% error!)"
2. 40% Deviation Intermediate Band
   "Data bandwidth . . . within 1 dB (author's note: 10% error!)"
3. 40% Deviation IRIG Lowband
   "Data bandwidth/ . . . within 1 dB (author's note: 10% error!)"
4. 30% Deviation Wideband Group II
   "Data bandwidth . . . +1, −2 dB (author's note: +10%–20% errors)"
   and, "Data bandwidth . . . +!, −3 dB (author's note: +10% error, −30% error)"

Metrum, at least occasionally, admits that their tape recorders are not flat anywhere in their published passband for certain configurations. This is the mark of a conservative manufacturer. Their hardware will beat these specifications every time—but these are the errors which are allowed by a recorder operating within specification.

But notice the confusion. The specification is not even consistent among different configurations of the same instrument from the same manufacturer. And nowhere is there a statement about terminal slopes. We haven't even mentioned direct record/reproduce yet! Metrum gets a D on this specification form.

2.3.4 Metrum Model RSR 512 Digital Storage Recorder

Metrum recently came out with their long-awaited, compact digital tape recorder/reproducer, the Model RSR 512 shown in Figures 2.13 and 2.14. It's digital—it must be better! Right? We'll see.

In Figure 2.13 you can see the effect of the marketeers. The recorder has "DC to 80 KHz record/reproduce." It also has—are you ready?—"Flat frequency response." This is utterly useless noninformation from a design perspective. You simply cannot design with this noninformation. You have to turn the page to Figure 2.14.

Buried in the section on available antialiasing filters are the words "Selectable flat amplitude (4 Pole Chebyschev) or transient response (linear phase) characteristics." No statement whatsoever about −3 dB points or terminal slopes. Metrum also flunks the course here and gets an D on this specification form. And remember—this confusion is coming from *reputable* vendors of good hardware.

2.3.5 What is the Message?

The message is that there is neither stardards nor consistency in the field about how frequency response is specified. Be very careful when you talk about this issue with vendors, or even among your design team. Be sure you know what you are talking about on this

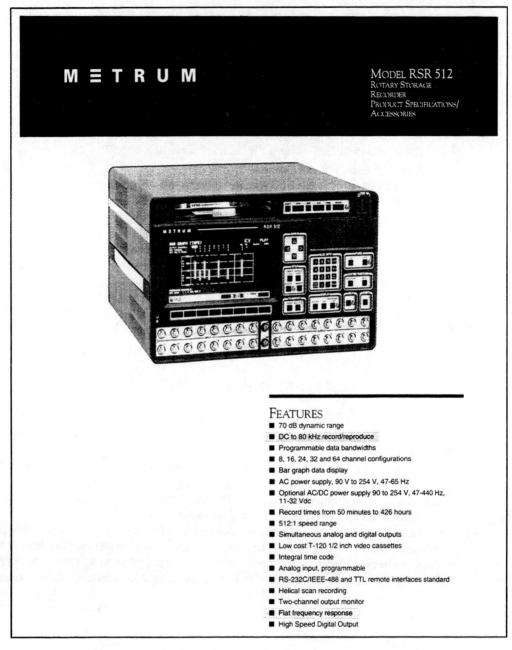

Figure 2.13 Metrum Model RSR512 Rotary Storage Recorder specifications (Printed through the courtesy of Metrum)

STANDARD OPTIONS

- Rack mount kit
- Remote control unit
- Vibration mount
- Voice kit, including microphone/headset
- Auxilary housing
- High Speed Digital Output (HSDO)

GENERAL

The RSR 512 has combined the characteristics of FM recording with the accuracy of digital recording. It accepts data in analog form, converts to digital for recording on the tape cassette and provides analog and digital outputs.

Size ....10.5 in (273 mm)
  H x 17.75 inch (450 mm) W x 22.5 inch (572 mm) D

Weight..Average 16 x 16
  channel record/reproduce system approx. 70 pounds (32 kg)

Power requirements
AC only ....90 to 254 V,
  47-63 Hz

Optional
AC/DC..90 to 254 V,
  47-440 Hz, 11-32 Vdc

Power consumption ....250 W typical

TAPE TRANSPORT

Recording/reproduce method ....PCM recording

Inputs ....Analog

Outputs ....Analog and digital

Tape ....EIAJ standard
  magnetic T-120 1/2 inch video cassette tape. METRUM T-120 certified tape cassette required to satisfy specifications and warranty.

Recording format ....Helical scan

Record/reproduce head.............................Quad-head, azimuth
                                                                     system

Erase head ....Full width

Tape speed 512:1 (80.8mm/sec
  to .16 mm/sec) tape speed automatically selected for optimum recording time; or can be manually set by user, extending time compression and expansion in the reproduce mode.

Recording time...50 minutes to over
  426 hours. Recording time based upon number of data channels, data bandwidth, and sample density.

Start/stop time..Approx. 5 sec from
                                                                        launch point

Fast/forward/rewind time ....Approx. 170 sec

Controls...F. FWD, REW, PLAY,
  STOP, REC, EJECT

SHUTTLE

Shuttle and search ....Shuttle and Search
  from: Test Name, Record Number, Date, Time, Stopwatch Time, Event Number, or Block Number.

DATA RECORD/REPRODUCE SYSTEM

Quantization..12 bit

Error correction ....Double Encoded
  Reed-Solomon combined with data interleaving.

Available filters..80 kHz, 40 kHz,
  20 kHz, 10 kHz, 5 kHz, 2.5 kHz, 1.25 kHz, 625 Hz, 312 Hz, 156 Hz, 78 Hz, 39 Hz, 20 Hz. Selectable flat amplitude (4 Pole Chebyschev) or transient response (linear phase) characteristics. Filters are for input anti-aliasing and output smoothing. User programmable and selectable.

Sample density ....4, 8, or 16 samples
  per bandedge cycle, user programmable. The sum of bandwidth x sample density for an 8-channel system < = 640 k and < = 1280 k for 16 or more channels.

Number of channels....................................8, 16, 24, 32 or 64

Data ....Channels are in
  modules of 8. The system accepts up to 4 modules. A 2-channel reproduce monitor is supplied as standard equipment.

Input/output..Front BNC; rear
  multipin connector. (Mating Connector Cannon DB-25P-K87, 16825952-003.)

Analog Input
Input voltage ranges ....Seven ranges, user
  programmable. ±0.1 V, 0.2 V, 0.5 V, 1.0 V, 2.0 V, 5.0 V, 10.0 V peak.

Max safe input voltage ....± 25 Vdc or peak ac

Gain accuracy...± 0.5% of full scale
                                                                      maximum

Over-range...± 102.4% of full
                                                                    scale input

Over-range warning ....Warning at approx.
                                                                       ± 94% of full scale

Input impedance ....1M ohm shunted by
                                                                      < 100 pF Differential

Source resistance ....Less than 1 k ohm

Analog Output
Output voltage range ....Two ranges, user
  programmable, ±1 V, or 2 V peak into 50 ohms.

Output current...50 mA max (short
                                                                     circuit protected)

Output impedance ....< 0.2 ohm (BNC)

Signal rms to noise rms ratio ....> 70 dB

Linearity ....± 0.2%

Frequency response ....< 1 dB (typical 0.3 dB)

Total harmonic distortion............................< 0.3%

Drift ....± 0.3% per 10° C
  change. (After 15 minute warmup)

Phase difference between
  channels (Skew)......................................< 3° (typical using
  same bandwidth filters)

Note: All specifications apply at 8 x sample rate, and excluding (±0.1 V) input range.

Figure 2.14 Metrum Model RSR512 Rotary Storage Recorder Specifications (Printed through the courtesy of Metrum)

issue. It will become clear very quickly if the other person knows what they are talking about. Most vendors do not adhere to the most meaningful way of specifying frequency response—not even the vendors with the best reputations. There is much confusion here— BE CAREFUL.

The proper questions are:

- What are the −3 dB frequencies and is there a plus and minus tolerance?
- What are the terminal roll-off slopes outside the passband?
- What is the overall and complete transfer function?

2.4 PHASE PORTION OF A MEASUREMENT SYSTEM'S TRANSFER FUNCTION

The second portion of a measurement system's transfer function is the phase response. How does a measurement system handle the phases of various frequencies in its input wave-shape?

2.4.1 Time Delay

It is impossible for information in a system's input waveshape to flow through the system to its output in zero time. This would imply that the speed of the electrical impulses within the system was infinite, which is physical nonsense. It would also imply that the measurement system had extrasensory perception—and that is not yet available as an option from the vendor. The last thing you want a measurement system to have is extrasensory perception! You don't even want a system to have an opinion. You want it to only report faithfully what happened.

What would have to happen for the waveshape at the output to be the same as the waveshape at the input? That, after all, is the fundamental job of most measurement systems. For the purposes of this discussion, assume that the system's amplitude portion of the transfer function is flat over the frequency range of interest, therefore adding negligible errors.

What would have to happen is that *all frequencies in the input waveshape would have to be delayed through the measurement system by the same amount of time*. You can see that if this amount of delay were a variable function of frequency, the phase relationships among these frequencies would change from the input to the output and the waveshape would change (become invalid).

In Figure 2.15 this relationship is drawn for just one of the frequencies in the input waveshape—the one a frequency f. Both the input sine wave and the output sine wave are shown. They are shown plotted against the time and phase axes. The period of the sine wave is T seconds, where $T = 1/f$. The period, on the phase axis, is 360 degrees or 2π radians. There is a time delay shown for the output sine wave, t. The delay in the phase axis

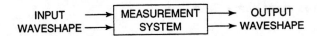

FOR A SINGLE FREQUENCY, f:

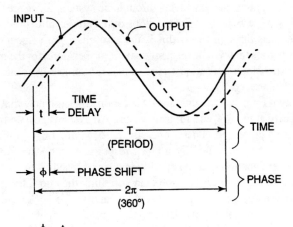

$$\frac{\phi}{2\pi} = \frac{t}{T} = tf$$

$$\boxed{\phi = 2\pi t f = (2\pi t)f} = (\text{CONSTANT}) \times (\text{FREQUENCY})$$

LINEAR PHASE SHIFT LAW

Figure 2.15   Phase shift versus frequency-linear phase shift law

is an angle, Phi. You can further see that the ratio of the delay to the total length of one period of the sine wave in either axis can be expressed on either axis and must be the same number.

Ratio on phase axis = Ratio on time axis
Phi/2π = t/T

Since the period on the time axis, *T*, is equal to 1/*f*, the expression can be rewritten in this manner:

Phi/2π = tf

Solving for the phase delay angle, Phi, you get

Phi = (2π) (t)f
Phase delay = Phi = (constant) (f)

Since we want the time delay, *t*, to be the same for all frequencies, (2π)(t) becomes a constant. You are left with an expression which says that *you want the phase shift at any frequency to be a constant times that frequency*. This is the linear phase shift law.

2.4.2 Implications of the Linear Phase Shift Law

In Figure 2.16 you can see the phase shift versus frequency characteristic for some candidate measurement system. Note that over some frequency range this characteristic is linear. The slope in this linear region is 2π times the time delay through the system. Since the time delay through the system cannot be zero, neither can the slope of the phase shift curve. But it can come very close. It may pass through zero—this is the zero phase shift frequency, which has some interesting properties that will be discussed later. Furthermore, for the most part you don't even care what this slope is. You just require that it be linear over the frequency range in which you are interested.

Figure 2.16 has been drawn for a system with non-DC response. Please note that the phase shift deviates from linearity at the low end for this system. All systems deviate from phase linearity at the high frequency end. Phase linearity here would imply infinite bandwidth. Be aware that, in my experience, nonlinear phase response errors occur further into the systems's overall transfer function than do amplitude errors caused by amplitude distortion. They occur further into the transfer function than you think they do. This must be taken into account in the system design and will be discussed later in this book.

The bottom line is this: a system that displays linear phase shift with input frequency over a frequency range of interest will not display phase distortion over that frequency range. Any nonlinearity between phase shift and frequency will cause the output waveshape to be distorted if there is more than a single frequency present. This effect is called phase distortion.

2.4.3 A Better Way to Look at Phase Shift

Phase shift curves should be plotted in a linear-linear fashion for interpretation. Such a curve plotted linear-log, or worse log-log, is impossible to interpret. So, how do many manufacturers present their phase shift data? Manufacturers who do not understand the importance of this information and how it is used present it with logarithmic frequency. You,

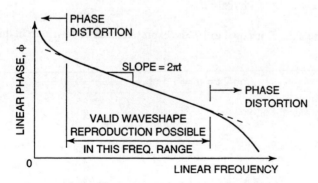

ONLY WHERE PHASE SHIFT IS A LINEAR FUNCTION OF
FREQUENCY IS VALID WAVESHAPE REPRODUCTION
POSSIBLE.

Figure 2.16   Linear phase shift

then, have to replot it linear-linear before you can make any judgments about it. This is a pain in the neck. A disadvantage of having to look at phase linearly, however, is that it now cannot be overlayed on the amplitude-frequency curves which are almost always logarithmic. So you need two pictures.

There is a better way—and here the electrical engineers have had the best of it over the mechanical engineers. If you differentiate the phase as a function of frequency characteristic, and form dPhi/*df*, it will have units of time, seconds. This is shown in Figure 2.17. These seconds are the *time delay* through the system. Where the phase-frequency characteristic is linear, the delay will be constant. If you are interested in direct waveshape reproduction, that is what you are looking for—constant time delay.

This characteristic may now be plotted logarithmically and compared directly to the amplitude portion of the transfer function on the same plot. You can see data presented in this fashion in Figure 2.18, which is data from Frequency Devices, Inc., for their LP01 elliptic filter. You can see here that this filter exhibits highly nonconstant delay within its passband and would, therefore, highly distort a waveshape passing through it with frequencies above 20% of its upper frequency limit.

Figure 2.19 shows the results when you get all done—with the amplitude and phase (in the form of time delay) overlayed on the same logarithmic plot for comparison. This is the most perceptive and, for once, easiest way to look at these characteristics. These data are for Frequency Devices LP03 Constant Time Delay filters. Note that the time delay for these filters is constant, showing absolute phase linearity, *past* the filter's upper frequency limit. This is superlative performance if waveshape reproduction is the criteria. This particular filter choice is the present optimum choice for waveshape filtering and most anti-

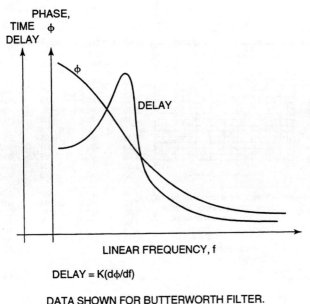

DELAY = K(dφ/df)

DATA SHOWN FOR BUTTERWORTH FILTER.   **Figure 2.17**   Time delay

FREQUENCY DEVICES™

Model LP01

8-Pole 6-Zero Elliptic Low-Pass Filter

NOTE

Description

The LP01 is an 8-pole 6-zero elliptic low-pass filter with a theoretical passband ripple of ± 0.035 dB. Its transfer function is a modified Cauer elliptic function that has been designed by FDI to minimize section Qs, delay and delay variation in the passband while maintaining an 80 dB .035 dB shape factor of 1.77, an 82 dB stopband floor and a 2-pole monotonic rolloff at high frequency.

Specifications

Transfer Function	8-pole 6-zero Elliptic Low-pass
Passband Ripple (0 - f_r) (theoretical)	± 0.035 dB
Stop Band Attenuation	80 dB
Cutoff Frequency f_r	± 2%
Amplitude	-0.035 dB
Phase	-323.5°

Phase Match [1]
0 - 0.8 fr	± 2° max
	± 1° typ
0.8 fr - 1.0 fr	± 4° max
	± 2° typ

Filter Attenuation (theoretical)
0.035 dB	1.00 f_r
3.01 dB	1.13 f_r
60.0 dB	1.67 f_r
80.0 dB	1.77 f_r

Amplitude Match [1]
0 - 0.8 fc	± 0.2 dB max
	± 0.1 dB typ
0.8 fc - 1.0 fc	± 0.4 dB max
	± 0.2 dB typ

Total Harmonic Distortion [2]	-90 dB typ
Broad Band Noise [2]	200 μV_{rms} typ
Narrow Band Noise [2]	10nV / $\sqrt{Hz}$ typ

Note 1: Channel to channel match for the same transfer function set to the same frequency and operating configuration.

Note 2: See text for definition and description.

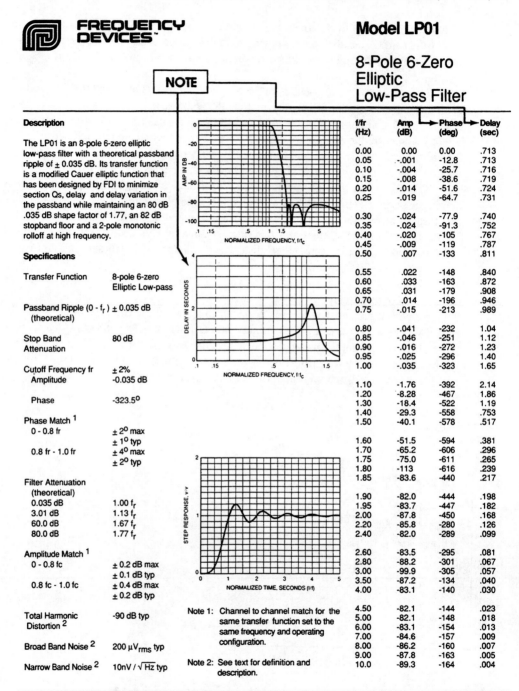

f/fr (Hz)	Amp (dB)	Phase (deg)	Delay (sec)
0.00	0.00	0.00	.713
0.05	-.001	-12.8	.713
0.10	-.004	-25.7	.716
0.15	-.008	-38.6	.719
0.20	-.014	-51.6	.724
0.25	-.019	-64.7	.731
0.30	-.024	-77.9	.740
0.35	-.024	-91.3	.752
0.40	-.020	-105	.767
0.45	-.009	-119	.787
0.50	.007	-133	.811
0.55	.022	-148	.840
0.60	.033	-163	.872
0.65	.031	-179	.908
0.70	.014	-196	.946
0.75	-.015	-213	.989
0.80	-.041	-232	1.04
0.85	-.046	-251	1.12
0.90	-.016	-272	1.23
0.95	-.025	-296	1.40
1.00	-.035	-323	1.65
1.10	-1.76	-392	2.14
1.20	-8.28	-467	1.86
1.30	-18.4	-522	1.19
1.40	-29.3	-558	.753
1.50	-40.1	-578	.517
1.60	-51.5	-594	.381
1.70	-65.2	-606	.296
1.75	-75.0	-611	.265
1.80	-113	-616	.239
1.85	-83.6	-440	.217
1.90	-82.0	-444	.198
1.95	-83.7	-447	.182
2.00	-87.8	-450	.168
2.20	-85.8	-280	.126
2.40	-82.0	-289	.099
2.60	-83.5	-295	.081
2.80	-88.2	-301	.067
3.00	-99.9	-305	.057
3.50	-87.2	-134	.040
4.00	-83.1	-140	.030
4.50	-82.1	-144	.023
5.00	-82.1	-148	.018
6.00	-83.1	-154	.013
7.00	-84.6	-157	.009
8.00	-86.2	-160	.007
9.00	-87.8	-163	.005
10.0	-89.3	-164	.004

Figure 2.18 LP01 Elliptic Filter Specifications (Printed through the courtesy of Frequency Devices, Inc.)

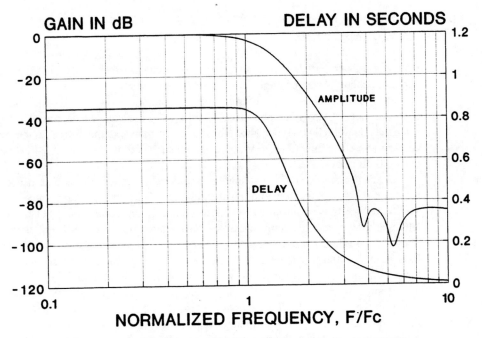

Figure 2.19   8-P, 6-Z constant delay filter (Frequency Devices, LP-03)

aliasing applications. It is the standard filter transfer function for critical and optimum filtering applications in the Measurements Engineering Department.

2.4.4 The Ultimate Zinger

Earlier, I discussed the amplitude portion of the transfer function as if there were no phase shift problems. Here, we've discussed phase shift as if there were no amplitude problems. I have discussed them as if they were separable.

In truth, we've got a chicken or the egg problem. Departures from phase linearity are caused by nonzero slopes in the amplitude response. The reverse is also true. You cannot have one without the other. You cannot say, "I'll just change the amplitude response on this thing a little bit by tweaking this resistor over here. The phase is OK right where it is." That design decision is not open to you. You cannot change one without changing the other. They are simply different mathematical pictures of the same phenomenon. It is for this reason that it is difficult to demonstrate an example of phase distortion in the analog world without accompanying amplitude distortion. I have never seen such a demonstration. If you know of a good one—let me know!

It is also a fairly difficult design criteria to work with since there are no standards as to how phase distortion errors should be quantified in an error budget. It is also harder to conceptualize phase-induced errors in a waveshape. So, it is a sort of local option—like winter rules in golf. But, if you are interested in waveshape, it must be taken into account.

2.5 TRANSFER FUNCTION OF A TYPICAL MEASUREMENT SYSTEM COMPONENT

How does this amplitude and phase information come together for a real measurement system component? As an example, we'll take a commercial passband filter, the Krohn-Hite Model 3700 filter set, as a model for an entire measurement system. This filter has two filter positions called "Max Flat" and "RC." Neither of these names is technically appropriate since they do not tell you much about the real performance of the device. In the "Max Flat" configuration, the filter has a terminal slope on each end of +/-4, or +/-24 dB per octave, or +/-80 dB per decade.

Let's use this filter as an example of a real measurement system and inspect the ramifications of its high and low end transfer function on its ability to reproduce waveshapes. To do this, we'll use the following scenario. A customer comes into your laboratory and tells you they would like some vibration data played back from FM/FM magnetic tape. They are not interested in frequencies lower than 50 Hz, nor higher than 5000 Hz. Please filter them out and provide waveshape data accurate to 1% for the frequencies between. Simple enough? By now you should be able to see this one coming.

I am now going to lead you down a very well traveled primrose path. Your response is to pull out the Model 3700 bandpass filter set, because it's the only one sitting on the shelf, and set its lower −3 dB frequency at 50 Hz and its upper −3 dB frequency at 5000 Hz. After all, that's just what the customer asked for, right? The amplitude portion of the filter transfer function shows in Figure 2.20. The terminal roll-offs show slopes of four. Note that at the −3 dB frequencies, you're already making 30% uncorrected amplitude errors! If you

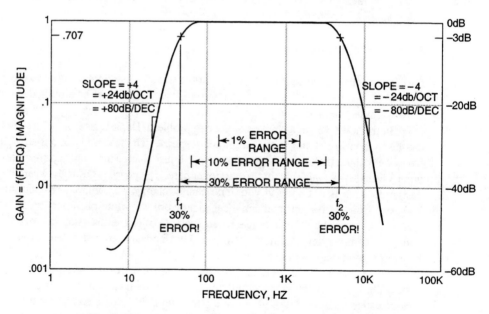

Figure 2.20   Krohn-Hite model 3700 filter set "maxflat" 50–5KHz

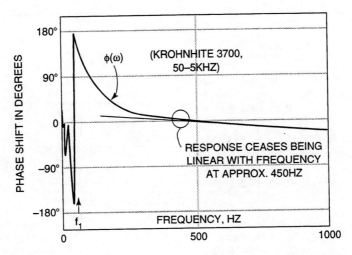

Figure 2.21   Filter phase response low end (Krohn-Hite 3700, 50–5KHz)

could live with 10% amplitude errors, you would be down to a frequency range of 60–4600 Hz. To provide the customer's 1% amplitude error budget, you're down to a frequency range of 200–1500 Hz! Amplitude errors larger than your 1% budget are occurring everywhere else.

Let's look at phase. Figure 2.21 shows the lower end of the filter's phase response. Note that it deviates from phase linearity at 450 Hz. Below this frequency the phase is nonlinear, preventing valid waveshape reproduction. Figure 2.22 shows the phase response at the high end. Phase nonlinearity begins at 3200 Hz. The overall picture is that the phase

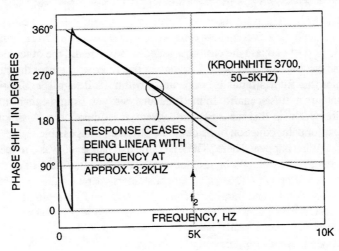

Figure 2.22   Frequency phase response high end (Krohn-Hite 3700, 50–5KHz)

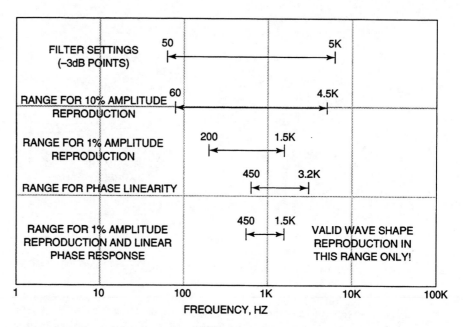

Figure 2.23   Frequency range for waveshape reproduction (Krohn-Hite 3700 filter set)

is linear only from 450–3200 Hz for this filter with the frequency limit dials set at 50–5000 Hz!

Figure 2.23 sums up our predicament. At the top are our filter frequency limit settings, 50 and 5000 Hz. Lower down is a line representing the frequency range over which we can hold the 1% amplitude error tolerance, 200–1500 Hz. Below that is the phase linearity line, from 450 to 3200 Hz. Only where we have amplitude errors of less than 1% *and* phase linearity can we meet the customer's request: valid waveshape reproduction. These two criteria only overlap between 450 and 1500 Hz, or a 1050 Hz bandpass. That represents 21% of the bandpass the customer asked for! Where did the other 79% go? It went into the trash can because over these frequencies this filter won't meet the customer's criteria. This is not this filter's fault. It is operating within its design specifications. It is our fault as amateurs in this scenario. In the final analysis, you simply cannot meet the customer's requirement with this measurement system. It is impossible. A simple, straightforward request for data reduction turns out to be physically impossible.

What happened here? The first thing that happened was a request for filtering from an uneducated customer who did not understand the ramifications of the request. After all, a filter is a filter, isn't it? Most of your customers operate at this level of measurement sophistication. The second thing that happened was a fatally simplistic response to an unspecific request. Neither party in this interaction operated in a professional manner responsive to what the customer really needed. The problem is that this level of interaction happens 50,000 times a day in this country alone.

The most important aspect of this scenario is a realization that the effects of the transfer function *outside* a measurement system's passband extend much further *inside* the passband than 99% of the users in the world think. The message is that you usually need much more bandwidth than your customer thinks you do.

2.6 SYSTEM RESPONSES TO STEP INPUTS: THE INDICIAL RESPONSE

There is another area of concern regarding the measurement system's transfer function. This consideration is of particular importance when specifying filters for use in suppressing aliasing frequency components prior to digitization, or merely performing filtering on the data for mechanical reasons.

This particular concern does not leap off the page at the measurement system designer or user when viewing the two parts of the system's transfer function—the amplitude and phase as functions of frequency. This concern is the system's response to a step change in level at its input. This is termed the system's *indicial response* and shows in Figure 2.24.

The primary importance of this response is when the data of interest resemble pulse trains, or have rapid steplike rising or falling edges. Examples of interest might be investigations of the firing pulse forms from ordnance channels on a deployment test, investigations into a telemetry waveform, or shock propagation studies.

It is also of importance in sampled (digital) systems because the indicial response is needed if data is to be corrected in the time domain for a system's imperfect transfer function. This is a way to extend an otherwise inextendable transfer function. If you can't design in enough low frequency response to capture the low frequency portion of the input wave-

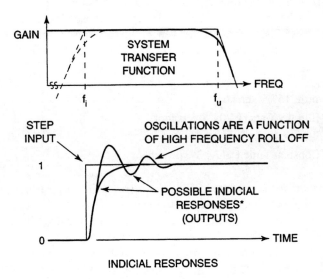

INDICIAL RESPONSES

* SECOND ORDER SYSTEMS SHOWN

Figure 2.24 System response to a step input: the indicial response

form or you get surprised by that waveform, you may be able to correct the problem in the digital domain by performing a time domain deconvolution on the data using the system's indicial response. Likewise, if you can't buy enough high frequency response or if you get surprised, you may be able to extend this range by correcting the data in the same manner using, again, the indicial response.

2.6.1 Powerful Methodology

When a step is input to a measurement system, the system's output defines the electrical transfer function entirely. Both the low frequency and high frequency portions of the transfer function are defined. The low frequency portion is defined by the system's reaction to the DC level in the step after the rise from zero. The high frequency portion is defined by the way the system reacts to the high frequencies causing the rise from zero to the fixed level.

This is a very simple yet powerful and perceptive methodology. Using this method, we can both calibrate and define the measurement system's electrical transfer function on-line, at the same time. The only caveat is that the installed transducer's mechanical transfer function is generally not verified by this method. The method is used in certain of our measurement systems such as the digital Transportation Measurement System used to monitor spacecraft motions during transit to the launch site. We use it all the time during system checkout and trouble shooting.

2.6.2 Examples of System Indicial Responses

Filter indicial responses.   Figure 2.25 shows indicial responses from four Frequency Devices, Inc.[1] low-pass filters. The data shown here is characteristic for these filter types. If you bought one of these filters from another vendor, this response must look exactly the same or the vendor is cooking the technical books! The overshoot results are as follows:

- LP00 8 Pole Butterworth: 16% overshoot
- LP01 8 Pole, 6 Zero Elliptic (Cauer): 20% overshoot
- LP02 8 Pole Bessel Filters: .4% overshoot
- LP03 8 Pole, 6 Zero Constant Time Delay: 3% overshoot

Remember, there is no overshoot in the input step. This *error* is created in this measurement system component. These overshoot numbers, varying from almost zero to 20%, *represent pure error* in the output waveform—presuming such error is not corrected digitally after the fact. This correction is possible as shown at the end of this section, but should probably be left to the professional with lots of time and money on his hands. As can be seen from these data, the filters of choice based *on this characteristic alone* would be the constant delay filter followed by the Bessel filter—again presuming no posttest data correction will

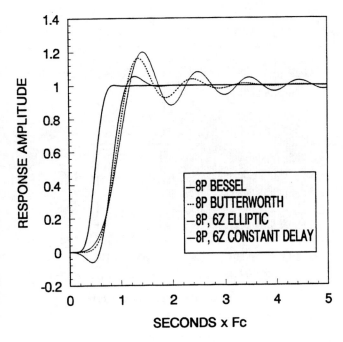

Figure 2.25   Filter step (indicial) responses

be done (i.e., do it right the first time). Both of these have the added advantage of linear phase responses. Both filter types are in daily use in our measurement systems. They do have the disadvantage of having less amplitude rejection above the cutoff frequency than the Butterworth or elliptic filters. The simple truth is that there is no *best* filter type. There is only a best filter type for today's application.

Pluse train calibrations.   Figure 2.26 shows a digital recording of a square wave pulse train used for automated, internal calibration and transfer function verification in a digital transportation measurement system. A square wave is merely a series of plus and minus step inputs. Here a 1 Hz square wave of nominal amplitude ±1Gpk is used to document system span and verify the transfer function at the same time.

An indicial response that flunked.   We recently procured a 16-channel state-of-the-science digital dynamic analyzer for use in doing analyses associated with dynamic testing. This device went out for competitive bidding among eight vendors.

Figure 2.27 is the response from one of the losing, but reputable, vendors. Buried in his technical response was the statement that his hardware used 7th order Chebyschev antialiasing filters. No filter specifications were provided. Chebyschev filters are a form of elliptical filters with known nonlinear phase and lousy indicial responses. We asked the vendor to provide us with the filter's indicial response. Figure 2.27 is it. Note that it shows a *17% overshoot* to a step. The vendor lost this $150K procurement on this specification—which, by the way, had never been asked for by any other potential client!

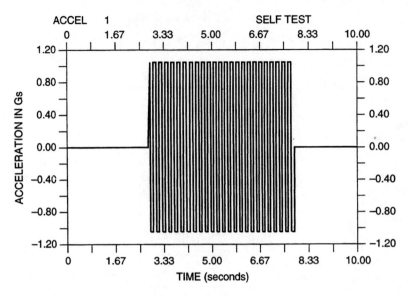

Figure 2.26   Pulse train calibrations for a transportation measurement system

2.6.3 Effects of Filter Indicial Responses on Actual Data

Figure 2.28 is a valid acceleration waveform obtained during a separation test of a large payload fairing in vacuum. The waveform is shown with a full scale of ±200Gpk and a time scale of 300 milliseconds. These data are valid from 2 to 300 Hz. For this test, that was considered wide band data covering the radial, rigid body acceleration of the fairing and its first 30 flexible body modes.

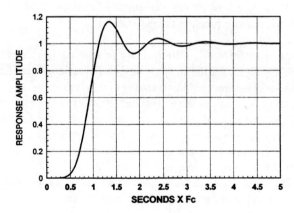

Figure 2.27   7th order Chebyshev filter indicial response

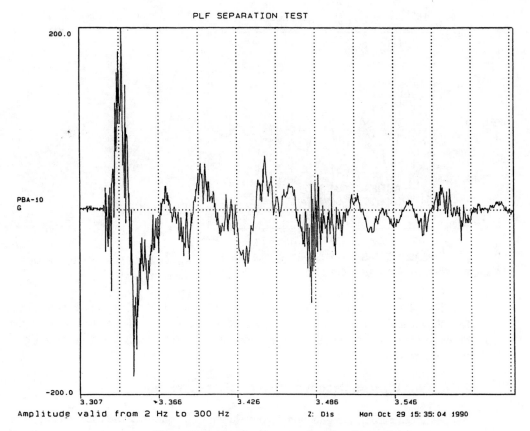

Figure 2.28   Acceleration waveform from a separation test

Constant time delay filters were used to recover the low level, rigid body acceleration which is buried in this waveform. They were used because of their optimum transfer functions (flat frequency response, linear phase response) for waveshape reproduction and their acceptable indicial response (3% overshoot). Figure 2.29 is a waveform result from trying to recover the valid lowest 15 Hz of this waveform. Please note the artifact on the filtered waveform. This artifact is put there by the use of the constant time delay filter and is not in the valid wide band data. The measurements engineer must know this if the data are to be validated.

In reviewing Figure 2.25, you can see that this artifact exactly mimics the indicial response of the 8 Pole, 6 Zero constant time delay filter in the lower right hand corner (look at zero time plus two divisions). This response has been caused by the very rapid rise of the data out of the system prefiring noise levels with the onset of motion. This looks to the measurement system just like a step input. The filter responds in the only way it can, by adding its own indicial response to the filtered data.

Figure 2.30 is a low-pass filtered waveform of the firing pulse from this separation

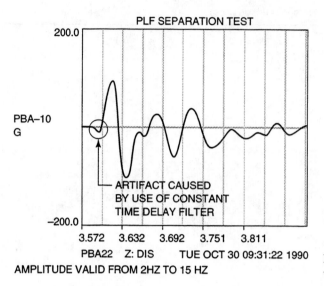

PLF SEPARATION TEST

ARTIFACT CAUSED
BY USE OF CONSTANT
TIME DELAY FILTER

PBA22   Z: DIS   TUE OCT 30 09:31:22 1990
AMPLITUDE VALID FROM 2HZ TO 15 HZ

Figure 2.29 Filtered acceleration waveform

test. This plot was run to investigate a suspected "glitch" in the firing pulse, which is, in fact, shown about one-third of the time after the pulse comes up. Please note, however, that the use of the same filter as above has installed artifacts on the leading and trailing edges of this pulse waveform.

2.6.4 Time Domain Deconvolution for Data Correction

As mentioned earlier, time domain deconvolution techniques may be used in certain circumstances to correct data for inadequacies in the measurement system's transfer function.

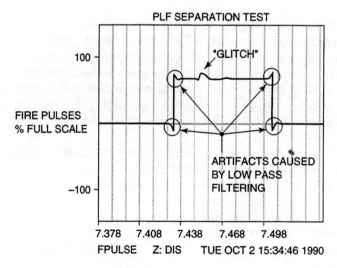

PLF SEPARATION TEST

"GLITCH"

FIRE PULSES
% FULL SCALE

ARTIFACTS CAUSED
BY LOW PASS
FILTERING

FPULSE   Z: DIS   TUE OCT 2 15:34:46 1990   **Figure 2.30** Filtered firing pulse

This is illustrated well in Bickle's[2] body of work from Sandia Laboratories in the early 1970s. To my knowledge, this body of work has not been surpassed since.

Figure 2.31 shows the indicial response from a system whose transfer function clearly shows a first order response at the low frequency end. It shows the exponential decay characteristic of first order systems with low frequency slopes of +1. This system would not respond to DC.

Figure 2.32 shows the system's response to a step input (true input) with noise thrown in for good measure, and its actual output (response plus noise). Note that, predictably, the system is incapable of handling the input's DC level due to its inadequate low frequency response, and it decays exponentially as it must.

By deconvolving the system output (response plus noise) with the system's indicial response (Figure 2.31), a corrected response can be developed which is also shown in Figure 2.32. A second example of this powerful correction method is shown in Figure 2.33 for a ramp input plus noise.

Although the transfer function problems defined in these examples occurred at the low frequency end, the method is generic and works (with certain general caveats) for high frequency inadequacies as well. The key to this method of data correction is the capture of the measurement system's indicial response as the basis for the correction.

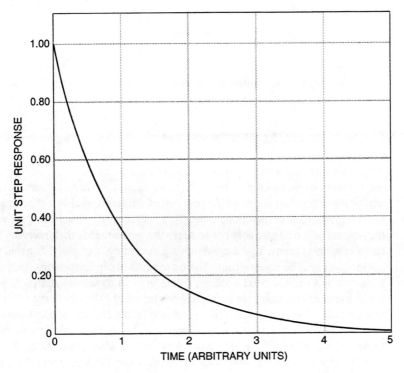

Figure 2.31 Indicial response of 1st order system with low frequency slope of +1

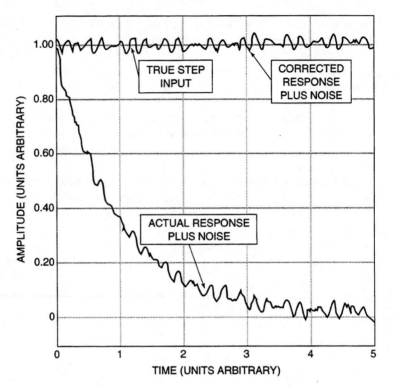

Figure 2.32   Corrected response to step using time domain deconvolution

2.7 RESONANT MEASUREMENT SYSTEMS

All the previous information in this chapter on transfer functions and linearity has been for nonresonant, or monotonic, systems. Monotonic systems have frequency responses whose amplitude never rises above the flat portion of the response. Most mechanical systems are, however, resonant or nonmonotonic. That includes the structures you will test and the transducers you will probably use to sense the measureands that interest you. In almost all cases cabling systems, signal conditioning, recording, sampling if included, and analysis components will be nonresonant. There are some exceptions to this such as galvanometer recorders and narrow band tracking filters used in dynamics analysis. So few electrical components of measurement systems show resonant behavior at the macroscopic level of interest to the user, that the rest of this section focuses on transducer behavior alone.

Resonant systems are inherently discriminatory to certain frequencies. At very low frequencies they have frequency responses that are almost flat. At other frequencies they show large amplification (resonance) or suppression (antiresonance) of input frequencies. In short, they ain't flat.

Figure 2.34 illustrates this. At the top is the amplitude portion of the transfer function

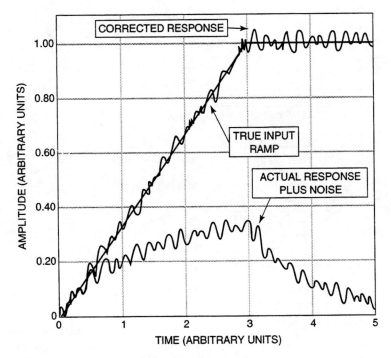

Figure 2.33 Corrected response to ramp using time domain deconvolution

of a typical nonresonant measurement system or system component. Instrumentation amplifiers, signal conditioners, FM/FM tape recorders and fiber optics oscillographs act like
this at reasonable measurements engineering frequencies. At the bottom is the amplitude
portion of the transfer function of a single degree of freedom (one mass, one spring, and one
damper in its mechanical engineering equivalent) resonant system. As the system excitation
or data frequency increases from some nominally low value, the system starts out being flat
(e.g., having constant gain). As the frequency approaches the first resonant frequency, the
gain increases rapidly to a peak at the damped natural frequency, f_d, and then decreases a
slope of -2. This figure shows only a single degree of freedom system.

In reality, engineering sturctures are distributed, multidegree of freedom systems
having multiple resonances that are not shown in this figure. This is the way the mechanical
world works—including the test structures you'll be working with and your mechanical
transducers for the measurement of motion, load, pressure, strain, deflection and its first
two derivatives, rotation, rotational rate, torque and so on. This behavior is ubiquitous
in transducers. It can occur at 0.1 Hz in certain large structures, at 30 Hz in gimballed
rate gyros, or at 1,000,000 Hz in certain piezoresistive shock acceleration transducers. A
thorough understanding of the behavior of the single degree of freedom system is the
foundation on which rests the entire field of dynamics and the measurement of dynamic
phenomena.

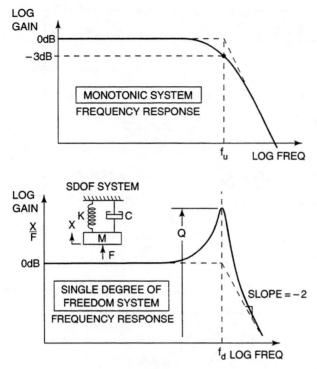

Figure 2.34 Comparison of frequency responses of monotonic and single degree of freedom resonant systems

2.7.1 Very Little Theory

Remember the single degree of freedom systems from your first vibrations class, if you were lucky enough to have one? This is a vibratory system with a single mass moving in one direction, with a single restoring spring and a single damper. There is no such thing as a single degree of freedom (DOF) system in the measurements world. If you really want one, you've got to go back to the classroom. In spite of this caveat, all the examples in the literature begin with a single degree of freedom system. I will too.

A single DOF system shows at the bottom of Figure 2.34. Figures 2.35 and 2.36 show the amplitude and phase portions of the transfer function shown above for various values of damping (as a percent, h, of critical damping). Don't get confused about all the curves from the different values of damping. There are only two values the measurement system designer needs to focus on. Let's inspect the amplitude curves first in Figure 2.35.

Assume you have very little damping in your single DOF system—the damper is barely there and the system is highly resonant. At very low frequencies the magnification factor, the system's gain, is 1, or 0 dB. This makes sense if you think of a weight hanging from a long rubber band that you hold in your hand. If you move your hand up and down slowly enough, the weight will move up and down the same amount—no magnification of motion, gain equals one. If you move your hand up and down faster, the weight will begin to move up and down more than your hand and maybe clunk you in the eye if you're not

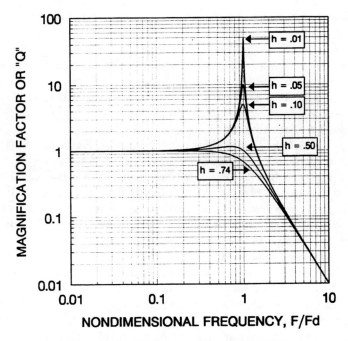

Figure 2.35   Single degree of freedom system magnification factors for various values of damping ratio (h)

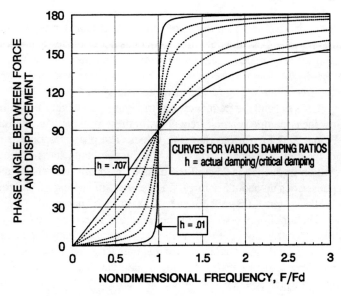

Figure 2.36   Phase angle between response displacement and excitation force for a single degree of freedom system with viscous damping

careful! At the resonant frequency it will be moving like gangbusters. That corresponds to the curve labeled $h = .01$ in Figure 2.35. Move your hand even faster and the weight will begin to move less and less, until it finally will appear not to move at all. You are now above the resonant frequency and the magnification factor, Q, is less than one. As you increase the damping in your SDOF system the maximum magnification factor, Q, decreases until about $h = .60$ where the system no longer resonates at all and Q loses meaning. When the damping ratio increases to about $h = .74$ the system gains some very interesting properties discussed below.

The SDOF system's phase response shows in Figure 2.36 as a function of the ratio of the excitation frequency F over the damped natural frequency F_d. At resonance, this ratio is 1. For low damping (h~.01) and at low frequencies the phase angle between your hand and the weight is zero. They go up and down together, in phase. As the frequency increases the phase stays near zero until you approach the resonance very closely (frequency ratio ~ 1). It quickly hops up through 90 degrees to –90 degrees and stays there. This characteristic is very interesting to the measurement system designer.

A second value of damping that interests the system designer is stated in the literature as 70.7% of critical damping (.707 of critical). Transducer elements by themselves are usually highly resonant with very low damping for the first couple of resonant frequencies. Manufacturers sometimes add a damping mechanism to the transducer so that the overall damping approaches .707 of critical. This is done by adding close tolerances and oil or gas damping to the transducer's flexural elements. This is done in many piezoresistive acceleration transducers and almost all servo force-balance accelerometers.

2.7.2 Why Do This?

Note the curve marked for the value of .707 of critical damping in the amplitude response. It does not rise above the flat portion—it does not resonate! This is a nonresonant resonator. The response is much flatter to higher frequencies than the undamped case. Further, notice that below the resonant frequency the phase is linear with frequency in Figure 2.36. This, again, is very interesting for the measurement system designer.

Let's expand the frequency axis in Figure 2.36 to get a better look at what is going on below the undamped natural frequency for these systems. The amplitude response is shown in Figure 2.37 for various damping values. The phase response is shown in Figure 2.38. Note the highly resonant case for $h = 0$. Its amplitude starts off at a gain of 1. When the excitation frequency is 25% of the damped resonant frequency (frequency ratio = 1), the response goes up about 5% to 1.05. Here, you are making a 5% error in amplitude, getting out of the system 1.05 units for every 1.00 unit you put in.

The amplitude response of the literature's favorite .707 of critical system is flat to within 5% out to about 50% of the system's damped natural frequency (as opposed to the 25% for the undamped case). Notice that it never goes above a gain of 1. Certain references in the literature state that this value of damping (.707 of critical) provides linear phase response. It, in fact, does not do that nor does any other value of damping. But it comes close for design purposes. Figure 2.39 shows the *deviation* from phase linearity for a single degree of freedom damped system for several values damping around .707 as a function of nondimensional frequency. You can see that no value actually provides linearity, which

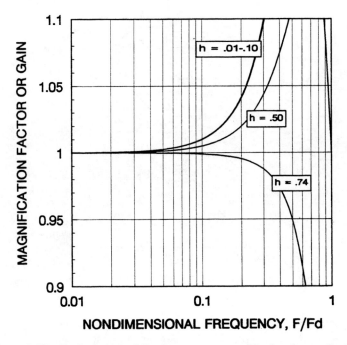

Figure 2.37 Single degree of freedom system magnification factors for frequency ratios less than one

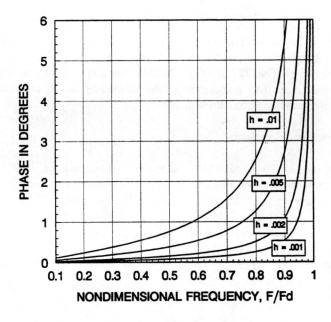

Figure 2.38 Phase response for a single degree of freedom system with low viscous damping

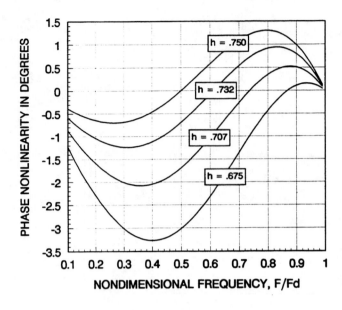

Figure 2.39 Phase nonlinearity for a single degree of freedom system with viscous damping around a damping factor of .707 of critical

would plot as a horizontal line at zero on this plot. As a matter of fact, the minimum phase nonlinearity case is provided at a damping ratio of about .74—not .707. This case provides phase nonlinearity of about +/− 1 one degree from zero frequency to the resonant frequency. For measurement system design purposes, this is as linear phase as you need for waveshape reproduction.

2.7.3 Necessary Resonant System Properties

Either system, very low damping or .707–.750 of critical, may be used to make valid waveshape measurements, but over different frequency ranges. The highly underdamped system (h very low) may be used to about 25% of its damped natural frequency and is limited by its amplitude response. The .707–.750 of critical damping system may be used to about 50% of the damped natural frequency, and is also limited by its amplitude response. Both damping values provide almost linear phase.

2.7.4 How do Real Systems Respond?

The amplitude portions of the transfer function for some real systems show in Figure 2.40. These data are for shock accelerometers (high acceleration levels, high resonant frequency) from Sandia National Laboratories in Albuquerque courtesy of Dr. Pat Walter[3]. These data were generated using an impulse shock acceleration technique. The frequencies here are specific to the application. If you are working with a low frequency system, a transducer with a 30 Hz resonance might be just fine.

Note that these responses resemble the theoretical ones in Figure 2.35. There are some responses below the primary resonant frequency in one of the plots and some above

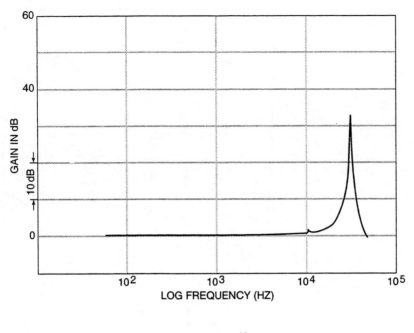

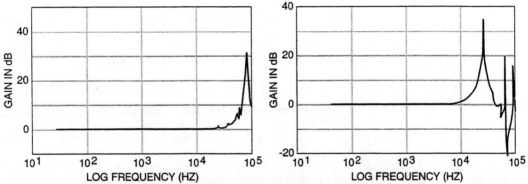

Figure 2.40   Examples of "real" resonant systems

the resonant frequency in another. The point is that the models are approximations of a nonexistent system—the ubiquitous damped single degree of freedom system. These responses are from real distributed systems having multiple degrees of freedom and multiple responses.

Figure 2.41 shows the amplitude and phase response for another real resonant system. This response is from a high frequency pressure transducer used in blast and shock propagation studies at Sandia. Again, the frequencies are unique to the application. Multiple resonances show in the amplitude response above the primary response. Below the resonance, it resembles the theoretical model. Note that the phase plot here violates the rules

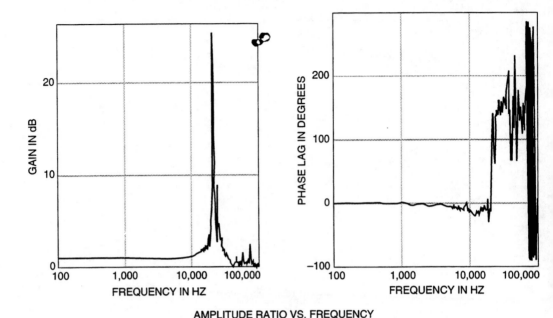

AMPLITUDE RATIO VS. FREQUENCY
FREQUENCY IN HZ.

Figure 2.41   Example of a "real" resonant system

by being plotted versus logarithmic frequency. The phase response starts at zero where it should be, and stays near zero until the 21KHz resonance occurs. It then immediately bounces up to 180 degrees followed by mass confusion! The confusion comes from other modes within the transducer above the first resonance. A linear plot of this response would straighten out the response so that it would appear linear until the resonance frequency.

The main points to remember about using resonant transducers or system components are:

1. Resonant components may be used to make valid waveshape measurements.
2. Two damping values are optimal.
   • very low (highly resonant)
   • .707–.750 of critical damping
3. Highly resonant transducers and systems may be used to make valid waveshape measurements to about 25% of their installed first resonant frequency.
   • flat amplitude response within 5%
   • almost linear phase response
4. Transducers damped between .707 and .750 of critical can be used to about 50% of the undamped first resonant frequency.
   • flat amplitude response within 5%
   • almost linear phase response

2.8 SYSTEM OUTPUT/INPUT LINEARITY

The third critical performance characteristic we will consider in this chapter is output/input linearity. Why do I call it output/input linearity and not the other way around—input/output? The system's gain is the ratio of the output over the input, output/input, and the inconstancy of this ratio defines the linearity problem.

In truth, the output/input linearity characteristic is not part of the measurement system's transfer function. The discussion appears here because this linearity notion is as important in waveshape and frequency content reproduction as are the amplitude and phase portions of the transfer function. This is the natural place to have the discussion.

A measurement system is linear if the ratio of output units per input unit is constant over the range of input amplitudes defined by the measurement window. The output/input characteristic for these systems is a straight line. The characteristic is usually defined at zero frequency, or DC for systems that respond to that frequency. For systems that exhibit both high and low frequency roll-offs, the characteristic has meaning only at the frequency of zero phase shift—that the manufacturer never defines. So, try to find that frequency in the specifications! You have to find it on the bench.

A measurement system is nonlinear if the ratio of output units per input unit is a function of the input level. Here, the output/input characteristic curves.

2.8.1 Here is the Big Zinger

- All material properties of all transducers are nonlinear.
- Therefore, all transducers are nonlinear.
- All other instruments in the measurement system chain are also nonlinear to some extent.
- Guess what? All measurement systems are nonlinear.

If you are willing to look closely enough, all measurement systems operate in a nonlinear fashion—even when in their *linear* range. The universe is nonlinear. Einstein merely discovered the fact.

This inherent nonlinearity is usually not a problem in a properly designed and properly used measurement system. The errors caused here are negligibly small compared to other error sources such as dynamic calibration uncertainties. Most of our measurement systems at TRW are consciously operated in their linear range. We take some pains in this regard because the price for nonlinear operation may be too expensive to pay.[4]

Do you ever get surprised? The Nobel Prize winning test conductor or analyst predicts that we'll see 300 engineering units maximum on a test—and the data channel is last seen passing 1,000! This happens more often than you might think. Is it an example of stupidity or incompetence? Usually not. People just get surprised sometimes. If they never got surprised, there would be no reason for testing. So, I guess we should celebrate the surprises.

What happens in this case? To make any statement about the resulting "data," it is vitally important that you understand how nonlinear systems work. This is of particular importance if your test has any frequency content in it.

2.8.2 The Mythical Linear System

Figure 2.42 shows a system having an output/input characteristic that is, over some range, linear—a straight line. Bowing to physics, we admit that at some high level input amplitude to the system will become nonlinear enough to matter. The system in Figure 2.42 is operating at its zero phase shift frequency in its linear range. This curve might be what you would get if you plotted your metrology data for a typical transducer.

  At the bottom you see a sine wave with an amplitude equal to A units. The output is a sine wave with an amplitude KA (gain of K times the input amplitude). This system is acting properly. It has not distorted the input waveform and the output data is valid.

2.8.3 A Real System

Figure 2.43 shows a real system that admits being nonlinear everywhere to some extent. Again, the input is sine wave of amplitude A. The amplitude is large enough to drive the system into nonlinear operation. The output waveform is severely distorted and the amplitude is less than KA—an error. This *amplitude distortion* is caused by the nonlinear operation.

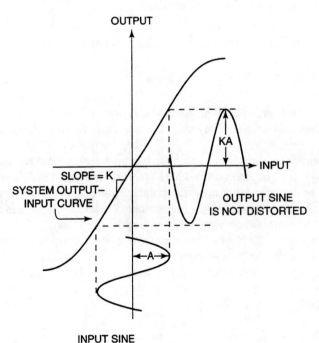

Figure 2.42 System operating in its linear range—output data is valid

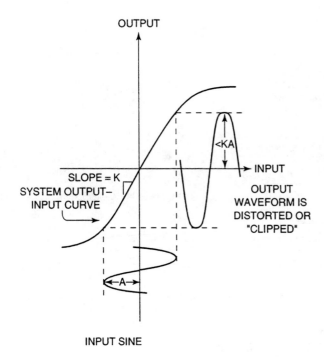

OUTPUT

SLOPE = K
SYSTEM OUTPUT–
INPUT CURVE

<KA

INPUT

OUTPUT
WAVEFORM IS
DISTORTED OR
"CLIPPED"

←A→

INPUT SINE

Figure 2.43 System is operating in its nonlinear range—output data is distorted (invalid) "amplitude distortion"

Be advised that this shape of nonlinearity curve is typical in measurement system components. There are others. For instance, if the positive going portion of the curve in the first quadrant had curved up instead of down, the output waveform would have been zero shifted (a DC error) as well as amplitude distorted. Since different components of measurement system have different nonlinearity curves, a general approach is needed to appreciate the problem.

2.8.4 One Each General Approach

What is really going on when a system operates in its nonlinear region and what are the general implications? The answers require some general background. Take a system with generic nonlinear output/input curve. Don't even specify its shape—any shape will do. The general case shows in Figure 2.44. Here, the input is X, the output is Y. We can describe the curve as a polynomial. There is a nonlinear output/input curve which crosses the Y axis at some value S. This is the DC component of the input X. The output is described as:

$$Y = S + AX + BX^2 + CX^3 + DX^4 + \ldots \ldots$$

If the values of B, C, D, and so on are zero, $Y = S + AX$ remains, the linear system description. So far, so good. The input to this system will be a simple unit sine wave, $N = (1)$ Sinωt, where ω is radian frequency and t is time. We are going to input a single frequency, ω, with unit amplitude and zero DC level.

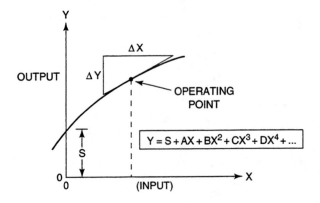

$$Y = S + AX + BX^2 + CX^3 + DX^4 + ...$$

Figure 2.44 Generic nonlinear system

Here is what comes out of the system:

$$Y = S + A\mathrm{Sin}\omega t + (1/2)B(1{-}\mathrm{Cos}2\omega t) +$$
$$(1/4)(3\mathrm{Sin}\omega t - 3\mathrm{Sin}3\omega t) +$$
$$(1/8)D(3 - 4\,\mathrm{Cos}2\omega t + \mathrm{Cos}4\omega t) + \ldots \ldots$$

Collecting terms:

$Y =$ $(S + (B/2) + (3D/8)) +$	*DC term screwed up*
$(A + (3C/4))\mathrm{Sin}\omega t -$	*The only frequency you input is screwed up*
$((B/2) + (D/2))\mathrm{Cos}2\omega t -$	*Error created at 2 times the input frequency*
$(3C/4)\mathrm{Sin}3\omega t +$	*Error created at 3 times the input frequency*
$(D/8)\mathrm{Cos}4\omega t +$	*Error created at 4 times the input frequency*
. . . . .	*And so on.*

You can see terms in the output equation at the DC level, the only frequency in the input, and at all harmonics of the input frequency. The zero frequency term is now in error because of the nonzero values of the coefficients B and D. The frequency component at the input frequency, ω, is now in error because of the nonzero value of the coefficient C. This error term would pass right through the best analog or digital tracking filter in the world— so filtering cannot be used to eliminate this error. The system creates new frequency components at the second, third, and fourth harmonic of the input frequency, and so on. These components are pure 100% error since there is zero input at these frequencies.

The measurement system, because of its nonlinear operation, has created frequencies in its output that were not there in its input. This is pure, unadulterated invalidity, not to mention bad news! This is the formal definition and test of a nonlinear system: nonlinear systems are frequency creative.

Let me put this discussion in context for you by assigning some numbers to this phenomenon. After all, how bad can it be? If there is a standard transducer used all over the country for the measurement of pyrotechnic, high level shock accelerations, it is the Endevco 2225 piezoelectric accelerometer. We use them at TRW for this purpose. This transducer has a ± 20,000 Gpk full scale range per the manufacturer. Endevco's linearity specification states "0.5% per 2000G, from zero to 20,000G." There is no mention of what happens between 0 and −20,000G. If you multiply that out you get a sensitivity at full scale that is 10% higher than at zero! *The accelerometer can be as much as 10% nonlinear while operating within the manufacturer's specifications.* I give Endevco credit for being a conservative manufacturer. Their specification is a guaranteed maximum value—but you could get an accelerometer with this specification.

2.9 SUMMARY ON LINEARITY

1. All measurement systems are nonlinear. To some extent, they are even nonlinear where the manufacturers say they're not.
2. Measurement systems are surely nonlinear when overdriven by plan or mistake.
3. Measurement systems operating in a nonlinear fashion are frequency creative. There will be frequencies in the output that are not in the input.
4. These errors can be substantial and can be *below*, *at*, and *above* the input frequency range.
5. In most cases, these errors are irrecoverable once they have occurred.
6. Never use measurement systems known to have anything but minimal nonlinearity on a dynamic, frequency-rich test.
7. There is only one frequency where measurement system nonlinearity lacks effect. That frequency is zero - DC. Nonlinear systems can, with care, be used for static measurements and we do it all the time.

NOTES

1. Frequency Devices, Inc., 25 Locust St., Haverhill, MA 01832.
2. L. W. Bickle, *Time Domain Deconvolution Technique for Correction of Transient Measurements,* Sandia National Laboratories, Albuquerque, NM; SC-RR-71 0658; November 1971.
3. Patrick Walter, *Limitations and Corrections in Measuring Dynamic Characteristics of Structural Systems,* lecture notes for the Measurement Systems Dynamics Short Course, Stein Engineering Services, Phoenix, AZ, 1979.
4. C. P. Wright, *Dynamic Data Invalidity Due to Measurement System Nonlinearity,* Proc. of the Western Regional Strain Gage Committee of the Society for Experimental Mechanics, Phoenix, AZ, 1985.

Chapter 3

Frequency Content
or Waveshape Reproduction?

3.1 THE MOST BASIC DESIGN QUESTION

The designer's job is to create a measurement system that will provide demonstrably valid data from some low frequency to some higher frequency, and from some algebraically low amplitude to a higher amplitude. In practice, this sounds like "measure valid dynamic loads for me from 5 to 5000 Hz, and from −1000 to +1000 pounds force." Here, the measurement window is 4995 Hz wide and 2000 pounds force tall. A specification like this is the most basic and succinct one for any measurement system design.

If the designer has done the job well, he must say there are negligible errors over that entire window throughout the measurement system. He is prepared to prove it—the proof being the difference between a professional and an amateur. This is not a simple task given the complex physics that must be controlled at that boundary.

In order for valid data to be presented at the measurement system's output, three criteria govern the handling of the input waveshape. These criteria concern:

- Amplitude portion of the entire system transfer function
- System output/input linearity
- Phase portion of the entire system transfer function

There are two fundamental classes of associated measurements problems. In the first class only the waveform's frequency content is to be validly reproduced. Examples of this measurement class are cases where the data is to be frequency analyzed (Fourier spectra, power or auto spectral density, filtered sine analyses, octave and one-third octave analyses). This is a large class in any widely based measurements organization. My department delivered 22,000 of these analyses alone in a recent year!

In the second class, the requirement is the valid measurement of the system's input waveshape. This means that the time history of the input measureand must be valid. This is a tougher job to for any measurement system to perform. Examples include any time history on any recording device from an oscilloscope screen to a high speed digital data acquisition system.

These two classes have different measurement requirements and I will discuss them separately.

3.1.1 Class I: Reproduction of Frequency Content

This is the easier design job of the two classes. In order for a measurement system to reproduce the frequency content of the input waveform, the first two criteria must be met:

1. All frequencies in the measurement window must lie within the *flat portion* of the amplitude portion of the transfer function.
2. All amplitudes in the measurement window must lie within the *linear range* of the system's output/input characteristic.

These are both *design* issues. These criteria must be met by the system designer. The system must be designed around this measurement window. (Editorial: To work the problem in the other direction is to sell your measurement soul. Selling your soul sounds like, "Well, my system works from here to here in frequency, and from here to here in amplitude—is that good enough?" to a person whose only tool is a hammer, all problems look like nails. Craftsmen carry *many* tools and use the right one for the job. The analogy is apt.

What about the third criteria—phase response? You can get away with a slight fast one here and it is a good thing you can. By the way, knowing what you can get away with, and what you can't, is another mark of a professional in this business. The types of analyses you would be doing in this measurement class (Fourier spectra, power or auto spectral density, filtered sine analyses, or octave/one-third octave analyses) are not sensitive to the phase relationships among the input waveshape's frequency components. These analyses mathematically throw away the phase information in most analysis systems. These analyses cannot answer the question, "Phase measured with respect to what?" You could change all the phase relationships among the frequency components and a commercial Fourier or PSD analyzer would never know the difference. You would get the same answer. The same thing is true for any RMS measuring instrument.

3.1.2 Class II: Reproduction of a Waveshape

This is a more severe problem for a designer. Unfortunately, it is the most general requirement that must be met. In the absence of knowledge about the final analysis requirements for the data, this case must be designed for even those cases where only a DC level is required fall into this class. The time history must be valid if there is to be a guarantee that the DC level is valid. Although it is possible that a sampled DC level could be valid while the waveshape is invalid, it is unlikely.

For a measurement system to reproduce the waveshape of its input, it must meet both of the criteria noted above. In addition, the third criteria must be met.

    3. All frequencies within the measurement window must lie within the *linear range* of the system's phase portion of the transfer function.

This last criteria is the hardest one to meet in a system design and there are several reasons. First, many vendors do not understand the importance of the phase response of their products and do not include the information in their specifications. Second, it usually costs money to meet this criterion. Third, this criterion is a crucial one for choosing filters to install in your system. In meeting this criterion, you may have to give up some performance in another area and that makes any designer nervous. For example, you may have to trade sampling rate for phase linearity in an anti-aliasing filter. Or, you may have to trade bandwidth for phase linearity in your signal conditioning. There will be more on these design trades later in the book.

These simple criteria are some of the most powerful in measurement engineering. Successful measurement engineers carry these ideas branded on their brains. You can easily tell them when they repeatedly say, "But tell me about the phase." You find out they have the entire transfer functions of 12 kinds of filters memorized!

In order to *reproduce the frequency content* of an input within the measurement window, a measurement system must have:

    1. Flat amplitude response
    2. Output/Input linearity

In order to *reproduce the waveshape*, the system must *additionally* have

    3. Linear phase response

3.2 SOME THOUGHTS ON FILTERING

The purpose of this section is twofold: (1) to supply information on the transfer functions of filters commonly used in experimental test laboratories so you can use your filters more intelligently; and (2) show you how much damage you can do to your data by uninformed filtering. This discussion fits in this section because of the havoc an inappropriate filter's transfer function can have on your hard won data. Do not filter casually! Filter with extreme care—and if you have any vote, not at all! If there is a generic problem in this business, it is the crime of improper filtering performed by the uninformed.

The reason this is so important is that most data customers have no idea what they are asking when they say, "Oh, just filter everything above 100 Hz." Filter how? Elliptic? Bessel? Chebyschev? Butterworth? Salman-Key? Passive? Active? Constant time delay?

High pass? Low pass? Bandpass? Band reject? What terminal slope please? – 1, 2, 3, 4, 5, 6, 7, 8, or more? What dynamic range please? If you ask these questions of an average data customer, you will see their eyes roll quickly up in their heads followed by an attempt to change the subject. It's a very simple question to pose, but a difficult question to answer. It generally must be answered in the measurement engineer's head.

The methodology for determining filter characteristics is generic. My choices of filters to show you does not drive the method. It can be used with any filter. I've chosen filters in common use in my department for various purposes. Some are buried in signal conditioning, some are stand-alone filter sets.

3.2.1 The Dynamics 7600/7860 Signal
Conditioner/Amplifier

This instrument is a laboratory quality Wheatstone bridge signal conditioner and amplifier. It has low-pass filters built in for both the primary and secondary outputs. The phase portion of the amplifier's frequency response with the filter "in" shows in Figure 3.1. The frequency shows nondimensionally so that the filter –3 dB setting, whatever it is, is 1 on this plot. You see that the phase response departs from linearity at a frequency ratio of 0.7. For frequencies below 0.7 times the filter setting, the phase is linear. Phase linearity extends to DC since this amplifier is DC coupled. Figure 3.2 shows the amplitude portion of the transfer function plotted versus normalized frequency, as before. These filters roll-off with a terminal slope of –2, or –40 dB/decade. The filters are 3 dB down (30% error) at a frequency ratio of one, 1 dB down (10% error) at 0.65, and .5 dB down (5% error) at 0.5.

For this instrument, neglecting for a moment the transducer's characteristics, input waveforms will be reproduced in a valid fashion from DC up to *only 50% of the filter setting* with a 5% amplitude error occurring first.

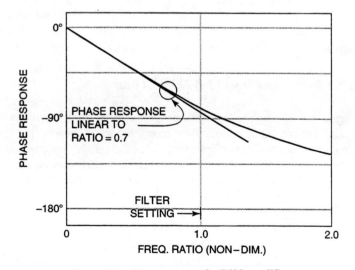

Figure 3.1   Phase response for 7600 amplifier

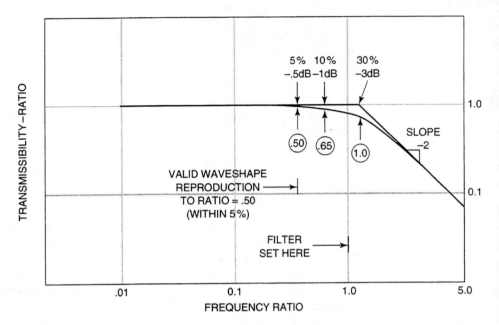

Figure 3.2 Amplitude portion of transfer function for Dynamics 7600 Conditioner

3.2.2 The Krohn-Hite Model 3342 Filter Set

This is a laboratory quality bandpass filter set. The filter's transfer function shows in Figure 3.3 in the low-pass mode. The filter has two characteristics to choose from—low-pass RC, and Max Flat both having terminal slopes of −8. The phase response for these two characteristics shows in Figure 3.4. Note that the phase response is plotted versus logarithmic frequency, rendering the plot almost useless for determining system performance.

3.2.3 Precision Filters LP-1 Elliptic Filter

Figure 3.5 shows the specifications for this filter from a manufacturer that understands filters. The amplitude portion of the transfer function shows in the upper left graph. This is a 6 pole, 6 zero low-pass elliptic filter. It's called elliptic because the equations that govern its behavior are elliptic functions. This filter's transfer function drops outside the pass band like a golf ball rolling slowly off a table top—straight down! Its nonlinear phase response shows in the upper right graph.

The phase stops being linear at 50% of the cutoff frequency. In this graph, "phase distortion" is the deviation from phase linearity. Valid waveshape reproduction is not possible past 50% of the bandwidth with this filter. The filter is flat, however, to within .5 dB (5%) out to 95% of the upper frequency limit. This filter is obviously designed for use in

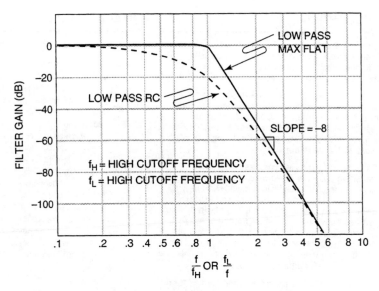

Figure 3.3 Amplitude portion of transfer function for Krohn-Hite 3340 Filter (Printed courtesy of Krohn-Hite, Inc.)

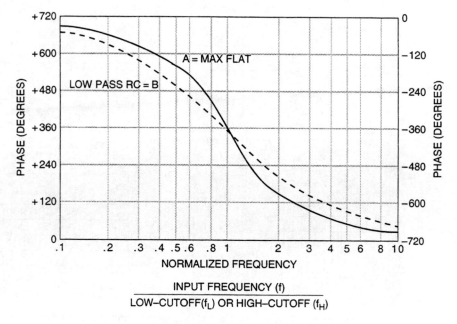

Figure 3.4 Phase portion of transfer function of Krohn-Hite 3340 filter (Printed courtesy of Krohn-Hite, Inc.)

DESCRIPTION

The LP1 6 pole, 6 zero Elliptic (Cauer) Low-Pass Filter has 0.1 dB p-p pass-band ripple, 80.07 dB minimum stop-band attenuation, and 80 dB/octave attenuation slope.

Of all filters with the same pass-band ripple, stop-band attenuation, and complexity (number of poles and zeros), the Cauer filter has the fastest transition from pass-band to stop-band making it ideal for rejecting interfering signals close to the pass-band.

For anti-alias applications, the LP1's fast transition provides more usable bandwidth, more attenuation of aliases, and a lower sampling frequency than can be obtained with more conventional filters. With sampling frequency set to three times the highest frequency of interest and the cutoff frequency set to 1.081 times the highest frequency of interest, the LP1 will provide 80 dB attenuation of all aliases with no more than 0.1 dB attenuation of the signal.

The LP1 can be cascaded with the HP1 to form a sharp band-pass filter, or it can be used with an HP1 to form a band-reject filter by driving their inputs in parallel and adding their outputs.

SPECIFICATIONS

Filter Type:	6 pole, 6 zero elliptic (Cauer) low-pass
Cutoff Frequency Amplitude:	−3.01 dB
Cutoff Frequency Phase:	−358.5°
Pass-Band Ripple:	0.1 dB p-p
Pass-Band Frequency:	0.925 F_c
Stop-Band Attenuation:	80.07 dB
Stop-Band Frequency:	1.971 F_c
Attenuation Slope:	80 dB in one octave
DC Gain:	−0.1 dB (0.9896)
High Frequency Gain:	−80.07 dB (99.197 E−6)
Zero Frequency Time Delay:	0.687/F_c Seconds
Overshoot:	19.2%
1% Settling Time:	4.64/F_c SEC
0.1% Settling Time:	8.45/F_c SEC
−0.1 dB Frequency:	0.925 F_c
−45 dB Frequency:	1.552 F_c
−60 dB Frequency:	1.776 F_c
−80 dB Frequency:	1.970 F_c

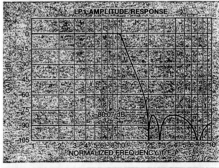
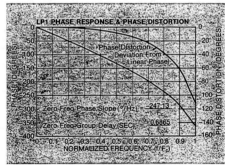
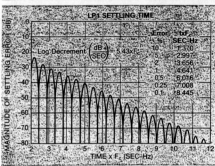
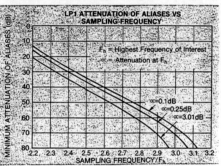

Figure 3.5 LP1 elliptic filter specifications (Printed through the courtesy of Precision Filters, Inc.)

systems where the wideband reproduction of frequency content only is desired. In fact, this filter is designed to work at the input of a digital frequency analyzer that does not know anything about phase.

3.2.4 Precision Filter's TD-1 Constant Delay (Linear Phase) Filter

Figure 3.6 shows the specifications for Precision Filter's TD-1 constant time delay filter in the same format as the previous filter. The amplitude plot in the upper left corner is not quite as flat as the elliptical filter, but it is still very good. The phase response, upper right, is dead linear to 130% of the upper frequency limit! The phase distortion (deviation from linearity) is less than +/- 0.5 degrees over this range. This filter is obviously designed for waveshape reproduction within its passband, and will do so ±.2dB to 80% of its upper frequency limit. We use them as antialiasing filters in digital data acquisition systems for pyrotechnic shock data, which is one of the most difficult waveshape reproduction tasks in our laboratories.

As of 5 years ago, this was the best combination on the market, in my opinion, of broad amplitude response, linear phase response, and graceful step response. There are now even better filters on the market. You can buy 8 pole, 6 zero constant time delay filters from several manufacturers, as you'll see in a section on antialiasing filters in a later chapter on sampled measurement systems.

3.2.5 Filter Step Response Comparisons

It is instructive to compare the responses of these filters to a standard, consistent input, a 1000 Hz repetition rate, zero-based square wave of unit amplitude. All of the data in this section was taken for this constant input, with all filter low-pass breakpoints at 10KHz. This mimics a measurement system trying to pass the square wave with errors of less than 1%. This corresponds to about the ninth-eleventh harmonic, thus the 10KHz.

Figure 3.7 shows the response of the Dynamics 7600 signal conditioner amplifier with a terminal slope of −2. The response is well mannered with an overshoot of about 1%.

Figures 3.8, 3.9, and 3.10 show the responses for the Krohn-Hite 3342 and 3700 filter sets, and the Precision Filter's TD-1 for the same input. Figure 3.11 shows all responses overlayed for comparison. These responses are all for the same input with the same filter conditions.

Which answer would you like from Figure 3.11?

The message is clear. These filters are just stand-ins for complete measurement systems. They are set, in this example, in the same manner seeing the same input. Their outputs are wildly different. Which one is correct? None of them are "correct," nor are any "wrong." All are operating within their specifications. Some are, however, certainly closer to reality than others. You have to choose the filter depending on what you want to do with it. The worst thing you can do is choose blindly.

DESCRIPTION

The TD1 6 pole, 6 zero Time-Delay Filter has a pass-band ripple of +0.21 to −0.13 dB, a minimum stop-band attenuation of 70 dB, an attenuation slope of 41.4 dB/octave, and a cutoff frequency amplitude of −3.00 dB.

An ideal filter would have no attenuation in the pass-band, infinite attenuation in the stop-band, and a linear phase response to preserve wave shape by delaying signal components equally. Elliptic filters approximate the ideal amplitude response, but they have a non-linear phase response. Bessel and equi-ripple delay filters approximate the ideal phase response, but their gaussian amplitude response unevenly attenuates signal components across the pass-band altering the signal's wave shape.

The TD1 uses 6 poles and 6 zeros to approximate the ideal filter. Six poles provide an equi-ripple approximation to linear phase ±2.5°, a pair of zeros flatten the pass-band, and four zeros sharpen the pass-band to stop-band transition. The TD1 has less attenuation in the pass-band, a sharper transition to stop-band than an 8 pole Bessel filter and it has good conformity to linear phase.

SPECIFICATIONS

Filter Type:	**6 pole, 6 zero equi-ripple phase, low-pass**
Cutoff Frequency Amplitude:	**−3.00 dB**
Cutoff Frequency Phase:	**−305.56°**
Amplitude Ripple:	**−0.13 dB to + 0.21 dB**
Phase Ripple:	**±2.5° p-p from linear**
Pass-Band Frequency:	**0.769 F_c (−0.13 dB)**
Stop-Band Attenuation:	**70 dB**
Stop-Band Frequency:	**3.225 F_c**
Attenuation Slope:	**41.4 dB/octave**
DC Gain:	**0 dB (x1)**
Zero Frequency Group Delay:	**0.849/F_c**
Overshoot:	**6.16%**
1% Settling Time:	**2.92/F_c SEC**
0.1% Settling Time:	**3.81/F_c SEC**
−0.1 dB Frequency:	**0.765 F_c**
−40 dB Frequency:	**2.200 F_c**
−60 dB Frequency:	**2.954 F_c**

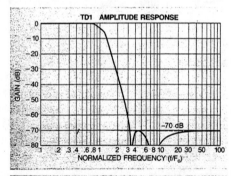

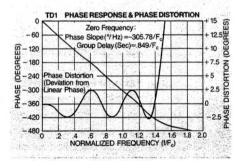

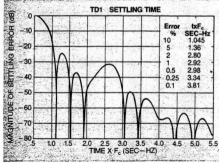

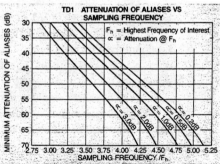

Figure 3.6   TD1 constant delay filter specifications (Printed through the courtesy of Precision Filters, Inc.)

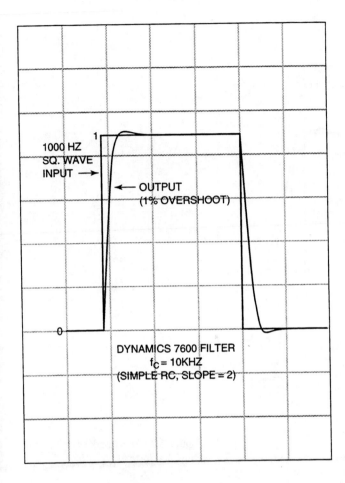

1000 HZ
SQ. WAVE
INPUT →

← OUTPUT
(1% OVERSHOOT)

DYNAMICS 7600 FILTER
$f_c = 10KHZ$
(SIMPLE RC, SLOPE = 2)

Figure 3.7   Step response of Dynamics 7600/7860 conditions, $f_c = 10KHz$

The details of a measurement system's entire transfer function and linearity characteristics must be understood before the system's performance can be judged. The designer's
job is just the opposite. He or she has to understand the entire transfer function as a part of
the design before the data ever arrives.

Remember the guy who went into the laboratory and picked up a thermocouple to
shove into the exhaust of the gas turbine? Well, the same guy went into the same laboratory
and looked for a "filter" to "filter" some data. *Which answer do you think he got?*

3.3 WHAT HAPPENS WHEN YOU VIOLATE THE RULES?

What happens when a measurement system, through improper design or surprise, violates
the rules? In this case, there are frequencies over too wide a spectrum or range of amplitudes for the system to handle properly. What happens?

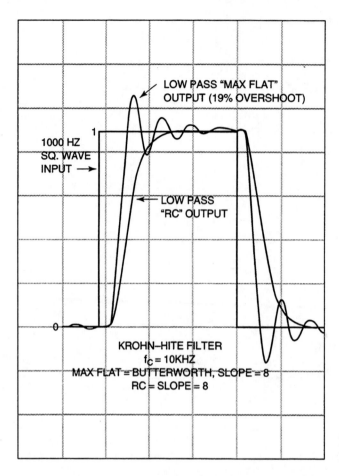

Figure 3.8 Step response of Krohn-Hite 3340 filter set, f_c = 10KHz

3.3.1 Rule 1: Flat Amplitude Response

I noted earlier that the amplitude and phase portions of the frequency response are both views of the same mathematical expression for the system—the transfer function. You cannot do something squirrely with one without having the evidence show up in the other. So, demonstrating problems with amplitude assuming phase is OK, and vice versa, is very difficult since they couple mathematically. The best method to show what violations do separately is via an analytical example. These examples are somewhat like Schroedinger's cat in relativistic physics. Something is happening to Schroedinger's cat hidden in a freely falling box. Well, phase and amplitude are a little like that hidden cat.

Let us use the example of the handy-dandy square wave input and its Fourier series shown earlier. Assume it is a zero centered square wave, so the constant term is zero. Further, only concern ourselves with the first two terms at frequencies f, and $3f$—the first and

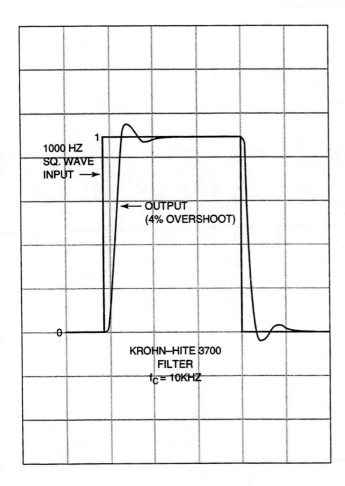

1000 HZ
SQ. WAVE
INPUT →

OUTPUT
(4% OVERSHOOT)

KROHN–HITE 3700
FILTER
f_c = 10KHZ

Figure 3.9   Step response of Krohn-Hite 3700 filter sets, f_c = 10KHz

third harmonics. Figure 3.12 shows these first two harmonics and their sum. You can see the first harmonic with an amplitude of 1 with one period from 0 to 360 degrees. You also can see the third harmonic, with amplitude 1/3, at frequency $3f$ with three periods from 0 to 360 degrees. The third harmonic is out of phase with the first harmonic as predicted in the Fourier series. The third harmonic is down at 90 degrees when the first harmonic is up at the same phase. The valid sum of the first two harmonics is shown. This is what a perfectly operating measuring system with infinite flat bandwidth will output for this input. Remember this waveshape. It is the standard for the next two cases. Take a flat amplitude violation first.

   Assume that the measurement system's phase response is perfect—no phase distortion. Figure 3.13 shows a measurement system's amplitude portion of the frequency response with an upper frequency limit at f_u. The first harmonic is within the flat region and the system passes it with full amplitude. The third harmonic exists at a frequency that is being rolled-off. A portion of that frequency component's amplitude is suppressed by the

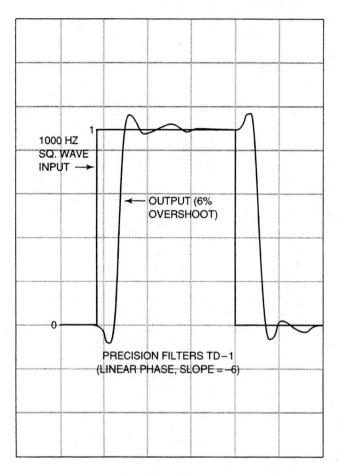

1000 HZ
SQ. WAVE
INPUT →

1

←OUTPUT (6%
OVERSHOOT)

0

PRECISION FILTERS TD−1
(LINEAR PHASE, SLOPE = −6)

Figure 3.10 Step response of Precision Filter's TD-1 constant delay filter, f_c = 10KHz

high frequency roll-off. The output must be distorted—and it is. Figure 3.14 shows the valid first harmonic and the invalid third—its amplitude is down to .15 versus and original .33 (a reduction of about 6 dB). You can see that the sum of the two is corrupted significantly compared to the valid sum in Figure 3.12. This is an example of frequency distortion caused by violation of the flat amplitude response rule. The system just does not have enough bandwidth to do this job.

3.3.2 Rule 3: Linear Phase Response

Now, look at the converse case. Assume the measurement system's amplitude portion of the frequency response is perfect—dead flat everywhere. Figure 3.15 shows the phase portion of the frequency response plotted in linear-linear form as it must be to make sense of phase plots. You can see the first harmonic booming through the system in its linear frequency range. The third harmonic is at a frequency where the phase response is nonlinear.

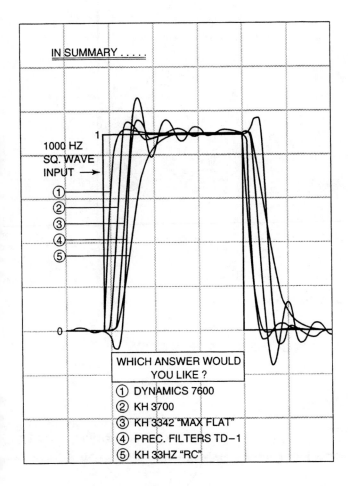

IN SUMMARY . . . . .

1000 HZ
SQ. WAVE
INPUT →

①
②
③
④
⑤

WHICH ANSWER WOULD
YOU LIKE ?
① DYNAMICS 7600
② KH 3700
③ KH 3342 "MAX FLAT"
④ PREC. FILTERS TD–1
⑤ KH 33HZ "RC"

Figure 3.11   Which answer would
you like?

Its phase has been shifted artificially with respect to its mate, the first harmonic. The frequencies in the output waveform cannot have the same phase relations as they did when they went in. And they don't.

Figure 3.16 shows the result. The valid first harmonic shows with the invalid third. The third harmonic's amplitude is, again, correct (.33), but its phase shifts due to the system's nonlinear phase response. The invalid sum also shows. This is called phase distortion. This spectacular waveshape corruption is due to a relative phase shift of 150 degrees. This level of relative phase shift is easy to get when using some high performance, but poorly chosen filters. This amount of phase shift is not overly large in reality.

Figure 3.17 shows the three sums together for comparison. The heavy lined valid sum shows with the phase and frequency distorted outputs. Note that at 90 degrees the answers vary over a factor of two! The waveforms agree at only a single phase value—about 185 degrees. The waveforms do not agree for the other 359 degrees on the figure! So, you were right 0.27% of the time! If your system's job was to reproduce just these first two frequency

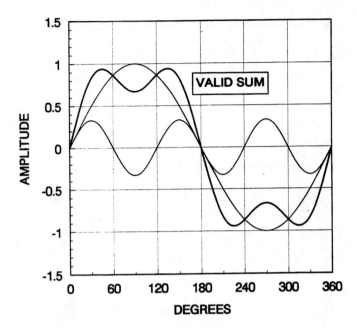

Figure 3.12   Valid 1st + valid 3rd = valid sum (no distortion)

AMPLITUDE PORTION OF THE TRANSFER FUNCTION

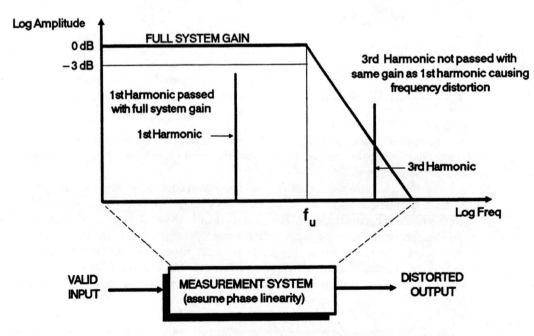

Figure 3.13   Frequency distortion

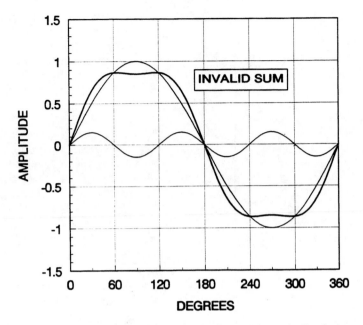

Figure 3.14   Valid 1st + invalid 3rd = invalid sum (violate rule #1—frequency distortion)

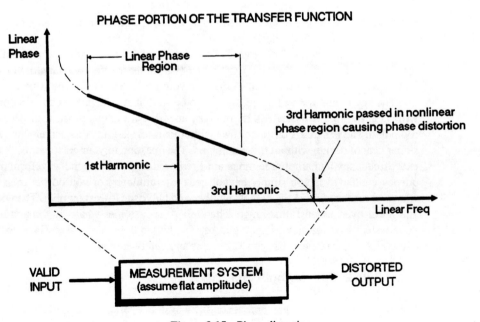

Figure 3.15   Phase distortion

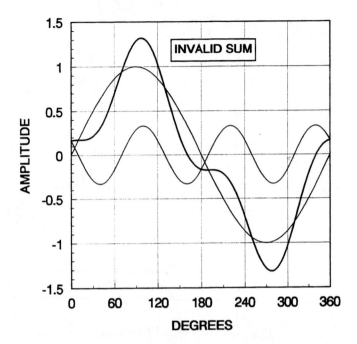

Figure 3.16   Valid 1st + invalid 3rd = invalid sum (violate rule #3—phase distortion

components of this simple square wave and your system produced anything other than the valid sum, you flunked the design course.

In these examples, I have artificially separated the frequency and phase distortions for the sake of simplicity and clarity. In your testing, these effects occur simultaneously in your data. You have to separate the effects in your brain. So, what is the system designer to do? The key to answering this question is to be sure about the data frequency content that will occur during the target test. If your customer doesn't know, or isn't sure, lock her in a room until she is sure. A customer owes you this information. A measurement system designer cannot design without these numbers. The question to your customer is, "Over what exact frequency and amplitude range am I required to reproduce the waveform or the frequency content?" If you simply cannot get these numbers, or you do not trust them for whatever reason, double the highest number and halve the lowest number. Then design flat and linear over, at least, that range. Then tell your customer what you did and how good your system's performance is going to be. The ball is then in her court since you've done everything you can and what you've done is absolutely ethical.

3.3.3 Rule 2: Output/Input Linearity

We discussed in the last chapter that nonlinear system operation was evil and to be avoided. It's evil because it creates frequency components that add or subtract from valid frequency

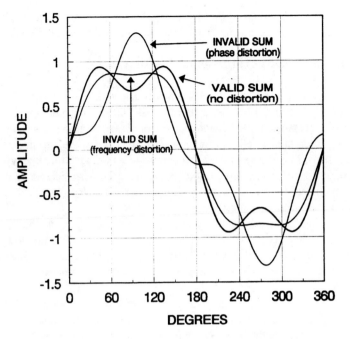

Figure 3.17   This is what happens if you don't follow the rules!

components in your data. Measurement systems are not supposed to add or subtract anything to your valid data. They are supposed to recreate that valid input at the output. A measurement system that is adding stuff to your data is writing an editorial for you, not reporting the news objectively.

It is highly unlikely that you will know which frequencies are corrupted in this frequency creation process, nor do you know how they were corrupted since their phase relationships are unknown. Your nonlinear measurement system has created an omelet from your carefully put together basket of valid data eggs.

Nonlinear operation can spread these invalid frequency components all over the spectrum in a very messy manner. The errors can occur below, at, and above your input frequency spectrum. This is important because there are people who think that nonlinear operation only creates frequencies above their input data. So, they can use low-pass filtering to suppress the invalid high frequency components. Wrong.

I heard a gentleman from a noted aerospace company at a technical meeting deliver a paper on the process by which he recovered the "data" from some highly nonlinear shock data from a missile launch sequence. They had ranged for 1000G and gotten 5000G on several telemetered data channels. He had spent literally hundreds of manhours "recovering" this data with all kinds of fancy analog and digital filtering algorithms. When it was pointed out to him that nonlinear operation causes irrecoverable errors within, above, and below the real input spectrum, and that his data was still invalid garbage after wasting tens of thousands of dollars on sexy filtering, he was (no pun intended) shocked. This is a dangerous error to make. Nonlinear operations should be avoided like the plague except at DC.

3.3.4 Examples of Frequency Creation

Example of frequency creation above the input spectrum.   This is the typical case. Figure 3.18 shows a 10-volt peak 100Hz sinusoidal input to a measurement system concocted of typical laboratory instrumentation amplifiers. This case used Dynamics 7514 differential amplifiers operating at a gain of one to represent the measurement system. Figure 3.18 also shows the FFT spectrum of its output while operating in its linear range. You see a frequency line at 100 Hz 10 volts tall—perfect. Some third harmonic comes through because this input had some third harmonic distortion. This third harmonic should be there since it was in the input.

Figure 3.19 shows this system operating in a slightly nonlinear fashion. You can see the waveshape is slightly clipped at the peaks and that there are now responses at the second, third, fourth, fifth, sixth, seventh, ninth, eleventh, and so on, harmonics. The measurement system *created* these frequencies—editorials, not news. You don't want your system's opinion—you want the facts.

Figure 3.20 shows the system operating in a highly nonlinear fashion. The fundamental clips at 13 volts as you can see in the time history. You also can see the odd harmonics (third, fifth, seventh, ninth, eleventh) dominating the spectrum. The waveform is approach-

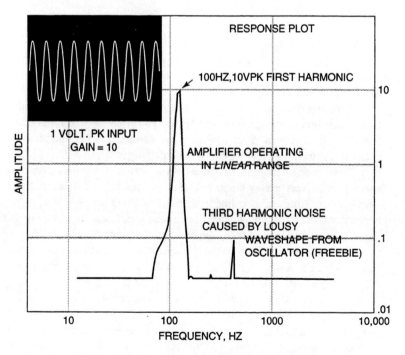

Figure 3.18 Conditioner operating in its linear range

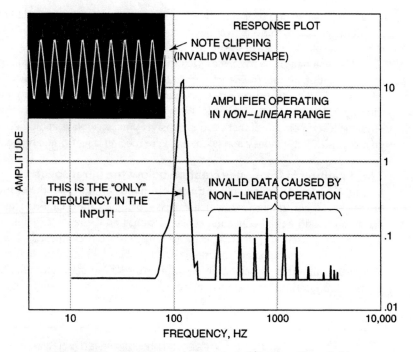

Figure 3.19 Conditioner operating in a slightly nonlinear fashion

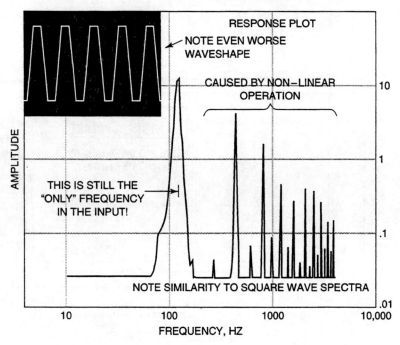

Figure 3.20 Conditioner operating in a highly nonlinear fashion

ing a square wave and so is the spectrum. A square wave has only odd harmonics, whose amplitude goes down as one over the harmonic number. This plots as a log-log straight line with a slope of −1. You can see the slope of the peaks is approaching −1. It all makes sense.

Can you imagine a frequency-rich shock event driving a measurement system nonlinear and creating all this frequency-rich stuff right on top of the input frequency spectrum? All of which adds together in some inexplicable fashion. The solution to this problem is not filtering. It is much too late for that. What you do is rerange the measurement channel and rerun the test if you can. If that's too expensive, you can always apologize! But don't waste your organization's money and talent trying to recover data from garbage like this.

Example of frequency creation below the input spectrum.   I waited a long time to see a definitive example of this effect in the literature. All the examples are like the one above, higher frequencies, which are easy to demonstrate. Getting frustrated while preparing our own internal education course, I went into the laboratory and created this one. This example used a piezoelectric measurement system composed of an Endevco 2225 piezoelectric accelerometer, an Unholz-Dickie charge line driver, and an Unholz-Dickie D11 charge amplifier. I used an Endevco insertion "T" to insert the signals into the system. In this configuration, the system does not know that the charge input is not coming directly

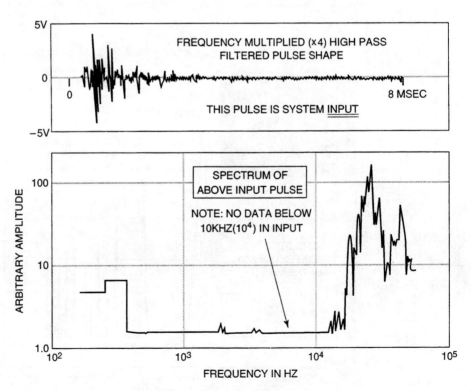

Figure 3.21   Input waveform and spectrum

from the piezoelectric accelerometer. More information about the very useful T-insertion method for piezoelectric systems will come later.

Figure 3.21 shows both the transient input waveform and its spectrum. I created this from a specially chosen shock waveform by frequency multiplying it by a factor of four, then high-pass filtering it to remove frequency components below 10KHz. When this is input to the measurement system, any components that come out below 10KHz must have been created in the system since there is zero input below 10KHz.

Figure 3.22 shows the system output for linear operation. The output looks like the input, the waveform, and the spectrum are valid. This measurement system is doing the job. It is reporting the *data news*. There is no editorial here.

Figure 3.23 shows the response of the system when driven into its nonlinear range. You can see the waveform is corrupted and there are significant frequency components below 10KHz. Please note the very high low end. This spectrum analyzer is trying to react here to the DC created in the invalid waveform. This DC component was also predicted by the nonlinear model. These frequency components have been created in the system by its nonlinear operation and would be indistinguishable from valid components in this frequency range. You would not have the slightest idea what came into your system if you saw

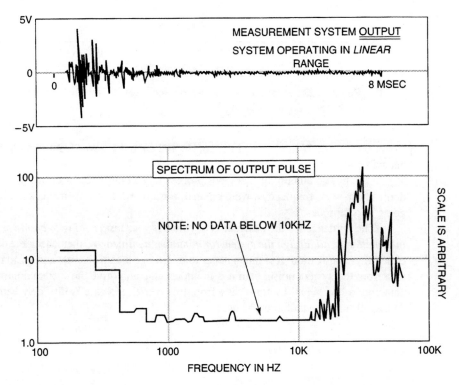

Figure 3.22 System output waveform and spectrum for linear operation (compare to Figure 3.21)

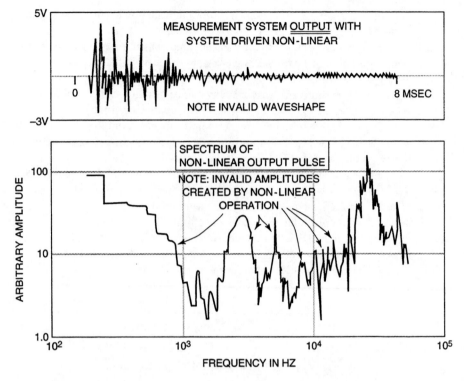

Figure 3.23   System output waveform and spectrum for nonlinear operation
(compare to Figures 3.21 and 3.22)

this at the output. You have had the course and cannot recover from this type of system behavior.

Usually, no amount of fancy postprocessing can dig you out of this hole. Since you don't know what frequencies were created, nor do you know their phases, what are you going to postprocess?

The guidance for the designer is obvious. *If your test is going to create any dynamic information at all, design the system for minimum nonlinearity,* then check the data for this effect after acquisition. It is much better to overrange a data channel and try to recover the data from the system noise, than it is to underrange and drive the system nonlinear. Valid data with a low signal-to-noise ratio from the overranged case is infinitely better than no data at all from the underranged one.

Chapter 4

Non-self-generating Transducers

The subject of transducers is so broad it is hard to know where to start. In the limited space here, only a cursory, almost sideways glance can be given. I'll note the generic issues you need to focus on when choosing transducers as a part of your design process, and when operating them in your systems. This approach is independent of what you want to measure today. Tomorrow you'll want to measure something else anyway.

Many books in this field have chapter headings like "Pressure Transducers," "Torque Transducers," "Acceleration Transducers," or "Load Cells." These headings imply that there are meaningful differences between transducer types. They imply that there is a unique body of technical knowledge associated with pressure transducers that is not applicable to torque transducers. I say—*balderdash*. There is a unified and powerful body of engineering knowledge that is applicable to all transducers[1]. This is the knowledge I'll share with you in this section—the very high spots at least.

The issues I'll discuss are:

1. Energy considerations at the transducer/process boundary
2. Boundary conditions of use
3. How to interpret the specifications

Why these three? A measurement system is a complex device that processes information and energy from a phenomenon. Everybody writes about the information—that is the prize. Very few write about energy flow—that is the booby prize. You must understand both to control both. You must control both by design to make valid measurements.

Boundary conditions of use differentiate your transducer in your test setup from the conditions extant when the transducer was calibrated. These differences make the idea of calibration traceability much less useful than your customer thinks it is.[2] Transducer and other component specifications can be confusing and overwhelming. You'll hear yourself say, "What does all this stuff mean?" I'll give you a generic method for organizing and understanding these specifications regardless of transducer type.

A transducer is a device that processes information and energy from a phenomenon under investigation. The formal definition includes the sense of energy change. In this definition a load cell is a transducer because it changes mechanical energy (force × displacement) into electrical energy (voltage × charge). By this formal definition, an amplifier is not a transducer because it does not change the form of the energy—although it can be modeled as a six terminal transducer using consistent mathematics.

In Measurements Engineering,[3] however, every component in the measurement chain is, and is modeled as, a transducer. This is a useful distinction from the formal definition since each component in the chain modifies the information as it passes through. Further, the information also may be modified at each succeeding boundary in the system. The system designer must be in control of these boundaries and transducers or evil things will occur to his data. Often, such evil things are irrevocable, and had best be recognized. I'll give you a concrete example of what not being in control looks like. Stein[4] relates this anecdote from Karl Anderson of NASA's Dryden Research Center:

> Anderson . . . relates the story of a test under the cognizance of a large airframe company who made sure that the boundary conditions at the input were duplicated for calibration and test conditions and who calibrated through the entire frequency range of interest before the test. During the test, at intervals which were not documented and due to requests from attending "Brass" observers, a number of additional instruments were connected to the system output. A CRO (author's note: cathode ray oscilloscope) was added for real-time quick look; a filter was added; an analog tape recorder; a plotter; a frequency analyzer; and so on. Frequency response calibrations after the test showed the upper frequency limit of the system to have lowered substantially enough due to these added isolation ratio or loading factors . . . that the *high frequency data were useless and test data costing over half a million of 1950s dollars had to be discarded.*

I wonder what half a million 1950s dollars translates to in today's dollars? Could you afford that on your test budget? This anecdote is 40 years old. There have been no improvements in measurement methodology over those intervening 40 years that make this startling example any less meaningful today than it was then. There have been improvements in technology (such as increased input impedance), but almost none in method.

There are only two kinds of transducers—*self-generating and non-self-generating.* There is confusion in the literature about other terms for transducer types. Some authors use the terms active and passive. IEEE defines a thermocouple as a passive transducer, while ISA defines it as an active transducer. By using the more specific terminologies of self- and non-self-generating, we'll avoid this semantic mine field.

4.1 NON-SELF-GENERATING TRANSDUCERS

A non-self-generating transducer is one in which the information about some external process is carried on a change of the transducer's internal impedance. This impedance change can be resistive, capacitive, inductive, or even optical. Strain gages and strain gage-based transducers are examples of resistive impedance change. Servo accelerometers, capacitive strain gages, and condenser microphones are examples of capacitive impedance change. Linear variable differential transformers (LVDTs) are examples of inductive impedance change, while photoelastic coatings for stress analysis operate with optical impedance change. Here, the information is carried on a strain-induced change in the optical impedance (speed of light) in the coating—called birefringence. This results in the beautiful colored patterns seen with this technology.

Process-induced information and certain noise components in this class of transducers wait in the transducer for you, carried on the internal impedance change. It is "latent." To get the information out of the transducer, you must supply a secondary energy source to the transducer. This secondary energy source interrogates the transducer for its internal information. So, it is called the interrogating, or secondary, or minor input, or the excitation. These terms are synonymous. For most of the transducers used in the experimental mechanics measurements business, this minor input is electrical since the impedance change is an electrical quantity.

4.2 THE INITIAL NON-SELF-GENERATING TRANSDUCER MODEL

Figure 4.1 shows the initial model for a non-self-generating transducer. A more detailed model will be presented in a later chapter. You can see a transducer receiving energy and information from some process, a minor input (the excitation), and an output. This model doesn't care what the process is, nor what the measurand is. In this crucial sense, there are no methodological differences between pressure transducers, load cells, accelerometers, potentiometers, strain gages, LVDTs, etc. They all can be fundamentally understood with this model.

Two quantities show at each of these boundaries. The product of these two quantities must be energy, or energy related, at each boundary. The measureands are either components of the energy, or time, space, or time and space derivatives of the energy quantities. At the major input, the quantity you want to measure today is Q_1 and the necessary pair quantity is Q_2. For a load cell, Q_1 is force and Q_2 is displacement. For a deflection measuring transducer, Q_1 is deflection and Q_2 is force. For a temperature measuring transducer, Q_1 is temperature and Q_2 is entropy. You can refer to the more complete energy pair list and discussion in Chapter 3. The minor input quantities, Q_3 and Q_4, are generally voltage and charge, as are the output quantities Q_5 and Q_6. Shown also is a crucial quantity called noise. This is everything else in the environment surrounding the transducer's space and time that you do not want to measure today, but that affect the measurement system anyway.

What you see in this figure is the electrical engineer's classic six terminal model.

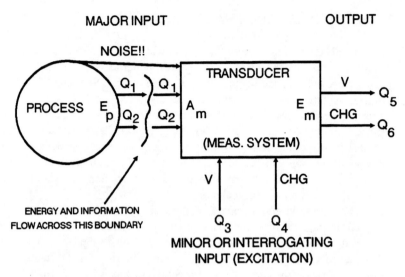

Figure 4.1   Initial non-self-generating transducer model (electrical example)

Transducers can be effectively described with the mathematics associated with the six terminal model. Underneath this top level, back-of-an-envelope view is a sound and complete theoretical model for transducer behavior including the complete Taylor series expansions for the partial differential equations that govern its behavior. This model has been developed over the last 20 years or so. If you need to delve into the theory of the six terminal transducer model to increase your understanding, I've noted two excellent references.[4,5]

4.3 CONSIDERATIONS AT THE PROCESS/TRANSDUCER BOUNDARY—THE MAJOR INPUT

Notice that from the process, there are two quantities, Q_1 and Q_2, available to you—say force and deflection. Any energy that flows across this boundary affects the process. This energy flow we do not want. You want the measurement system to be "isolated" from the process. You do not want the process to know the measurement system is there.

The process is, in a sense, emitting information about itself. It is making information available. Its ability to do that is a function of the ratio two quantities at the process output assuming the measurement system is not there. We call this the process *emission ratio*.[7]

$$\text{EMISSION RATIO} = E_p = Q_1/Q_2 \text{ (at the process)}$$

Only in certain cases is the process emission ratio directly measurable with no error. These special cases will be discussed later. In all other cases it must be inferred from the measured value, which is in error by definition, and corrected.

The measurement system is accepting information from the process. Its ability to do that is a function of the ratio of the two quantities, Q_1 and Q_2, at its input assuming the process is not there. This ratio is the measurement system's *acceptance ratio*. Both cases assume total isolation.

$$\text{ACCEPTANCE RATIO} = A_m = Q_1/Q_2 \text{ (at the measuring system)}$$

The questions are: "Given a process and a measuring system, how much of the information available from the process about the quantity we want to measure, Q_1, actually gets into the measurement system across this boundary? And, how much are we losing because of the dreaded energy flow?" The answers lie in the degree to which the process and the measuring system are actually "isolated" from each other at calibration and test time. If the measurement system is perfectly isolated from the process, all available information about Q_1 will enter the measurement system and there will be no invalidity. To the degree the measurement system is not isolated, energy will flow and error will result.

The isolation ratio defines the degree of this isolation between the process and the measurement system. The definition is:

$$\text{ISOLATION RATIO} = \frac{A_m}{E_p + A_m}$$

The isolation ratio defines how the information will flow across the boundary in the presence of energy flow. This is a very useful ratio for the system designer to be aware of. It is easy to interpret. If the emission ratio of the process, E_p, is zero, then the isolation ratio is one, and all the information flows across the boundary with no error at this boundary. If the acceptance ratio is infinite (possible) the isolation ratio is one. In both cases the systems are perfectly, mathematically isolated and perfect information transfer occurs.

If the acceptance ratio of the measurement system is very high compared to the emission ratio of the process, the isolation ratio approaches one and insignificant errors occur at this boundary. Unfortunately, you usually have neither case. The process usually has a finite emission ratio and the measurement system usually has a finite acceptance ratio. So how useful is this whole idea?

You want an isolation ratio between the process and the measurement system to be as close to one as possible and reasonable to minimize the flow of energy and reduce the errors caused by that flow. How do you do that? Choose a transducer that has an acceptance ratio that is so large, compared to the process emission ratio, that errors are negligible. The larger the ratio of the measurement system's acceptance ratio to the process' isolation ratio, the smaller the error. All else being equal, always use transducers with as high an acceptance ratio as is possible and reasonable.

How do you use this information as a part of your design process or on your test floor? What is the acceptance ratio for several transducers you might use for various measurements and how would you figure it out?

4.3.1 Example 1: Load Cell

For a load cell (or, more formally, a load transducer), you want to measure Q_1 = load and Q_2 = deflection. The acceptance ratio for a load cell is Q_1/Q_2, or force/deflection. This is the load cell's stiffness at full scale in pounds/inch. If you are lucky, the manufacturer will tell you the stiffness in the specifications. If not, you have to figure it out. And the answer is, the stiffer (higher acceptance ratio) the better! If you have a 1000-pound full scale load cell with a full scale deflection of .004″, the acceptance ratio would be:

$$A_m = Q_1/Q_2 = 1000 \text{ lb}/.004 \text{ inches} = 25,000 \text{ pounds/inch}$$

In the load cell world, you always want to use the stiffest transducer you can for any application. The design problem is this—generally, the stiffer the cell for a certain flexure material and general design, the lower the sensitivity. That stiffness versus sensitivity issue is a design trade you must make.

4.3.2 Example 2: A Linear Variable Differential Transformer (LVDT)

A linear variable differential transformer, an LVDT, is a transducer designed to measure deflection. An example is the Schaevitz Engineering Co. (pronounced Shay-vits) GCA-121–500 LVDT gage head. This is a transducer with a full scale of +/− .500 inches. A gage head LVDT has a spring loaded tip and is used for precision deflection measurements during structural testing. Since the tip is spring loaded, the transducer exerts some force back on the process—energy flow. The acceptance ratio for this LVDT is Q_1/Q_2, or deflection /force. This is the reciprocal of the transducer's stiffness, Q_2/Q_1. Remember, Q_1 and Q_2 are redefined for this transducer. If you deflect this transducer from zero to full scale the load will change by .09 pounds. So

$$A_m = Q_1/Q_2 = .500 \text{ inches}/.09 \text{ pounds} = 5.55 \text{ inches/pound}$$

4.3.3 Example 3: Pressure Transducer

This information can be difficult to get from the specifications. Most pressure transducer manufacturers don't supply it because they don't appreciate its importance. For a pressure transducer to work, the diaphragm must be deflected by the impinging pressure so that the internal flexures are stressed. This diaphragm deflection causes a change in volume in front of the diaphragm to occur. The acceptance ratio for a pressure transducer is pressure change at full scale/volume change at full scale. A Dynisco DHF 50 psia transducer has a volume change of .005 in.^{3} for full scale pressure. The transducer's acceptance ratio is:

$$A_m = Q_1/Q_2 = 50 \text{ psia}/.005 \text{ in.}^3 = 10,000 \text{ psia/in.}^3$$

That is how you can calculate the acceptance ratios for the transducers you are considering for your design. It is very convenient to say, "Use transducers with high acceptance ratios." But how high is high enough?

Figure 4.2 shows data that will support your answer to that design question. Here, the magnitude of the error created by the interaction between the process and the measurement system (lack of perfect isolation) shows versus the ratio of the process emission ratio to the measurement system's acceptance ratio. That sounds like gobbledygook. Let me simplify.

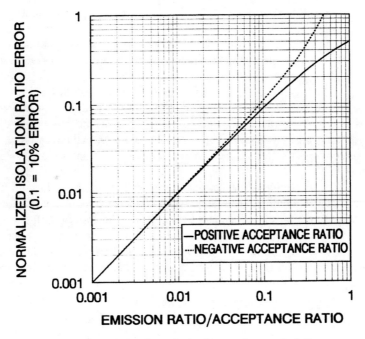

Figure 4.2   Errors due to lack of system/process isolation

I've said use transducers with as high an acceptance ratio as possible. As the measurement system's acceptance ratio goes up, the ratio of emission/acceptance must come down. So, the smaller this ratio is, the closer the isolation ratio is to one, and the better off you are. The errors show vertically. If the process emission ratio is one-hundredth (.01) of the acceptance ratio, a 1% amplitude error is made. The measurement system reads .99 units when the process, in the system's absence, would put out 1.00 unit. If the emission ratio is one-tenth of the acceptance ratio, the error climbs to 10%. If the ratio drops to one-fifth, the error is 34%! Errors for both plus and minus acceptance ratios show.

A good design rule of thumb is if the measurement system's acceptance ratio is 100 times the process' emission ratio, a 1% error will occur, which is probably negligible in most experimental mechanics measurements. A correction for this error is possible in certain cases.

4.4 SIMPLE EXPERIMENTAL EXAMPLE OF THE USE OF THE ISOLATION RATIO

A simple example is in order to show how the isolation ratio tells you what is going on at the boundary between the process and the measurement system. You can play this game two ways. If your customer for the data is on his or her toes, they will have a feel for the emission ratio of the process but won't know what the term means. Frankly, not enough people do! If the customer has such an idea, or better yet real data, you simply use a measurement system with an acceptance ratio one hundred times the magnitude of the emission ratio knowing that errors of 1% will occur—or the ratio of your design choice. You might even correct these errors in data reduction by making some assumptions. You may choose to neglect errors at 1% level.

The second case is more usual. No one has any idea about the local value of the process' emission ratio. What do you do here? You use a system with as high an acceptance ratio as possible, run the test, and note the measured emission ratio of the process. Then you assess the errors you know you've made. Again, by making certain assumptions, you can correct the data after the fact. Of course, the corrections are only as good as the assumptions!

This example is for the second case and shows in Figures 4.3 through 4.6. Assume you have an aluminum cantilever beam 30 inches long, 2 inches wide and .50 inch thick. There is nothing special about these numbers. The answer we'll get is just specific to this example. You'll get a different answer for your design problem. Such a beam is not trivial and would weigh about 3 pounds. It is cantilevered by holding it horizontal and welding the end into an infinitely stiff, vertical wall shown in Figure 4.3. Here is the problem:

> Neglecting the effects of gravity, what is the vertical spring constant (in pounds/inch) of the beam for dead weight loads up to a maximum of ten pounds, applied at the free end of the beam?

This is a simple enough problem. Mechanical engineering students would get a problem to analyze like this in the junior measurements course. University engineering students might

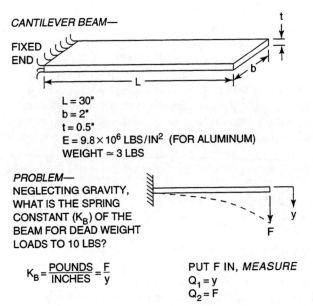

CANTILEVER BEAM—

FIXED END

$L = 30"$
$b = 2"$
$t = 0.5"$
$E = 9.8 \times 10^6$ LBS/IN^2 (FOR ALUMINUM)
WEIGHT $\simeq$ 3 LBS

PROBLEM—
NEGLECTING GRAVITY, WHAT IS THE SPRING CONSTANT (K_B) OF THE BEAM FOR DEAD WEIGHT LOADS TO 10 LBS?

$$K_B = \frac{\text{POUNDS}}{\text{INCHES}} = \frac{F}{y}$$

PUT F IN, *MEASURE*
$Q_1 = y$
$Q_2 = F$

Figure 4.3 Experimental problem

(emphasize might) see an experiment like this in graduate school. Problems like this appear in structural test laboratories all over the world every day. Simple problem—put in dead weight load, F, measure the deflection at the free end, Y, and calculate the stiffness, F/Y. Simple.

We measure the deflection Y. So Y is our quantity of interest, Q_1. The test set up shows in Figure 4.4. The beam is dead weight loaded with a weight, F, and we are measuring with the same gage head LVDT as before, the Schaevitz GCA-121–500. The experimental data shows at the bottom of Figure 4.4. The beam stiffness has been measured as 22.10 pounds/inch. How do you know that this is the right answer? Well, the LVDT was calibrated just before its use and the dead weights had little green calibration stickers on them . . . so, the answer must be right. Right? Maybe.

The calibrated LVDT and weights are necessary for the measurement, but not sufficient. How much has the measurement system affected the process? What would the measurement have been if the measurement system had not been there? This is the real question. And remember, this is not the metrologist's question. This is the measurement engineer's and the customer's question.

This LVDT has an internal spring whose purpose is to keep the core plunger in contact with the beam tip no matter where it goes. A spring you say? Springs exert force and that force will push the beam away an amount that is dependent on the beam's position. The

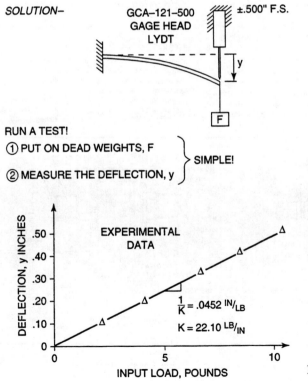

Figure 4.4   Solution

Type DHF—High Frequency Pressure Transducer

0-20 through 0-5000 psi Capacities

Rugged, small size, all welded, needs no mounting brackets

High frequency, negligible response to vibration and static acceleration

Diaphragm sensing element, simple cleanout

For gases and fluids compatible with 17-4PH stainless steel

Diffuser protects against high velocity solids and thermal shock

Integral shunt for system calibration over wide temperature range

CAPACITIES and FREQUENCY RESPONSE

Rated pressures	Natural frequency, approx.	Rated pressures	Natural frequency, approx.
0 to 50 psia	3,000 Hz	0 to 500 psi	14,000 Hz
0 to 100 psia	3,000 Hz	0 to 1000 psi	20,000 Hz
0 to 200 psia	9,000 Hz	0 to 2000 psi	27,000 Hz
0 to 350 psi	12,500 Hz	0 to 3500 psi	32,000 Hz
		0 to 5000 psi	36,000 Hz

PERFORMANCE

Rated output (R. O.)	3.0 mv/V
Calibration accuracy	0.25% R. O.
Nonlinearity, 20—50 psi:	0.25% R. O.
100—5000 psi:	0.15% R. O.
Repeatability,	0.03% R. O.
Hysteresis, 20—200 psi:	0.10% R. O.
350—5000 psi:	0.05% R. O.

The built-in shunt resistors simulate approximately 50% of full scale output. (Exact output recorded on calibration certificate.) Shunt resistor output will not be affected by temperature changes within the calibration tolerance.

TEMPERATURE

Temperature range, safe	−65 to 250 F*
compensated	0 to 150 F
Temp. effect on zero balance	0.25% per 100 F
Temp. effect on R. O.	0.35% per 100 F

*Temperature errors on either span or zero will not exceed 1% per 100°F with operation in safe temperature range.

ELECTRICAL

Excitation, recommended	10 V, AC or DC
maximum	15 V, AC or DC
Zero balance	1.0% R. O.
Terminal resistance, input	
20—100 psi:	350 ohms, min
200—5000 psi:	350 ±3.5 ohms
output	350 ±5 ohms
Termination	2 options; see drawing below
Insulation resistance,	
bridge to ground	5000 megohms
shield to ground	2000 megohms

OVERLOAD RATINGS

Safe	200%
Ultimate	400%

PHYSICAL

Internal volume at zero pressure	0.05 cu. in.
Change in internal volume caused by applied pressure	0.005 cu. in.
Material in contact with the pressure medium	17-4PH stainless steel
Pressure fitting	AND 10050-4 tube fitting 7/16-20 UNF-3B
Weight	12 oz.

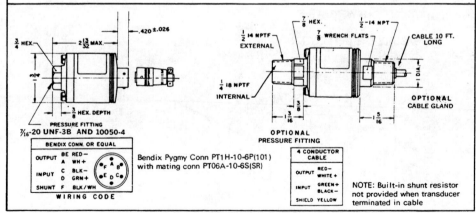

Figure 4.5 DHF pressure transducer specifications (Printed through the courtesy of Dynisco)

Type GP-H

Continuous operation to 425°F, no cooling required
Fast rise time on step input
One-piece sensing tube with integral threads, no gaskets
All stainless steel construction

CAPACITIES AND FREQUENCY RESPONSE

Rated pressures, gage or absolute	Natural frequency, approx.	Rated pressures, gage or absolute	Natural frequency, approx.
0 to 100 psi	5,000 Hz	0 to 1,500 psi	25,000 Hz
0 to 200 psi	8,000 Hz	0 to 2,000 psi	29,000 Hz
0 to 350 psi	11,000 Hz	0 to 3,000 psi	33,000 Hz
0 to 500 psi	14,000 Hz	0 to 5,000 psi	40,000 Hz
0 to 750 psi	17,000 Hz	0 to 10,000 psi	45,000 Hz
0 to 1,000 psi	20,000 Hz	0 to 20,000 psi	50,000 Hz

PERFORMANCE

Rated output (R. O.) .... 3.0 mv/V
Calibration accuracy .... 0.25% R. O.
Nonlinearity,
    100—1,000 psi: .... 0.50% R. O.
    1,500—20,000 psi: .... 0.25% R. O.
Repeatability .... 0.10% R. O.
Hysteresis,
    100—1,000 psi: .... 0.35% R. O.
    1,500—20,000 psi: .... 0.25% R. O.
Rise Time .... Less than 100
                 microseconds to a step input

TEMPERATURE

Temperature range, safe .... —50 to 425 F
             compensated ..... 15 to 425 F
Temp. effect on zero balance .... 0.5% R. O. per 100 F
Temp. effect on R. O. .... 0.5% R. O. per 100 F

ELECTRICAL

Excitation, recommended ..... 12 V, AC or DC
           maximum .... 18 V, AC or DC
Zero balance .... 1.0% R. O.
Terminal resistance, input .... 350 ±3.5 ohms
           (375 ±8.0 ohms at 425 F)
         output .... 350 ±10.0 ohms
Termination .... 4-cond. cable, 10 ft.
Insulation resistance,
  bridge to ground .... 1000 megohms
  shield to ground .... 1000 megohms

OVERLOAD RATINGS

Safe, below 300 F,
      100—1,000 psi: 150%
      1,500—20,000 psi: 120%
      above 300 F .... none
Ultimate, below 300 F,
      100—1,000 psi: 300%
      1,500—20,000 psi: 200%
      above 300 F .... none

PHYSICAL

Internal volume at zero pressure 0.7 cu. in.
Change in internal volume caused
by applied pressure .... negligible
Material in contact with the
pressure medium .... 17-4PH stainless steel
Pressure fitting
  100—5000 psi: external .... ½-14 NPTF Dryseal thread
              internal .... ¼-18 NPTF Dryseal thread
  10,000 psi: external .... ½-14 NPTF Dryseal thread
              internal ...... ⅛-27 NPTF Dryseal thread
  20,000 psi: external ...... 1-14 NF2A thread with
                   copper washer
Weight
  With cable: .... 23 Oz.

Figure 4.6 GP-H pressure transducer specifications (Printed through the courtesy of Dynisco)

LVDT will affect the position of the beam. We have caused energy to flow across the process/transducer boundary and errors result. That energy went into compressing, or extending, the spring.

How much have you influenced the process? We know the measured beam stiffness is 22.10 pounds/inch and we surmise that this number is in error due to lack of isolation.

$$\text{Beam stiffness} = 22.10 \text{ pounds/inch}$$

$$\text{Beam compliance} = 0.04525 \text{ inches/pound}$$

What is the acceptance ratio of the transducer? A simple test will show that it is 5.555 inches/pound. So

$$A_m = 5.555 \text{ inches/pound}$$

The next step takes a small leap of faith. Assume you did the job correctly and your beam stiffness was close to the emission ratio. Now you can use the isolation ratio to test the assumption.

The isolation ratio is then

$$I = A_m/(E_p + A_m) = 5.555/(0.04525 + 5.555) = 0.992$$

This example shows an isolation ratio of 0.992, or an error of .8%. You would have made an .8% error in the stiffness measurement that could be corrected. That error level might be perfectly acceptable for you. You probably would accept the 22.10 pound/inch number for beam stiffness. The result says the beam moved more with the LVDT in place than it would have without it by the ratio of 1/.992 or 1.008. If you choose, you could correct the experimental data to account for the lack of perfect isolation. The corrected value for beam stiffness would be 22.10/.992, or 22.28 pounds/inch.

4.5 HOW EASY IS IT TO FIND THE NECESSARY ISOLATION RATIO INFORMATION?

Unfortunately, it is sometimes not easy. It is not generally accepted that this concept exists, let alone is useful. Most people believe those little green stickers and answer the question, "What did the meter read?" The transducer manufacturer may or may not give you the information to calculate the acceptance ratio. If he does give it to you, it may be buried in the specifications so that you have to dig it out like a gold nugget. I'll show examples of each case.

Figure 4.5 shows specifications for a commercial strain gage-based DHF-50 pressure transducer from Dynisco, a reputable and quality transducer manufacturer. The important information is in the lower box I've marked—change in internal volume (in front of the diaphragm) for full scale pressure—very nice. This information, with the full scale pressure, allows you to calculate the acceptance ratio that we did in the earlier example. Now this is a little tricky. Note that the manufacturer specifies "change in internal volume"—that is the key. Right above it, however, is a red herring labeled "internal volume." For this exercise, who cares what the internal volume is? The internal volume is not the problem.

Figure 4.6 shows the specifications for another Dynisco pressure transducer, the GP-H series. In the similar box for the previous transducer, the manufacturer tells us that the change of internal volume for full scale pressure is . . . *negligible!* This is interesting but not useful. Whether, indeed, it is negligible depends totally on your design application, not on the manufacturer's opinion!

Look at similar data for load transducers. Figure 4.7 shows the data for a series of precision, low profile load cells from Interface, Inc. We use these transducers for our most exacting mass properties measurement systems. Figure 4.8 shows the cell's technical specifications with the column marked where the deflection at full scale load is noted by model number. Using these data and the full scale load ratings, you can easily calculate the acceptance ratios.

Figure 4.9 shows the specifications for Interface's line of SSM load cells. No deflection at full scale load is given at all. If this interests you, and it should, call their applications department and they will tell you. These numbers should clearly be in their specifications.

Figure 4.10 shows a page from the Lebow torque transducer catalog. We use several models of these transducers for static testing and torque and pure moment calibrations. The manufacturer gives you the acceptance ratio, torque/angle, directly in the column marked "rotary stiffness."

4.6 MORE BIG ZINGERS

There are, unfortunately, two large time bombs ticking in here. The entire discussion prior to this point has covertly assumed that both the emission ratio and the acceptance ratio were constants. *They aren't.*

Both properties are nonlinear. To be more specific, they each have their own linear input range. Some processes are simply nonlinear. Here, the ratio of Q_1 to Q_2 changes as a function of the level of Q_2. The universe is nonlinear. What can I say? Further, transducer properties are nonlinear. If you were to exceed the full scale range of a load cell, it will begin to yield and the ratio of load to deflection will change—a changing acceptance ratio. Also, the change is in the wrong direction! The cell gets less stiff when it yields, causing ever larger isolation ratio errors. All the assumptions you made at zero frequency and amplitude, whether you know you made them or not, go out the window.

It gets worse. *Both the process emission ratio and the transducer acceptance ratio are functions of frequency and have a complex magnitude and phase.* Try to find that information in the specifications!

I'll give you an example. The acceptance ratio is sort of an internal gain for the transducer. At zero frequency you can calculate the acceptance ratio knowing the full scale load and deflection for a load cell. As the frequency of loading increases, the transducer will begin to go into resonance, and the ratio of pounds to inches, its acceptance ratio, will change—again in the wrong direction. At resonance, for a lightly damped transducer, it takes almost no dynamic load at all to deflect the transducer. The same rules apply as in our

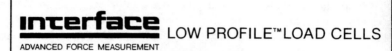

INTERFACE
ADVANCED FORCE MEASUREMENT

LOW PROFILE™ LOAD CELLS

Model 1220AF-25K
Shown with B103 Base

The INTERFACE load cell concept has established new dimensions in every aspect of precision performance:

- COMPACT SIZE
- RESISTANCE TO EXTRANEOUS FORCES
- HIGH OUTPUT
- TRUE LINEARITY
- LONG TERM STABILITY & FATIGUE LIFE
- LOW DEFLECTION
- THERMAL STABILITY
- BAROMETRIC COMPENSATION
- SYMMETRICAL OUTPUT
- METRIC VERSIONS AVAILABLE
- STAINLESS STEEL AVAILABLE

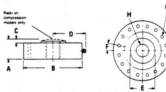

DIMENSIONAL SPECIFICATIONS

SERIES	Model No.	A(1)	B	C	D	E	F	G	H
LOW PROFILE UNIVERSAL TENSION AND COMPRESSION	1010	1.37	4.12	0.12	2.63	1.34	22.5°	9/32 Dia., 8 Holes, EQ. SP., 3.50 B.C.	5/8-18 UNF-3B
	1020	1.75	6.06	0.12	3.60	2.65(1)	15°	13/32 Dia., 12 Holes, EQ. SP., 5.125 B.C.	1 1/4-12 UNF-3B
	1032	2.50	8.00	0.25	4.57	3.76	11.25°	17/32 Dia., 16 Holes, EQ. SP., 6.500 B.C.	1 3/4-12 UN-3B
	1040	3.50	11.00	0.50	6.07	4.81	11.25°	11/16 Dia., 16 Holes, EQ. SP., 9.00 B.C.	2 3/4-8 UN-3B
	1110	1.37	4.12	0.12	2.63	1.34	22.5°	9/32 Dia., 8 Holes, EQ. SP., 3.50 B.C.	5/8-18 UNF-3B
	1120	1.75	6.06	0.12	3.60	2.65(2)	15°	13/32 Dia., 12 Holes, EQ. SP., 5.125 B.C.	1 1/4-12 UNF-3B
	1132	2.50	8.00	0.25	4.57	3.76	11.25°	17/32 Dia., 16 Holes, EQ. SP., 6.500 B.C.	1 3/4-12 UN-3B
	1140	3.50	11.00	0.50	6.07	4.81	11.25°	11/16 Dia., 16 Holes, EQ. SP., 9.00 B.C.	2 3/4-8 UN-3B
	1210	1.37	4.12	0.12	2.63	1.34	22.5°	9/32 Dia., 8 Holes, EQ. SP., 3.50 B.C.	5/8-18 UNF-3B
	1220	1.75	6.06	0.12	3.60	2.65(2)	15°	13/32 Dia., 12 Holes, EQ. SP., 5.125 B.C.	1 1/4-12 UNF-3B
	1232	2.50	8.00	0.25	4.57	3.76	11.25°	17/32 Dia., 16 Holes, EQ. SP., 6.500 B.C.	1 3/4-12 UN-3B
	1240	3.50	11.00	0.50	6.07	4.81	11.25°	11/16 Dia., 16 Holes, EQ. SP., 9.00 B.C.	2 3/4-8 UN-3B
LOW PROFILE COMPRESSION (ONLY)	1111	1.37	4.12	0.12	2.63	1.34	22.5°	9/32 Dia., 8 Holes, EQ. SP., 3.50 B.C.	--
	1121	1.75	6.06	0.12	3.57	2.65(2)	15°	13/32 Dia., 12 Holes, EQ. SP., 5.125 B.C.	--
	1211	1.37	4.12	0.12	2.63	1.34	22.5°	9/32 Dia., 8 Holes, EQ. SP., 3.50 B.C.	--
	1221	1.75	4.75	0.12	2.95	1.57	45°	11/32 Dia., 4 Holes, EQ. SP., 4.00 B.C.	--
	1231	2.25	7.50	0.25	4.32	3.12	15°	15/32 Dia., 12 Holes, EQ. SP., 6.250 B.C.	--
	1241	3.25	8.25	0.25	4.70	3.16	15°	11/16 Dia., 12 Holes, EQ. SP., 6.75 B.C.	--

Notes: (1) O.D. of hub 12 2.41 on 12.5K range.
(2) O.D. of hub is 2.41 on 25K range.
(3) All 1100 Series models sold with Base included (See Pg. 3 for base heights.)

Figure 4.7 Interface Low Profile Load Cell specifications (Printed through the courtesy of Interface, Inc.)

Interface Load Cell Specifications / SERIES	MODEL	RANGES (lbs.)	Output mV/V	Static Error Band ±% Full Scale (1)	Non-Linearity % Full Scale (2)	Hysteresis % Full Scale (2)	Non-Repeatability % Full Scale (2)	Tens. & Comp. Symmetry ±% Full Scale	Compensated Temp. Range °F	Weight in Pounds	Temp. Effect on Zero % Full Scale/100°F	Temp. Effect on Sensitivity % Reading/100°F	Sale Overload % Rated Range	Deflection at Rated Range in Inches (5)	Optional Base P/N
FATIGUE–UNIVERSAL Premium materials, special processing and critical inspection and the most advanced concept in design are combined to provide the ultimate fatigue load cell. It will withstand in excess of 100 million fully reversed load cycles without failure.	1010*	250	1	0.05	0.05	0.02	0.02	0.1	0° to +150	1½	0.08	0.08	300	0.0005	B101
	1010*	500, 1K	1	0.05		0.02				1½				0.0005	B101
	1010	2.5K, 5K	2	0.05		0.03				3¼				0.001	B102
	1020	12.5K, 25K	2	0.05		0.03				9½				0.001	B103
	1032	50K	2	0.05		0.03				26				0.002	B112
	1040	100K	2	0.07		0.03				68				0.003	B105
ULTRA PRECISION–UNIVERSAL The highest accuracy Ultra Precision Load Cell is ideal for calibration systems, thrust measurements, laboratory standards, or any precise requirement. (Includes base installed.)	1110*	300, 500	2	0.02	0.02	0.01	0.01	0.1	0° to +150	3¼	0.04	0.08	150	0.002	B101
	1110*	1K, 2K	2	0.02	0.02	0.01				3¼				0.002	B101
	1110	5K, 10K	4	0.03	0.03	0.03				7¼				0.004	B102
	1120	25K, 50K	4	0.04	0.03	0.04				21½				0.004	B103
	1132	100K	4	0.06	0.05	0.04				52				0.006	B112
	1140	200K	4	0.07	0.05	0.04				146				0.012	B105
ULTRA PRECISION–COMPRESSION - The highest accuracy Ultra Precision load cell in a compression only configuration. (Includes base installed.)	1111*	1K, 2K	2	0.02	0.02	0.01	0.01	—	0° to +150	3¼	0.04	0.08	150	0.002	B101
	1111	5K, 10K	4	0.03	0.03	0.03				7¼				0.004	B102
	1121	25K, 50K	4	0.03	0.03	0.03				21½				0.004	B103
PRECISION–UNIVERSAL The general purpose Precision series offers low profile, high accuracy, and immunity to extraneous forces for all your requirements.	1210*	300, 500	2	0.05	0.05	0.02	0.02	0.1	+15 to +115	1½	0.08	0.08	150	0.001	B101
	1210*	1K, 2K	2	0.05		0.02				1½				0.001	B101
	1210	5K, 10K	4	0.05		0.04				3¼				0.002	B102
	1220	25K, 50K	4	0.05		0.04				9½				0.002	B103
	1232	100K	4	0.06		0.04				26				0.003	B112
	1240	200K	4	0.07		0.04				68				0.006	B105
PRECISION–COMPRESSION The small, lightweight design makes installation in weighing applications such as scales, tanks, bins and hoppers, extremely easy.	1211*	1K, 2K	2	0.04	0.05	0.02	0.02	—	+15 to +115	1½	0.08	0.08	150	0.001	B101
	1211	5K, 10K	4	0.04		0.03				3¼				0.002	B102
	1221	25K, 50K	4	0.04		0.03				6¾				0.002	B106
	1231	100K	4	0.04		0.04				13½				0.003	B104
	1241	200K	4	0.06		0.04				40				0.004	B108

*Aluminum - All others are steel.

Note 1: Static Error Band (SEB), where shown, is the guaranteed performance specification. Static Error Band is calculated as the best straight line through zero including the effects of non-linearity, hysteresis and non-repeatability. (SMA Load Cell Terminology)

Note 2: Specifications noted are typical values where Static Error Band is shown.

Note 3: All universal low profile cells are normally calibrated in tension only. If compression only or tension and compression calibration is desired, please specify with your order.

Note 4: Output standardized to ±0.1% optionally available. Base required. Input resistance is 350Ω + 10%-1%.

Note 5: Total Deflection, load cell with base, nominally 2x load cell value.

Note 6: Recommended excitation voltage 10VDC. (Max. 20VDC)

Note 7: Useable temperature -65 to +200 range F° for all models.

Bridge Input and Output Resistance 350Ω ±1% all series except Note 4.
Zero Balance ±1% Full Scale all series.
Insulation Resistance Bridge to Ground 5000 Megohms all series.

RESISTANCE TO EXTRANEOUS LOADS

The INTERFACE low profile design provides optimum resistance to extraneous loads to insure maximum operation life and minimize reading errors. The following chart tabulates maximum allowable extraneous loads that may be applied singularly without electrical or mechanical damage to the cell and the maximum error that can be expected from side forces or bending moments. Several loads can be tolerated simultaneously if the total combined load is not more than 100% of the allowable maximum extraneous load.

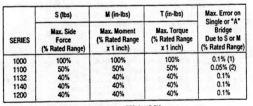

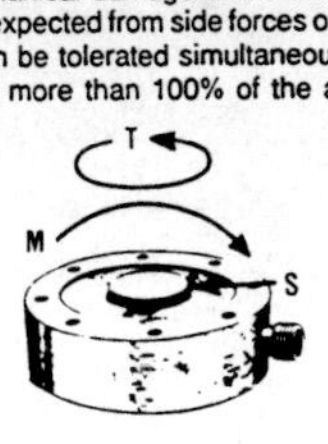

SERIES	S (lbs) Max. Side Force (% Rated Range)	M (in-lbs) Max. Moment (% Rated Range x 1 inch)	T (in-lbs) Max. Torque (% Rated Range x 1 inch)	Max. Error on Single or "A" Bridge Due to S or M (% Rated Range)
1000	100%	100%	100%	0.1% (1)
1100	50%	50%	50%	0.05% (2)
1132	40%	40%	40%	0.1%
1140	40%	40%	40%	0.1%
1200	40%	40%	40%	0.1%

(1) 0.25% for multiple bridge load cells ranges 250 thru 2.5K.
(2) 0.12% for multiple bridge load cells ranges 300 thru 5K.

Figure 4.8   Interface Low Profile Load Cell specifications (Printed through the courtesy of Interface, Inc.)

interface SEALED SUPER-MINI LOAD CELLS

ADVANCED FORCE MEASUREMENT

**Designed
for
Precision
Electronic
Force Measurement**

Model SSM-100   Model SSM-AF-500

FEATURES

NOTE THAT DEFLECTION
SPECIFICATION IS MISSING

- Ultra Precision
- Excellent Linearity
- High Repeatability
- Thermally Compensated
- Low Moment Sensitivity
- Low Cost
- Easily Installed

RATED CAPACITIES: 50, 100, 250, 500, 750, 1000, 2000, 3000, and 5000 pounds
(222N, 445N, 1112N, 2224N, 3336N, 4448N, 8896N, 13345N, 22241N)

The Sealed Super-Mini load cell is a precision strain gage load cell which is waterproof and barometrically insensitive. It is designed for testing, weighing & force measurements in tension & compression. Interface's application of proprietary advanced materials technology, in strain gage and flexure design, produces load cells with the highest accuracy in the industry yet priced competitively with lower performance units.

These rugged cells have no moving parts to wear out or get out of adjustment. The specifications listed below illustrate the superior performance of Interface Sealed SSM Series load cells and are a major factor in their worldwide acceptance in applications such as structural force testing, thrust measurement, steelyard rod conversions, conveyor scales, check weighers, counting and white scales, tensile testing and engine dynamometers.

For metric applications see Metric Sealed Super-Mini Series offering 200N, 500N, 1000N, 2000N, 5000N, 10kN, and 20kN capacities with metric mounting threads.

For applications not requiring waterproof sealed units, see the Super-Mini series of load cells with Moisture Resistant (MR) coating.

SPECIFICATIONS[1]

Non-Linearity—% Rated Output	±0.05
Hysteresis—% Rated Output	±0.03
Non-Repeatability—% Rated Output	±0.02
Temperature Range, Compensated—°F (-15° to 65°C)	0 to 150
Temperature Range, Operating—°F (-55° to 90°C)	-65 to 200
Temperature Effect on Rated Output—% of Reading/100°F (% of Reading/55.6°C)	±0.08
Temperature Effect on Zero—% of Reading/100°F (% of Reading/55.6°C)	±0.15
Creep, After 20 Min.—% Rated Output[2]	±0.03
Overload Ratings—% Rated Capacity	
Safe	±150
Ultimate	±500
Nominal Output—mV/V	3
Zero Balance—% Rated Output	±1
Input Resistance—Ohms	350+50/-3.5
Output Resistance-Ohms	350±3.5
Excitation Voltage	
Recommended—VDC	10
Insulation Resistance, Bridge to Case—Megohms	>5000

(1) Per SMA "Load Cell Terminology and Definitions".
(2) Creep specification is determined at rated capacity. Creep performance at reduced loads is proportional to the applied load.

Figure 4.9 Interface Sealed Super-Mini Load Cell specifications (Printed through the courtesy of Interface, Inc.)

- No maintenance of slip rings, bearings, or brushes
- Minimal friction error
- Low end sensitivity
- Reaction measurements eliminate speed limitations

Model No.	Capacity Oz. In.	Dimensions	Over-Load Oz. In.	Torsional Stiffness Lbs. In./Rad.	Max. Overhung Moment W x S Lbs. In.	Max. Shear W Lbs.	Max. Thrust P Lbs.
2120-50	50	See Figure 2	150	300	3.1	2.6	12
2120-100	100		300	890	6.25	3.6	35
2120-200	200		600	2,310	12.5	5.0	60
2120-500	500		1,000	2,560	31.25	10	120
2120-1000	1,000		1,500	5,130	62.5	16	140

Model No.	Capacity Lbs. In.	Over-Load Lbs. In.	Torsional Stiffness Lbs. In./Rad.	Max. Overhung Moment W x S Lbs. In.	Max. Shear W Lbs.	Max. Thrust P Lbs.	Key Size Square In.
2121-100	100	150	6,430	100	20	280	3/16
2121-200	200	300	17,000	200	26	400	3/16
2121-500	500	750	45,200	250	500	500	1/4
2121-1K	1,000	1,500	103,000	500	1,000	660	1/4
2121-2K	2,000	3,000	197,000	1,000	1,500	2,000	1/4
2122-5K	5,000	7,500	379,000	2,000	2,100	3,000	3/8
2122-10K	10,000	15,000	750,000	5,000	4,000	6,000	3/8
2124-20K	20,000	30,000	2,610,000	10,000	6,500	10,000	1/2
2125-50K	50,000	75,000	6,840,000	24,000	12,000	18,000	3/4
2125-100K	100,000	150,000	12,200,000	50,000	20,000	30,000	3/4
2126-200K	200,000	300,000	19,950,000	90,000	30,000	40,000	1
2126-500K**	500,000	750,000	25,250,000	150,000	42,000	60,000	1

*Calibration performed to 300,000 lbs. in. maximum.

* SAFETY CONSIDERATIONS: "It would be unsafe to operate Lebow® Torque Sensors and Load Cells beyond Static Overload or Ultimate Extraneous Load Limits as defined on page 111 in the Glossary of Terms or, when applicable, higher than maximum speed. When in doubt consult the factory. Eaton Corporation is not responsible for any property damage or personal injury which may result because of the misapplication of the Transducer."

Figure 4.10 Lebow 2100 series torque transducer specifications (Printed through the courtesy of the Lebow division of the Eaton Corp.)

previous discussion of resonant systems. The acceptance ratio for a resonant system is essentially constant up to 20% of the damped natural frequency. For a transducer with .70–.74 of critical damping, the ratio is essentially constant up to 50% of the damped natural frequency.

The point of this discussion is that the interaction of a transducer with a process is governed by some very complex physics. This is difficult stuff to work with. You may not be able to find any information, and both ratios are nonlinear functions of frequency with separate magnitudes and phases. The only data you are going to find to work these design problems will be at zero frequency—when you can find it at all. Remember, some manufacturers think this stuff is *"negligible."*

The situation is this. You are never going to have enough information to characterize totally the interaction between the measurement system and the process under investigation. The manufacturers simply don't know, and/or it would be too expensive to generate this data. Furthermore, sometimes no one knows the local emission ratio of the process before the test begins anyway, even to an order of magnitude. For instance, if you're measuring dynamic pressure inside the combustion chamber of a chemical laser, the information you want is the local, complex acoustic impedance of the combusting gases over the entire measurement frequency and amplitude window. My sense is that information might not be forthcoming right away.

You must, however, be aware of what is going on at the process/transducer boundary and use good practices at that boundary when you use a transducer. Errors made at this boundary because of dynamic interactions probably cannot be fixed by computational means. It would require you to be significantly smarter than you are likely to be and cost more money that you'll be willing to spend. Your design philosophy should be to place the frequencies of interaction (poor isolation ratio) so far outside the measurement window that errors are negligible. That may seem like a brute force, and possibly unacademic, approach. It is the only approach I know that works on a real test, with real measurements and a real budget and schedule.

4.7 BOUNDARY CONDITIONS OF USE

There are two very simple points to be made here. Notice that the test fixtures and apparatus and the electrical conditions that existed during a transducer's calibration in your metrology department are not those of use in your test. All that stuff, which totally defines the transducer's performance at calibration time, is not there in your test. Other stuff is there— other test fixturing, a totally different environment, and your signal conditioning that is probably different from that the metrologist used. This is a roundabout way of saying that the mechanical and electrical boundary conditions that exist during the test (isolation ratios) are generally not those which existed during the calibration. This is a major cause of data invalidity in measurements and the point is not well understood. Here are some examples from our work and from the literature. I could recite pages and pages of these examples.

1. Resistance temperature transducers were calibrated in a constant temperature bath where the primary heat transfer mechanism is *convection* from the fluid to the transducers. The transducer time constant is measured by dunking it into the constant temperature bath. They are used in thermal vacuum testing where the heat transfer mechanisms are *conduction and radiation*, with convection excluded by definition.

2. An LVDT is calibrated against an infinitely rigid jig and used to measure deflection on a real structure that is not rigid—like the beam in the example.

3. Several pressure transducers, calibrated to within 0.1% on a dead weight tester, read differently in a shock tube by 50% for the same input.

4. A Wheatstone bridge based load cell has fantastic thermal output characteristics at the vendor. The customer cannot recreate them at his facility.

5. A liquid cooled pressure transducer is used to measure chamber pressure in a high power fluorine chemical laser. When the lasing combustion begins, the transducer reads −160 psi, violating the laws of physics. Pressure cannot be negative.

6. A load cell is calibrated in metrology to 0.1% and is isolated from the excitation supply during the calibration (zero source impedance) and from the reading instrument (open circuit). In a test the load cell reads more than 1% in error.

7. An accelerometer is carefully calibrated on a small shaker in a uniaxial acceleration field—no lateral accelerations. It is used in an environment with triaxial motion. The transducer has nonzero transverse sensitivity and reads significantly high at certain frequencies.

Problems like these happen thousands of times every day in laboratories all over the world. What is the common denominator in these events? Transducers are calibrated in your metrology laboratory as isolated devices. Metrologists are experts at constructing the conditions of isolation at all three transducer interfaces—major input, interrogating input, and output. This is what they get paid to do. Your metrology engineers cannot dream up and account for all the boundary conditions and environments you will put on these transducers in your test. When the transducers are used, they are used in a measurement system that includes the process, the total environment, and your boundary conditions—none of which existed at calibration time unless you put them there.

Each of the problems listed above was caused by the conditions of calibration being necessarily different from the conditions of use. Real, predictable, and meaningful errors will accrue from this situation. This is what happened in each of the seven cases above. This also is what happened to the guy who made the 0.1% error measurements of a temperature in error by 220 degrees in the gas turbine! The isolation ratio approach can be used to account for these differences. This is a fundamental error if not accounted for in your design.

A second cause of problems occurs when the user does not recreate the manufacturer's recommended boundary conditions of use. This is a subset of the general boundary condition problem. I mention it because it can be handled very easily by a designer who is aware of the problem. Figure 4.11 shows one very reputable manufacturer's directions for

INSTALLATION INFORMATION

Mechanical installation of an **INTERFACE** load cell is easily accomplished due to its inherent immunity to side loads, bending moments, torques, and thermal gradients. The alignment to the load is not critical, as it is with column type cells, because it measures only the forces perpendicular to the mounting surface and will reject other components and bending moments, up to the limits listed.

The load cell should be mounted on a surface which is flat within 0.0002 T.I.R. for universal and for compression cells. This surface can be supplied with a standard INTERFACE base, mounting plate, or by precision grinding of your mounting plate. Certified drawings available for use in designing mountings and fixturing.

The cell should be mounted to the surface with grade 8 bolts. The bolts should be alternately and evenly tightened to the following torques:

Bolt Size	1/4-28	5/16-24	3/8-24	7/16-20	1/2-20	5/8-18
Install Torque (ft. lbs.)	5 (alum. L/C) 10 (steel L/C)	25	45	80	120	250

The mating thread for the load cell should be class 3 fit. The yield strength for a mating grade 8 bolt when engaged one diameter is as follows:

Thread Size	5/8-18	1 1/4-12	1 3/4-12	2 3/4-8
Yield strength x 1000 lbs.	35	125	250	620

ORDERING INFORMATION

The **INTERFACE** model numbering system is simple and easily indicates changes from the basic catalog unit. The model number is broken down as shown:

1210   AF   5K

Basic Model	Configuration code (AF represents the standard catalog unit with electrical connector)	Range in lbs. (K=000)

The configuration code is established by **INTERFACE**. Any change in a standard load cell must be identified by a new **INTERFACE** code. The code will guarantee the fabrication and shipment of exactly what was ordered both on the initial order and on follow-up orders.

Please consult our local representative or the factory for special features, quantity purchases, or blanket orders.

For further information on load cells with Metric threads and calibrations in Newtons, or on Stainless Steel load cells, request a copy of the Metric Low Profile Data Sheet or the Stainless Steel Low Profile Data Sheet respectively.

WHEN WRITING YOUR SPECIFICATIONS

To assure receiving an **INTERFACE** load cell, list the following specifications in your order:

Output: Table on page 5.
Size: Information from page 1.
Weight: Information from page 5.
Deflection: Table on page 5.
Side load: Information from page 5.
Bending moment: Information from page 5.
Torque: Information from page 5.
Thermal Gradient: Information from page 2.
Symmetrical Output: Table on page 5.
Fatigue Life: 10⁸ fully reversed cycles for fatigue series page 5.
Barometric Compensation: Information on page 2.

To best apply the **INTERFACE** load cell to your requirements be sure to specify the required adapters, options or special features such as:

Base
Load Buttons
Multiple Bridge Option
Positive Overload Option
Output Standardization
Mounting Plates

Please request certified drawings before designing mountings or fixtures since dimensions are subject to change.

Warranty: **INTERFACE, INC.**'s standard two-year warranty is applicable to the Low Profile load cell. **INTERFACE, INC.** certifies that its calibration measurements are traceable to the U.S. National Institute of Standards and Technology (NIST).

HOW TO ORDER

Minimum Order Charge: $50.00

Address your order to: **INTERFACE, INC.**
7401 East Butherus Drive, Scottsdale, Arizona 85260
or send to **INTERFACE, INC.** c/o our local representative.

Additional ordering information can be obtained by calling (602) 948-5555 collect, or Telex 825-882.

Standard terms: Net 30 days in U.S. dollars, F.O.B. Scottsdale, Arizona.

Prices and specifications subject to change without notice.

interface

ADVANCED FORCE MEASUREMENT   Copyright © 1984 by INTERFACE, INC.

INTERFACE, INC. • 7401 E. BUTHERUS DR. • SCOTTSDALE, ARIZONA 85260 USA • (602) 948-5555 • FAX (602) 948-1924 • TELEX 825-882

Printed in U.S.A. 15-14K 1292

Figure 4.11   Interface Low Profile Load Cell installation instructions (Printed through the courtesy of Interface, Inc.)

122

mounting his low profile load cells. We use these load cells in our mass properties measurement systems that require validity and system level tolerances of 0.1% full scale. What you are trying to do is more closely approximate in use the actual conditions of the calibration. Detailed information is given by the manufacturer including mounting surface flatness, bolt torques, and even the order of bolt torquing. What the manufacturer does not tell you is that if you do not heed his instructions, the transducer probably will not meet the published, tight specifications you paid for. The technical way to approach this argument is to say that transducers must have boundary conditions of use similar to those of design and calibration at the vendor if the transducers are to perform the way the vendor states they will. Vendors build into their transducers some incredible performance specifications. You can easily throw this performance away by improperly installing and operating the transducer.

It is your responsibility as a measurement system designer to be aware of and recreate these necessary boundary conditions at each interface in your system. In the absence of this level of concern, it is likely you may not be able to explain your data.

4.8 TRANSDUCER SPECIFICATIONS AND WHAT THEY MEAN

If you look at the specification sheet from a reputable transducer manufacturer (manufacturer of any other component in a measurement system for that matter) your impression might be where in the world do you start? What does it all mean? Is all this information really useful to you? Are you being had? Are you comparing apples with oranges?

The purpose of this section is to give you a road map to help you in traversing this sometimes confusing ground. Referring back to Figure 4.1, the initial transducer model, a non-self-generating transducer has a major input (what you want to measure today plus another quantity that comes with it), a minor or interrogating input (usually voltage and charge), and an output (usually voltage and charge). It has a transfer function that relates its output to its input—with no statement at all about conditions at the minor input. It is subject to noise levels in its environment that can even include noise levels caused by what you want to measure. This last statement seems, somehow, unfair. This is the conversation this last statement should bring up in your mind: "Noise levels caused by what I want to measure? I thought what I want to measure was data—not noise. I know everything else is noise—but the data too? Is the universe really that perverse?" Yes it is.

Any halfway decent technical specification can be grouped and compared in a logical manner. You may have to do the unscrambling because most manufacturers won't do it for you. This is the structure I suggest:

1. MAJOR INPUT SPECIFICATIONS
   - What goes on at the transducer input with regard to what you want to measure today?
2. MINOR INPUT SPECIFICATIONS
   - What goes on at the minor input, the interrogating input, the excitation?
3. OUTPUT SPECIFICATIONS
   - What goes on at the transducer output?

4. TRANSFER SPECIFICATIONS
- How does the output relate to the major and minor inputs?

5. NOISE LEVEL SPECIFICATIONS
- How does the transducer handle the rest of the environment about which you do not yet know?

This is the order in which I'll discuss them. To do that we'll use the specifications for the Model 206 pressure transducer from Taber. This is a standard production, high reliability, strain gage based pressure transducer from another reputable, conservative manufacturer. The entire specification shows in Figure 4.12. For our example we'll choose the 1000 psig full scale model. As you see, I've included the category number defined above for each specification.

4.8.1 Major Input Specifications 1

What do you need to know about the major input quantity you want to measure today, pressure, before you can make valid measurements with this transducer in your measurement system? You need some information about how you connect to the transducer pressure port and with what fluids the transducer is compatible. Does Taber give you this information? Yes, they do. However, they give you no information on allowable torques that can be applied to the interface, nor do they give you an optimum torque. Your mind should be going back to the example of the bolting torque regimen on the precision low profile load cell. Can you put in a million inch-pounds of torque on this fitting? Doubtful.

You need to know how much major input quantity, pressure, you can put in and still have the transducer meet its performance specifications. Taber tells you this under both the "range" and "proof pressure" headings. Be careful here. Taber tells you proof pressure will not cause a change in the transducer's performance. This is a unique way to define proof pressure. Most manufacturers define proof pressure as that pressure the transducer will stand and not fail catastrophically. Taber takes a much more conservative definition. There are ISA standards for specifications in this area, but not all manufacturers follow them. You, as a designer, have to know.

How about the mechanical input impedance information about the transducer acceptance ratio? Taber tells you about the volume of the pressure cavity (not what you want to know), but not about the change in this volume for full scale pressure (what you do want to know). The initial pressure cavity volume is, however, useful for calculating the frequency response of the pressurized measuring system. No data is provided to calculate the acceptance ratio. This is a major omission.

4.8.2 Minor Input Specifications 2

These specifications tell you what you need to know about the transducer excitation. What type of connector do you need? They tell you. How is the connector wired? There is no information in the specification regarding the wiring standard. This is a major omission.

Model 206

Bonded Strain Gage Pressure Transducer

Specifications:

Measurand Fluids	All fluids compatible with 347 and Carpenter 20 Stainless steel. 347 ① stainless steel diaphragm is replaceable.
Full Scale Output	3.00 ±0.015 mV open circuit per volt excitation. Calibrated at 10.00 ④ Vdc excitation.
Zero Balance	0.00 ±0.03 mV per volt at + 70°F (+21°C). ④
End Point Linearity	Within ±0.25% FSO. ④
Hysteresis	Within 0.25% FSO.
Repeatability	Within 0.10% FSO. ④
Resolution	Infinite.

Natural Frequency	Range (PSI)	Frequency (kHz) ④
	0-200	2.6
	0-300	3.2
	0-500	4.2
	0-750	5.5
	0-1000	6.7
	0-1500	8.8
	0-2000	11.0
	0-2500	13.0
	0-3000	15.0
	0-3500	16.0
	0-4000	18.0
	0-5000	21.0

Proof Pressure Rating	0-200 thru 0-750 PSI ranges: 3.0 times range. ① 0-1000 thru 0-2000 PSI ranges: 2.5 times range. 0-2500 thru 0-3500 PSI ranges: 2.0 times range. 0-4000 thru 0-5000 PSI ranges: 1.5 times range. Application of proof pressure will not cause any change in performance characteristics.
Compensated Temperature Range	-30°F to +170°F (-34°C to +77°C). ⑤
Operating Temperature Range	-100°F to +225°F (-73°C to +107°C). ⑤
Thermal Sensitivity Shift	Less than ±0.005% FSO per °F over compensated temperature ⑤ range (±0.009% FSO per °C).
Thermal Zero Shift	Less than ±0.010% FSO per °F over compensated temperature ⑤ range (±0.018% FSO per °C).
Triaxial Mechanical Shock	30 G's applied for 11 milliseconds will not cause change in ⑤ performance characteristics.

Acceleration Error	Along most sensitive axis:
	0-200 PSI range .... ±0.0145% FSO/G ⑤
	0-300 PSI range .... ±0.0130% FSO/G
	0-500 PSI range .... ±0.0100% FSO/G
	0-750 PSI range .... ±0.0070% FSO/G
	0-1000 PSI range .... ±0.0050% FSO/G
	0-1500 PSI range .... ±0.0035% FSO/G
	0-2000 PSI range .... ±0.0030% FSO/G
	0-2500 PSI range .... ±0.0025% FSO/G
	0-3000 PSI range .... ±0.0020% FSO/G
	0-3500 PSI range .... ±0.0015% FSO/G
	0-4000 PSI range .... ±0.0013% FSO/G
	0-5000 PSI range .... ±0.0010% FSO/G

Excitation	10 volts dc or ac rms recommended. 15 volts dc or ac maximum. ②
Input Resistance	350 ±3.5 ohms at +70°F (+21°C). Input circuitry symmetrical. ②
Output Resistance	350 ±5.0 ohms at +70°F (+21°C). ③
Insulation Resistance	Greater than 10k megohms at 50 Vdc between all terminals ② in parallel and case at +70°F (+21°C).

Pressure Connection	7/16-20 internal thread per MS33649-4 is standard. ① Options available.
Pressure Cavity Volume	0.08 in² (1.31 ml) excluding MS33649-4 fitting. ①
Electrical Receptacle	Stainless steel receptacle to mate with MS3106-14-2S. ② ③ Standard wiring: Excitation +A, -D; Signal +B, -C. Options available.
Enclosure	Entire housing and pressure cavity of stainless steel. All electrical ① components sealed against adverse environmental conditions.
Weight	Approximately 53 ounces (1.5 kg).

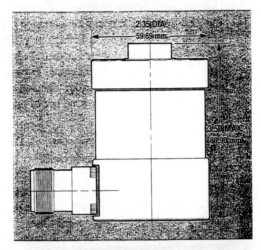

Warranty:

Teledyne Taber, herein after designated as the Company, warrants that any part or parts of the product which, under normal operating conditions in the plant of the original purchaser thereof, proves defective in material or workmanship within one year from the date of shipment by the Company, as determined by an inspection by the Company, will be repaired or replaced free of charge provided that the original purchaser promptly sends to the Company the defective material, transportation charges prepaid, with notice of the defect and establishes that the product has been properly installed, maintained and operated within the limits of rated and normal usage. Replacement parts will be shipped F.O.B. the Teledyne Taber plant. The terms of this Warranty do not in any way extend to part or parts of the product thereof which has a life, under normal usage, inherently shorter than the one year indicated above. Said Warranty in respect to replacement of defective parts and any such additional warranties express or implied, including any implied warranty of merchantability, or fitness for any particular purpose.

Warranty specifications and qualitative calibration data, as supplied with each transducer, are based on tests performed on and values obtained with N.I.S.T. traceable laboratory standards and test equipment of Teledyne Taber.

Teledyne Taber reserves the right to make changes without notice at any time in materials, specifications and models, and also to discontinue models.

⚛ TELEDYNE TABER

455 Bryant Street, N. Tonawanda, N.Y. 14120 U.S.A. • Phone: 716-694-4000
TWX: 710-262-1264 • Toll Free: 1-800-333-5300 • FAX: 716-694-1450

Figure 4.12 Model 206 pressure transducer specifications (Printed through the courtesy of Taber)

What types of excitation waveforms can be used and what levels are allowable? The "excitation" specification tells you DC up to 10 volts DC. In fact, the transducer is designed to operate at this excitation level—but the manufacturer does not tell you that. The transducer's thermal compensation has been optimized at this excitation. It may not even meet its performance specifications at other values. The specification states that up to 10 Vrms AC excitation can be used. What about sine waves, pulse trains, square waves? At what frequencies can these excitations be used? a maximum excitation level of 15 volts is recommended—15 volts DC or AC? What happens if you exceed this level? Does the transducer zero stability go to hell, or does it catch fire or melt? No statement is made. This excitation specification is significantly incomplete. At least the manufacturer has intimated to you that non-DC excitation may be used.

What is the electrical input impedance? You need to know that if you are designing signal conditioning or have to diagnose a problem. Under "input resistance" they tell you and give you a tolerance. The tolerance is used to determine if the transducer is operating properly, or has sustained damage or yield. They also tell you how this impedance, resistance in this case, distributes around the bridge. How about the resistance between the electrical circuitry in the transducer and its case? Problems in this specification cause zero shifts in the data. They tell you under "insulation resistance."

This is, generally, a good set of minor input specifications. It is as complete as you can expect.

4.8.3. Output Specifications 3

What happens at the transducer's output and how do you hook it up? The connector was defined in the minor input section. No data on what pins go where or how to handle the cabling shield is provided on this page.

What is the transducer's electrical output impedance? You need this to design or specify signal conditioning and for trouble shooting. They give you the necessary information under "output specifications." They give you the value and the tolerance. This is about all you have to know at the output.

4.8.4 Transfer Specifications 4

These are the specifications that tell you how the transducer *transfers* information from its major input to the output via the excitation at the minor input. For every specification in this section, values of all three (major input, minor input, output) must be stated if the specification is to have meaning.

The first thing you want to know is the basic sensitivity at zero frequency when operating within its linear range. This is stated as "full scale output." Notice the confusion—a clear transfer specification is stated as an output specification. No wonder specifications can drive otherwise sane engineers mad. This specification tells you how much output the transducer will provide at full scale pressure input per unit of excitation. Note that Taber says "3.000+-.015 millivolts open circuit per volt excitation, Calibrated at 10 VDC excitation." Since non-self-generating transducers carry their information on an internal impedance change (delta R/R here, or delta C/C, or delta L/L), the sensitivity must be specified in terms of change in output (millivolts) per unit of minor input (volts or amps) per full scale input (psi). This specification is in millivolts per volt at full scale, which is the standard way to present this information. Strain gage based transducers typically are designed for from 1.0 to 4.0 millivolts per volt at full scale input. The Model 206 fits nicely into this range. Note that Taber has told you that the sensitivity was measured "open circuit" at the output. They have specified for you the isolation ratio between their transducer and the reading instrument at calibration time. If you do not match this isolation ratio in your system, you should expect a slightly different answer. This is an excellent example of a transfer specification at zero frequency.

The next piece of information you need is an estimate of the internal noise level in the transducer for zero pressure input at recommended excitation. If you apply zero pressure, what comes out? This is a noise level associated with the major input. Taber tells you that for zero pressure input and 10 VDC excitation, the open circuit output will be less than ±.03 millivolts per volt (or +/−30 microvolts/volt!). This is equal to ±1% of the rated full scale sensitivity. Taber is telling you they trim this transducer model to within ±1% of full scale output. This is a fairly standard level. Note that the specification is at zero frequency. What about noise levels at other frequencies of interest? No information is given and I have never seen it anywhere else either.

When you buy such a transducer, you want linear behavior with zero hysteresis. These numbers show under "end point linearity, hysteresis, and repeatability." These errors are stated as percents of full scale output at rated excitation. These errors should be considered lower bounds on experimental uncertainties. Below these uncertainties, it is usually economically infeasible to operate. You can do it—but bring money and time!

How does the transducer react to major input frequencies other then zero? What is the transfer function? How does the basic sensitivity vary in amplitude and phase as a function of input frequency? These are all equivalent questions. Taber does not tell you nearly enough to answer this question. At least, they admit their transducer is resonant with a damped natural frequency of 6.7 KHz for this 1000 psig transducer. No statement is made of the transducer's damping ratio. a conservative assumption is that the damping is very low, and that the amplitude response is flat +/−5% to 20% of the damped natural frequency, or 1340 Hz. If the transducer damping gave you a mechanical gain of 100 at resonance, the acceptance ratio would be down by the same factor of 100 at that frequency. This means that 100 times as much energy would be drawn from the test process at that frequency! Since the transducer is a mutltidegree of freedom structure, the interaction between it and the natural multidegree of freedom dynamics of the process is very complex. In general, this is the best you are going to get from any manufacturer All this information deals with the amplitude portion of the frequency response.

What about the phase portion? No information is given. You must infer the phase response from your knowledge of underdamped second order systems. This information was included for just this reason in Chapter 2. Phase information is almost nonexistent in transducer specifications. You might get some phase information for an electrical component like a signal conditioner/amplifier.

4.8.5 Noise Level Specifications 5

How does the transducer react to the other parts of the environment that may be there at test time? Most manufacturers give you information about at most two noise levels. The most general is temperature. Some will give you response to motion. That is about all you should expect.

Over what temperature range may you operate the transducer? This is stated as "operating temperature range." Operate outside this temperature range and all bets are off on performance.

Has the manufacturer taken any pains (usually great pains!) to minimize the response over any particular range? Yes they have, and this is called "compensated temperature range." Within this range, how does the transducer perform? To know this you must specify the temperature effect on the system zero and on its sensitivity. "Thermal zero shift" tells you about how the zero varies with temperature. This specification means that for zero pressure input and rated excitation, if you vary the transducer temperature over the compensated temperature range, what you should expect at the output. Taber says less than 1% of full scale output (1% of the voltage equivalent of 1000 psig) comes out per 100°F of temperature change. No mention of the sign of this noise level. You have to find that out experimentally.

How does the sensitivity change with temperature? Look at "Thermal sensitivity shift." Taber tells you the sensitivity changes less that 0.5% of full scale output for full scale input at rated excitation per 100°F of temperature change. There is a problem with both specifications. The problem is a covert condition that existed when this performance was verified at the vendor. That covert condition was that the temperature change during these tests was *very, very slow and uniform*. This problem occurs with almost all manufacturers. What do the specifications tell you about the transducer's performance in the face of transient temperatures? They tell you nothing. What about the performance in the face of temperature gradients? After all, the pressurant medium is in contact with only one end of the transducer—the end with the diaphragm. If the pressurant medium's temperature changes at all, a temperature gradient exists in the transducer. What do the specifications say will happen in this, or any other, temperature gradient condition? They tell you nothing.

Transducers which are carefully designed and manufactured to give optimum thermal output performance in the presence of slow and uniform temperature changes can give wildly different and spectacularly invalid readings under other temperature conditions. This is another example of calibration conditions being different from use conditions.

How does this pressure transducer react to motion? How about shock? "Triaxial mechanical shock" tells you how the transducer reacts to one very specific shock pulse—30G applied for 11 milliseconds. Peak Gs or RMS? What waveform over the 11 milliseconds—half sine, triangular? The specification tells you nothing about how the transducer will react to shock pulses on your test. The specification is troublesome in another regard. It says that some acceleration level applied for 11 milliseconds will not change the performance characteristic. It tells you nothing about what comes out of the transducer during those 11 milliseconds during which you want to measure the pressure! What happens during the pulse when the transducer resonance gets rung like a bell at 6.7 KHz by the high frequencies in the pressure waveform? Further, no information about the axis of the shock input is given. The specification is, essentially, useless.

"Acceleration Error" tells you there will be an apparent output less than such and such percent of full scale for per unit of acceleration. In what axis is the acceleration applied? Was it in the axis of the transducer and normal to the diaphragm? Or was it lateral and in the plane of the diaphragm? You can bet this makes a meaningful difference in the response. What is said about the frequency of the acceleration? You can bet the output will be different at the resonant frequency than it will be at 10Hz! Over what levels of acceleration amplitude will the specification be valid? How about 100,000 Gs? This is another impressive, but largely useless, specification for the system designer.

4.9 SUMMARY REMARKS ON SPECIFICATIONS

You can see there is much more going on with respect to transducer specifications than generally meets the eye. These conclusions are clear from these specifications from Taber. And this is a good set of specifications from a reputable manufacturer! You can imagine what you get from manufacturers who do not operate at Taber's level of integrity. You can how see how specsmanship can be easily played with these numbers and how an uninformed designer or user can get into serious performance trouble.

There is no way out of this one. You must study specifications in detail to make informed design judgments about measurement system components. You must be willing to study some and dig for the answers you need for your design. The manufacturers can never tell you everything you need to know for your design purposes. There are several reasons for this: (1) they don't know what is important to you; (2) it's too expensive to test for anyway; and (3) they don't know how to run the test in a meaningful way. Remember, every specification a vendor publishes is a performance guarantee to you. Therefore, they probably test for every one they publish. If you need a different set of specifications for your design, be prepared to pay for the privilege.

I've presented a general method for looking at and organizing transducer, and other component, specifications. You can look at any transducer or measurement system component this way. Although the discussion has been for a particular model and type of transducer—a strain gage based pressure transducer—the structure, analysis method and the questions asked and answered are generic. Here is the simple structure:

1. Major input specifications
2. Minor input specifications
3. Output specifications
4. Transfer specifications
5. Noise level specifications

NOTES

1. Peter K. Stein, *The Unified Approach to the Engineering of Measurement Systems*; Stein Engineering Services, Phoenix, AZ, 1992.
2. ———, *Traceability—The Golden Calf;* Proc. 1967 Western Regional Conference of the American Society of Quality Control; Milwaukee WI, 1968.
3. See note 1.
4. Ibid.
5. ———, *A New Conceptual and Mathematical Model Application to Impedance Based Transducers Such as Strain Gages;* VDI Bericht Nr. 176, Dusseldorf FRG, 1972.
6. ———, *A New Conceptual Model for Components in Measurement and Control Systems;* Proc. 5th International Conference on Temperature, Its Measurement and Control in Science and Industry, ISA, Pittsburgh PA, 1973.
7. See note 5.

Chapter 5

Everything You Ever
Wanted to Know
about the Wheatstone Bridge

5.1 A SHORT HISTORY OF THE WHEATSTONE BRIDGE

Georg Simon Ohm discovered, in 1827, the basic laws of electrical current and resistance. He published these findings in Germany in *"Die galvanische Kette mathematische bearbeitet."* For his discoveries, he later had the fundamental unit of resistance, the Ohm, named after him.

In London in 1833, Samuel Hunter Christie, M.A., F.R.S., M.C.P.S., published his paper *"Experimental Determinations of the Laws of Magneto-electric Induction"* in the Philosophical Transactions of the Royal Society. In this paper he anticipated Wheatstone's bridge when discussing his experimental apparatus. About this paper, Sir Charles Wheatstone later wrote in 1846 that Christie "has described a differential arrangement of which the principle is the same as that . . . described in this section (of Wheatstone's later paper)". To Mr. Christie must, therefore, be attributed the first idea of this useful and accurate method of measuring resistance. Wheatstone had not read Christie's paper.

In 1843, Sir Charles Wheatstone, D.C.L., F.R.S., presented the annual Bakerian lecture to the Royal Society. His subject was *"On New Processes for Determining the Constants of a Voltaic Circuit."* In his lecture, he noted that "Slight differences in the lengths and even in the tensions of the wires are sufficient to disturb the equilibrium (of my circuit)." In the work he detailed a bridge arrangement which today carries his name. Wheatstone had the strain gage in his hand but did not realize it and considered it an annoying noise level. By all rights, we should be talking about the Christie bridge!

Thirteen years later, in 1856, Sir William Thomson, M. A., F. R. S., also presented the Bakerian lecture at the Royal Society. His subject was *"On the Electrodynamic Qualities of Metals."* During the work, he discovered that both iron and copper wires change resistance when strained. He also had the strain gage in his hand and did not realize it. In a footnote to his published lecture that was added later he said, "An hour before the meeting of the Royal Society at which this paper was read, I learned that a method of testing resistances was given by Mr. Wheatstone, which would probably be found to be the same . . . as that which I had described." Thomson had not read Wheatstone's paper. Thomson was later made Lord Kelvin by Queen Victoria for this and many other discoveries. A modification of this method became famous as the Kelvin bridge, still used more than a century and a half later, for precision resistance measurements.

The Wheatstone bridge is probably the most general measurement system component in use in the world. There are hundreds of thousands of Wheatstone bridge conditioning channels in use all over the world built by 500 manufacturers. There are 560 channels in our laboratories alone. As I write this, there are 7000 channels of Wheatstone bridge signal conditioning on a single aircraft structural test here in Southern California! If you had to get along with only one measurement method to get at all the measureands you needed, the Wheatstone bridge would be the method of choice. You can use it with appropriate transducers to measure almost anything of interest to the mechanical engineer from zero frequency to well above 100KHz.

When new signal conditioning arrangements for non-self-generating transducers appear in the literature, they are invariably compared with the performance of the Wheatstone bridge. The Wheatstone bridge is the benchmark for this type of signal conditioning, and will, I think, justifiably remain so for decades to come. Is it the only way to support non-self-generating transducers in measurement systems? Certainly not. Other options include the ballast circuit, the constant current loop and other bridge configurations. All are beyond the scope of this book. Since it is the benchmark signal conditioning method, understanding of this most useful measurement component is of fundamental importance to the measurement system designer.

5.2 WHEATSTONE BRIDGE BASICS

5.2.1 Bridge Equations

The basic Wheatstone bridge circuit for non-self-generating transducers shows in Figure 5.1. All Wheatstone bridges shown in this book carry the nomenclature standard defined by the Western Regional Strain Gage Committee of the Society for Experimental Mechanics and later adopted by the Instrument Society of America. Here, you can see a constant voltage source, V, which can have any waveform—DC, sine waves, pulse trains. The source has a series resistance R_s looking from the bridge back into the excitation source. The bridge is formed from arms $R_1, R_2, R_3,$ and R_4. The input impedance, looking into the bridge from the excitation source, is R_i. The bridge output impedance, looking back from the measuring instruments, is R_o. The measuring instrument, the readout device, has an input impedance R_m.

For this discussion, to a first order approximation and for most work, the impedances are discussed as though they are resistive only. This assumption will get us close enough for

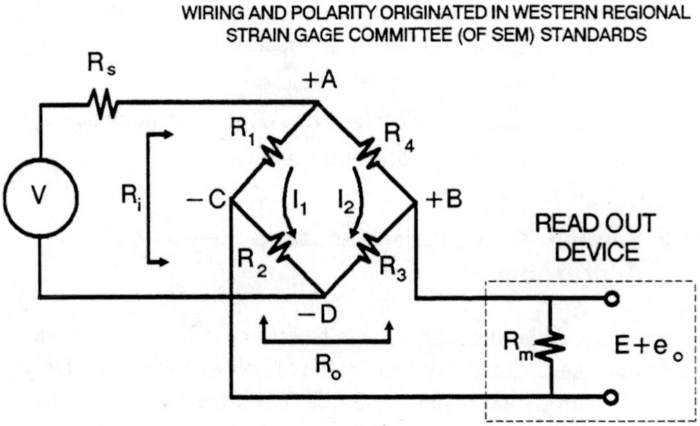

Figure 5.1   The basic Wheatstone bridge circuit

design purposes. In reality, these impedances are complex, and should be considered as such particularly if high frequency excitation or data waveforms are needed.

For a full and complete derivation of the bridge equations, consult Murray and Stein[1] and Shull.[2] The voltages from C to D, and B to D are

$$V_{CD} = \frac{R_2}{R_1 + R_2} \cdot V \qquad\qquad V_{BD} = \frac{R_3}{R_4 + R_3} \cdot V$$

The voltage from B to C is, therefore

$$V_{BC} = \left(\frac{R_3}{R_4 + R_3} - \frac{R_2}{R_1 + R_2} \right) V = \left(\frac{R_1 R_3 - R_2 R_4}{(R_1 + R_2)(R_3 + R_4)} \right) V$$

Now, $V_{BC} = E = 0$, if $R_1 R_3 = R_2 R_4$. This is the condition for a balanced bridge. Define $a = R_2/R_1 = R_3/R_4$ for a symmetric bridge. The open circuit bridge output voltage, e_o, is caused by $\Delta R/R$ in the bridge arms.

The overall bridge equation is, therefore

$$\frac{e_o}{V} = \frac{a}{(1 + R)^2}(r_1 - r_2 + r_3 - r_4)(1 - \eta)$$

where η is the bridge nonlinearity term. The total equation is then

$$e_o = V \left(\frac{1}{1 + R_s/R_I} \right) \left(\frac{1}{1 + R_o/R_m} \right) \left(\frac{a}{(1+a)^2} \right) \left[\frac{\Delta R_1}{R_1} - \frac{\Delta R_2}{R_2} + \frac{\Delta R_3}{R_3} - \frac{\Delta R_4}{R_4} \right] (1 - \eta)$$

$$\underset{①}{\phantom{\left(\frac{1}{1 + R_s/R_I} \right)}} \underset{②}{\phantom{\left(\frac{1}{1 + R_o/R_m} \right)}} \underset{③}{\phantom{\left(\frac{a}{(1+a)^2} \right)}} \underset{④}{} \underset{⑤}{}$$

Term 1 is is the bridge input loading term showing interaction between the excitation source impedance and the bridge input impedance.

Term 2 is the bridge output loading term showing interaction between the bridge output impedance and the input impedance of the measuring instrument.

Term 3 is the bridge symmetry factor.

Term 4 includes all the bridge arm ΔR/R contributions.

Term 5 is the bridge nonlinearity, if any.

5.2.2 Simplify Terms Where Reasonable—Useful Approximations

For most cases, these equations can be simplified. Most modern signal conditioners have source impedances on the order of 0.01Ω in constant voltage (DC or AC) excitation modes. A typical strain gage resistance is 350Ω. Then, the Term 1 loading is equal to .9999714, and can safely beconsidered to be equal to one. This is as close to open circuit conditions as you are ever going to get without using a potentiometric circuit. In creating this open circuit condition you have recreated the transducer boundary condition that likely existed at calibration time.

A modern and typical readout device, such as a digital multimeter, oscilloscope, or A/D converter, will have an input impedance of $R_m = 10^7\Omega = 10M\Omega$. With a bridge output impedance of 350Ω, the output loading Term 2 becomes .999965 ~ 1.000. This, again, is very close to open circuit conditions and recreates the likely calibration condition.

However, if the readout/recording device had been a direct or FM/FM tape record/reproducer with a relatively low input impedance of $50K\Omega$—not $10M\Omega$—the loading factor would have been .993, representing a loading error of almost 1%. This level would have to be taken into account operationally at calibration time or corrected after the fact.

Most of the bridges used in experimental mechanics are balanced. An exception might be the external half bridge completed with resistors of another value in the signal conditioning. For the general balanced bridge case, $R_1 = R_2 = R_3 = R_4$ approximately. Therefore, $a = 1.00$ approximately. The Term 3 bridge symmetry factor is, therefore, $a/(1 + a^2) = 1/4$.

Term 4, the bridge nonlinearity term η, is somewhat more complicated. If we define $r = \Delta R/R$, the nonlinearity term becomes

$$\eta = \frac{\left[\dfrac{(r_1 + ar_2)(r_1 - r_2)}{1 + a + r_1 + ar_2}\right] - \left[\dfrac{(r_4 + ar_3)(r_4 - r_3)}{1 + a + r_4 + ar_3}\right]}{(r_1 - r_2 + r_3 - r_4)}$$

Be aware that the nonlinearity term, η, is a function of the relationships among the ΔR/R terms for the four bridge arms. Some bridge configurations have $\eta = 0$ which is an advantage since corrections are not necessary. Other bridge configurations have η not equal to zero, although it is generally small. The general rule is that if the bridge configuration forces the current through each half of the bridge to be constant (for constant voltage excitation) then the output is a linear function of the sum of the ΔR/R terms in the bridge arms.

We can now simplify the overall bridge equation and write it as

$$\frac{e_o}{V} \approx \frac{1}{4}\left(\frac{\Delta R_1}{R_1} - \frac{\Delta R_2}{R_2} + \frac{\Delta R_3}{R_3} - \frac{\Delta R_4}{R_4}\right)(1 - \eta)$$

Note the form of the equation. The input is in terms of $\Delta\Omega/\Omega$, milliohms/ohm or micro-ohms/ohm. The output is in terms of $\Delta e/V$, or millivolts/volt, or microvolts/volt.

This is an important point. A Wheatstone bridge is a millivolt per volt device—not a millivolt device. Its basic sensitivity is driven by $\Delta R/R$ in the bridge arms. You can now see why resistance based, non-self-generating transducers are specified in terms of millivolts per volt per full scale major input. Some manufacturers of piezoresistive accelerometers and pressure transducers define their transducer sensitivity, incorrectly, in terms of millivolts per G or PSI of input. The excitation numbers are buried somewhere else in the specification. This is usually because the manufacturer comes from a self-generating piezo-electric transducer world where this form of specification is legitimate. In the non-self-generating, Wheatstone bridge world, this form is not correct. So, be careful when dealing in this particular area.

For most applications in the experimental mechanics world, $1-\eta\sim1$, and may be neglected as a significant source of error.

If currents I_1 and I_2 in the opposite sides of the bridge are equal for all values of $\Delta R/R$, then the bridge is, by definition, linear. The best reference for bridge linearity understanding and calculation is by Measurements Group.[3]

In summary, there are two main things to note. Note the minus signs in the final bridge equation. The minus signs for the $\Delta R_2/R_2$ and $\Delta R_4/R_4$ contributions are the key to using the Wheatstone bridge as an analog computer. In the absence of these minus signs, the bridge would not be one-tenth as useful.

Second, note that equal $\Delta R/R$ values in adjacent bridge arms cause no output because of the minus signs in the final bridge equation. Assume $\Delta R_3/R_3$ and $\Delta R_4/R_4$ are zero. Given these values, if $\Delta R_1/R_1$ equals $\Delta R_2/R_2$, no output occurs.

5.2.3 The Bridge as an Analog Computer

The two minus signs in the final bridge equation are the key to its usefulness. The minus signs in terms two and four allow the Wheatstone bridge to operate as an analog computer. The location of the computer also is optimal. It is right on the part in the case of strain gages, or right inside the transducer in that case. No other measurement hardware or software assets are necessary to use this computing power. Several case histories will illustrate the point.

Case #1—The Quarter Bridge.   Assume constant voltage minor input and a single strain gage wired in bridge arm 1 with fixed precision resistors completing the bridge. This is the classic quarter bridge configuration for strain measurement. Here

$$\Delta R_1/R_1 = (\text{gage factor})(\text{strain}) = K_\varepsilon$$

$$\Delta R_2/R_2 = \Delta R_3/R_3 = \Delta R_4/R_4 = 0 \text{ since they are fixed resistors}$$

$$e_o/V = (1/4)(\Delta R_1/R_1)(1-\eta) = K_\varepsilon P/4 = K_\varepsilon/4$$

where, P = bridge factor = 1 for a quarter bridge (i. e., one active arm). The bridge factor P is driven by how the $\Delta R/R$s are arranged in the bridge by the designer.

Note: I mentioned the words "fixed precision resistors" in the previous paragraph. The operative words are fixed and precision. The nature of the Wheatstone bridge is this: it has parts per million sensitivity—but is *stupid.* It will report $\Delta R/R$ from its bridge arms, according to their arrangement, regardless of the source of the $\Delta R/R$. It cannot tell the difference between $\Delta R/R$ coming from what you want to measure, like strain, and $\Delta R/R$ from time instability and temperature change in the completion resistors and intrabridge wiring. Your only defense from this insidious problem is the use of fixed, precision resistors in the .01% resistance tolerance, 25 ppm/year drift or less, .6 ppm/F temperature coefficient class. These are not cheap resistors. They'll cost you about $10 apiece and are worth every penny. The only thing more expensive than this investment is the labor dollars you'll waste trying to diagnose problems in bridge output caused by your *not* spending this money. *Never* use cheap wire wound or carbon resistors inside a Wheatstone bridge. Use them in your stereo system or your garage door opener—but not in a Wheatstone bridge.

Case #2—The Full Bridge.   Assume a beam is loaded with a complex set of inputs. Of these, you want to measure only strains due to bending because you want to calculate bending stresses. You want to measure nothing else. The part has been strain gaged with a full bending bridge as shown in Figure 5.2. The full bridge means that all four bridge arms are active, so the bridge factor $P = 4$.

The load F will cause a moment around the Y-axis, M_y. The stress caused by this M_y is what we want to measure. For this bending load

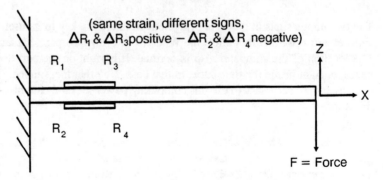

Figure 5.2   Cantilever beam gaged for bending

$$\Delta R_1/R_1 = \Delta R_3/R_3 = -\Delta R_2/R_2 = -\Delta R_4/R_4$$

since gages R_1 and R_3 see tension strains on the beam's top, and R_2 and R_4 see equal but compression strains on the beam's bottom, as the beam bends down due to load F. Therefore

$$e_o/V = (1/4)(\Delta R_1/R_1 - \Delta R_2/R_2 + \Delta R_3/R_3 - \Delta R_4/R_4)$$

$$= (1/4)(\Delta R_1/R_1 - (-\Delta R_1/R_1) + \Delta R_1/R_1 - (-\Delta R_1/R_1))$$

$$= (4/4)\,\Delta R_1/R_1 = \varepsilon K P/4 \text{ where } P = 4$$

So, the bridge works for bending about the Y-axis and give four times the output that a single gage would give. It has multiplied the normal output by four. You've installed an amplifier directly on the part—where it counts.

What about the other input loads which are noise levels to you since the bending strain due to M_y is the answer you want? Figure 5.3 illustrates the point.

Noise level: Axial load in the X-axis.   For any load F_x, the strains in the four gages will have the same magnitude and sign.

$$\Delta R_1/R_1 = \Delta R_2/R_2 = \Delta R_3/R_3 = \Delta R_4/R_4$$

$$e_o/V = (1/4)(\Delta R_1/R_1 - \Delta R_1/R_1 + (-\Delta R_1/R_1) - (-\Delta R_1/R_1)) = 0$$

The bridge is "compensated" for the axial load F_x.

Noise level: Lateral loads in the y-axis.   For any load F_Y, the strain in the gages will have the same magnitude strains assuming they are all the same distance from the neutral bending axis for M_y. R_1 and R_2 will be positive, while R_3 and R_4 will be negative.

$$\Delta R_1/R_1 = \Delta R_2/R_2 = -\Delta R_3/R_3 = -\Delta R_4/R_4$$

$$e_o/V = (1/4)(\Delta R_1/R_1 - \Delta R_1/R_1 + (-\Delta R_1/R_1) - (-\Delta R_1/R_1)) = 0$$

The bridge is "compensated" for F_Y and the resulting M_z.

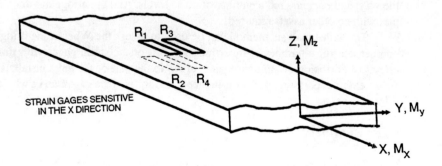

Figure 5.3   Full bending bridge measuring M_y

Noise level: Uniform Temperature Rise.   If we raised the temperature of the unstrained beam uniformly (same temperature everywhere), and all strain gages acted exactly alike to this stimulus, the $\Delta R/R$ for each gage would be similar. If they are all equal, $e_o/V = 0$ and the bridge is compensated for uniform temperature rise. We'll talk more about the transducer "alikeness" assumption in the upcoming noise level control and documentation chapter in the section on mutual compensation. The assumption is, in fact, somewhat dangerous but is made here for convenience in explaining the workings of the Wheatstone bridge.

Noise level: Temperature gradient in the X-axis.   This is the case of a temperature gradient along the beam length, its *X*-axis. Therefore, all gages installed at the same *x* station would experience the same temperature, T_1, and the $\Delta R/R$s would be the same, and $e_o/V = 0$ again. The beam is compensated for an *X*-axis temperature gradient.

Noise level: Temperature gradient in the Y-axis.   Here, the @ sign means "evaluated at." In this case R_1 and R_2 are at temperature T_1, and R_3 and R_4 are at temperature T_2. The equation reduces to

$$e_o/V = (1/4) (\Delta R_1/R_1 @T_1 - \Delta R_1/R_1 @T_1 + \Delta R_3/R_3 @T_2 - \Delta R_3/R_3 @T_2) = 0$$

The bridge is compensated for temperature gradients in the *Y*-axis.

Noise level: Temperature gradient in the Z-axis.   In this case R_1 and R_3 on the beam top are at one temperature T_1, while R_2 and R_4 on the beam bottom are at a different temperature, T_2.

Therefore,

$$e_o/V = (1/4) (\Delta R_1/R_1 @T_1 - \Delta R_2/R_2 @T_2 + \Delta R_3/R_3 @T_1 - \Delta R_4/R_4 @T_2)$$

$$= (1/4) (2\Delta R_1/R_1 @T_1 - 2\Delta R_2/R_2 @T_2)\ \text{NOT compensated for this case}$$

In summary, this bridge configuration is self-compensated for both orthogonal loads F_x and F_Y, two orthogonal moments M_z and M_x (although we did not do this analysis, the beam is also compensated for torque, M_x), uniform temperature rise, and temperature gradients in the *X* and *Y* axes. Not too bad. You can see that the Wheatstone bridge has allowed us to *compute in* what we want—bending strains due to F_z and multiplied the answer by four by the way, and *compute out* a number of noise levels. And the bridge has done it right on the part with no other assets required!

The same analysis method can be used to assess the Wheatstone bridge's ability to compensate for a number of other inputs—drift of the strain gages with time, long-term effects of corrosion or atomic oxygen, creep in the part or the gage installation, magnetic fields, external pressure, radiation-induced zero shifts in the gages, and a whole raft of other effects.

5.2.4 Summary

1. The Wheatstone bridge is an extremely sensitive analog computer at below the one part per million level. It exists right in your transducer or right on the test article itself.

2. It is used to: (a) "compute in" what you want to measure; (b) "compute out" what you do not want to measure; and (c) give you system gain of up to X4 right where you need it.

3. In certain configurations, the bridge output is a nonlinear function of $\Delta R/R$, regardless of the transducer linearity specifications.

4. Nonlinearities in the bridge output are usually negligable for experimental mechanics purposes (for a quarter strain gage bridge@1000 microstrain, $\eta=.001$ or 0.1% of reading). For precision work, they are not neglectable and must be corrected.

5. The bridge will behave in a fashion similar to DC constant voltage, for non-DC constant voltage waveforms like sine waves and pulse trains.

6. The Wheatstone bridge will behave differently for constant current excitation. This is the subject of the graduate course, however!

7. Bridge performance for at least span, output/input linearity and temperature compensation for zero and span will be compromised by the insertion of a "T" or "I" configuration balance network for zero control.[4] Performance with such a balance network must be understood but is beyond the scope of this book.

8. The stability of any component placed inside the bridge corners of a Wheatstone bridge is crucial. This applies, most particularly, to bridge completion resistors.

9. The Wheatstone bridge is very sensitive—but very stupid. It cannot distinguish among $\Delta R/R$s coming from its four bridge arms and the intrabridge wiring. That is the designer's problem.

5.3 METHODOLOGY FOR PROPER SETUP AND OPERATION OF WHEATSTONE BRIDGE BASED MEASUREMENT SYSTEMS

This section provides a valid and uniform method for the setup and operation of any measurement system using resistance based non-self-generating transducers and Wheatstone bridge signal conditioning. This general method is useful regardless of the actual measure and, transducer, signal conditioner/amplifier, or readout device. The steps should always be performed in the order shown, however, independent of the particular Wheatstone bridge system you are using. The following discussion assumes constant voltage excitation of any waveform (DC, sine waves, pulse trains), although the method is generic. If constant current excitation is used, only switch (#4) would be different as mentioned in the text.

5.3.1 Generic Setup Methodology

Verify the excitation level.   Prior to connecting the non-self-generating transducer to the signal conditioner input, verify that the excitation is approximately at the level you will want later in the process. You should know this level before you even begin setup. It is part of your system design.

Verification at this point will prevent inadvertent overvoltage on a transducer caused by the previous user not resetting it to a low value. This is of particular concern if your transducer has small components in it, or you are using strain gages of gage length less than .100 inches. In an overvoltage condition, the small gage area cannot dissipate its internal I^2R heating by conduction through the adhesive layer into the substrate, and its temperature will rise. This potential overheating can cause all kinds of problems with the later performance of the gage—if it is not destroyed outright in the overheating process. The key is to verify the excitation level before connecting the transducer cable to the rest of the system. After verification, connect the transducer and its extension cable to the signal conditioner/amplifier.

Further, always disconnect a transducer from a signal conditioner before pulling that conditioner out of its rack position. The act of pulling a powered signal conditioner/amplifier out of its case may cause voltage transients of 20–40 volts to flow out the cables to the transducer. This transient is high enough to fail small strain gages and strain-gage-based transducers. The correct order is: disconnect, pull signal conditioner/amplifier, replace signal conditioner/amplifier, verify excitation, reconnect transducer.

Establish the condition of zero output for zero input.   The measurement system in Figure 5.4 is a generic one. It is a model for your particular system. The features shown may or may not be in your hardware. If they are not there, I strongly recommend that you put them in and, certainly, into your next design. The system has: (1) a Wheatstone bridge based transducer measuring strain, load, pressure, acceleration, torque, moment, deflection, or anything else; (2) a signal conditioner providing excitation and housekeeping functions for the bridge output; (3) an instrumentation amplifier which may, or may not, be

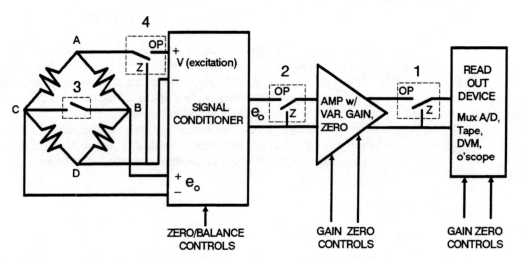

Figure 5.4   Wheatstone bridge signal conditioner with zero and noise level verifications switches

there; and (4) a readout device that could be a DVM/DMM, oscilloscope, meter, galvono-meter recorder, XY plotter, tape recorder, or multiplexer A/D converter.

There are two general initial conditions you want to establish in this measurement chain. The first is *zero voltage through the entire measurement system for zero major input*—zero load, strain, pressure, acceleration, torque, zero whatever. The second is *zero output for zero input, independent of system gain*. The second condition is as important as the first. You want the overall system zero to stay put if you have to change the system gain. If you do not create a gain independent zero, and you change the system gain at any time, you no longer know where zero is. You are lost within the measurement window and all DC information has gone "out the window!"

This methodology also allows almost instant problem solving when things go wrong. I've seen engineers and technicians scratch their heads for hours trying to determine where a certain unexpected DC level in a measurement system was coming from. Invariably, when you see this symptom, the measurement system was not set up for zero out for zero in everywhere. There were compensating DC offsets all over the place. When the pressure is on, this condition is deadly to efficient problem solving.

Setting system zero and span uses switches #1 and #2 in the figure. Switches #3 and #4 are used in noise level hunting and documentation, and will be discussed in a later chapter. The zeroing, or balancing process begins at the right end of the measurement system and proceeds to the left. Switch #1 shows as an actual electrical switch at the input to the voltage sensitive readout device. The switch may or may not be there in your system. You may have to pull the readout device inputs and electrically short the input terminals to cause a condition of zero voltage input. A short circuit is defined as zero volts. An open circuit is not allowable here—an open circuit is zero for current, not voltage. You cannot merely pull the inputs out and assume zero—you must short circuit the input terminals for a voltage sensitive device. Exceptions to this would be unbuffered, current sensitive galvanometers and piezoelectric transducers used with charge line drivers. In these cases, use open circuits for zero checks.

With the input to the readout device shorted (in a condition of zero voltage input), set the gain on the readout device to its lowest value, or its range to the highest value. Using the zero of balance control set the output to zero. Increase the gain to the highest value, or the range to the lowest value, and repeat the zeroing process. Run the device gain through its entire range, and verify that zero stays put on all ranges. When you achieve this condition, you have a balanced system with a gain independent zero. Return the switch to the *OP*erate position, or reinstall the input cable. If the readout device has gain control, set this control to an intermediate value.

Move to switch #2 at the amplifier input. Repeat the balancing procedure as before with the amplifier. With its input shorted (switch #2 in the Zero position), vary the zero balance controls until the amplifier output voltage is zero and independent of gain. Return the switch to the *OP*erate position.

Move to the signal conditioner zero or balance control. Create conditions in the transducer such that the measurand is zero. Adjust the balance or zero control until the signal conditioner output is zero. Verify that the voltage at the readout device input is still zero within your tolerances. If it is not, find the problem and correct it.

You have now created the following valuable conditions:

1. You have a measurement system whose electrical zero at any interface corresponds to the zero value of the input measureand; and
2. Your system's zero is independent of gain.

You now know that any voltages you see at any interface in the system are, except for noise levels, coming from the major input, the measureand, and not from some arbitrarily set zero control. This is a very comforting and convenient situation to have. You must know, at any point in the system, whether the voltage you see is the right answer or not. This can be extremely difficult with arbitrarily offset zeroes. Do not use them unless you absolutely have to. I do not recommend offsetting the mechanical zero on a transducer to positive or negative full scale and using the entire plus and minus full scale range. This can become confusing and is easy to forget.

Establish the system span.   There are two primary methods for setting system span. In method #1 you set the transducer excitation voltage to the manufacturer's recommended excitation level. You then apply the appropriate engineering unit input (dead weight load for a load cell, deflection for a deflection transducer) or a shunt calibration, and you vary the system's gain until you have the voltage you want at the readout device input. It may be the case that there is no amplifier in your system. Here the data will usually be read on some form of autoranging device such as a digital multimeter or a software-ranged multiplexer A/D converter.

In either case, after setting the system's full scale output, check linearity by applying lower equivalent value shunt calibrations or engineering unit inputs. If possible, check full scale and linearity in the opposite direction. Never run a test with a single value for shunt calibration in a single direction if you can possibly avoid it. Single value, single polarity span setting data proves *nothing* with regard to system linearity. Multiple value shunt calibrations in both directions, or multiple value engineering unit inputs are documentation, by definition, of system linearity.

Some people in the business seem to have trouble with this advice on the grounds that using multiple value bipolar levels for span setting and linearity documentation takes too much time. They say, "Time is money!" and it takes too long to do this. This position is penny wise and pound foolish. A little care in setting and documenting your system spans and linearity will save you hours when trying to explain or diagnose a measurement system problem. Given a successful test and valid data, your customers will quickly forget the short amount of time you spent checking this. They will never forget the serial time you waste trying to diagnose a zero or linearity problem after not having performed the check.

Method #2 is used when it is necessary, for some reason, to operate with fixed and trustable settings on your system gain controls. Here you want to operate with your variable gain controls in the detents or at the ends of their settings so you know they have not been changed inadvertently. In method #2, you set the gains at convenient but fixed gains, then vary the transducer excitation to get the overall sensitivity you want. When varying the excitation, do not vary it more than a factor of two away from the recommended value from the manufacturer. If you find that you're running with the excitation at 20% of the recommended value, decrease the overall system gain and raise the excitation level. Remember, in any Wheatstone bridge based measurement system you have several gain controls, with the excitation being one of them. It is a legitimate design parameter for the designer to use.

After the system span is set, recheck the system zero by reapplying zero measureand at the input. Readjust the signal conditioner balance control to get zero out/zero in conditions. You no longer have the right to adjust the amplifier or readout device balance controls. After rechecking system zero, recheck the span and adjust as required.

Whether you are using method #1 or #2, you must now record the excitation value used to run the test. Write it down.

5.3.2 General Rule for Setting Transducer Excitation Levels

Always set the excitation for a non-self-generating transducer as high as you can commensurate with overall measurement system zero stability. This is where too high an excitation value will show—the system will have an unstable zero. Use the manufacturer's rated excitation if you can. For strain gages, see vendor literature.[5] Maximizing the excitation in this manner maximizes the transducer's internal signal-to-noise ratio and will improve its performance. It allows you to operate down-stream amplifiers and readout devices at lower gains, which will generally increase their zero stability. Put the gain in the system as close to the transducer as you can—namely *in* the transducer by maximizing the excitation.

How do you effectively tell what the maximum excitation level should be? A good practice is to set the excitation at a low level, say 1 volt, and monitor the system zero output. If stable, double the excitation to two volts and monitor again. If stable, double again to four volts and monitor. Eventually, with increasing excitation levels, the transducer cannot dissipate enough internal I^2R heating, will become thermally unstable, and will exhibit zero drift with time. You have identified how much is too much! Decrease the excitation level until the output becomes stable again and run.

5.4 THREE WIRE CIRCUITS FOR STRAIN GAGES

The bonded resistance strain gage is the most common transducer used with the Wheatstone bridge in experimental mechanics work. Its purpose is the measurement of surface strain on a test article with the possibility of the calculation of surface stress. The strain gage is the perfect example to use in these discussions of the Wheatstone bridge. Its properties can be used to illustrate many significant issues in the use of the bridge.

As mentioned before, the Wheatstone bridge is an exquisitely sensitive, very stupid device. The bridge will react to $\Delta R/R$ in its bridge arms regardless of the cause. It cannot discriminate between $\Delta R/R$ caused by strain you might want to measure, and $\Delta R/R$ caused by an increase in magnetic fields or having the temperature of the lead wires change. The bridge simply does know where the $\Delta R/R$ comes from. It is your job as the system designer to assure yourself that the $\Delta R/R$ is coming from the measureand you want, and not from some noise level you don't want.

5.4.1 Two Wire Circuit

To explain how sensitive the bridge is, take an example of a 350Ω strain gage in the quarter bridge (single active gage, three fixed bridge completion resistors, $P = 1$) configuration with constant voltage excitation. This arrangement shows in Figure 5.5. Assume the gage is connected to the remaining bridge arms with 50 feet of 22 AWG twisted shielded cable in a two wire configuration. You can see that the lead resistance in the 50-foot cable appears as R_{L1} out to the gage and R_{L2} coming back. This length and wire gage would have values of $R_{L1} = R_{L2} = 1\Omega$, the total series resistance being 2Ω.

What happens in this bridge configuration if the temperature of the *lead wires* changes by just 1°F? The copper leads have a thermal coefficient of resistance of about $+1800 \times \text{x}10^6 \Omega/\Omega$ /F around room temperature. This means there will be a ΔR in the lead wires of .0036Ω caused by the 1°F temperature change. Not much, right? The bridge does not know where this ΔR is coming from. Here, it is a noise level—but the bridge is too stupid to know it and reports it at the bridge output. How much error is created in this 1°F benign scenario?

$$\Delta R_L = (1800 \times 10^{-6}\Omega/\Omega/F)(2\Omega)(1F) = .0036\Omega$$

$$\Delta R/R = (\text{strain})(\text{gage factor}) = \varepsilon K$$

$$\varepsilon = \Delta R/RK = (.0036)/(350)(2) = 5\ \mu\varepsilon!$$

The bridge temperature sensitivity in this configuration is then *5 microstrain per degree F* from the lead wires alone! This is clearly unacceptable even in a laboratory ambient environment where = +/– 10°F changes are routine. What you have with the two wire circuit is a thermometer!

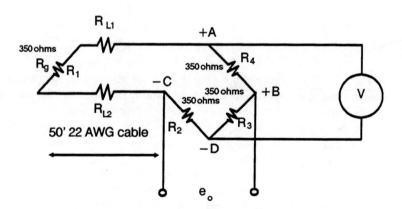

Figure 5.5 Two wire strain gage circuit

For this reason, it would always bother me when another contractor's hardware would arrive in our laboratories for structural tests wired with two wire strain gage circuits. The reason given by the contractor was that less labor was needed using two wire, not three wire hookups—about 5% less perhaps. Then we had to deal with the resulting continual zero shifts in the measurements as the structure's temperature wandered around in the test laboratory. This is a particularly difficult task if you are using the zero returns from a structural loading to detect yielding or slippage in the structure, and the article's temperature is eight degrees warmer at the test completion than when the test began hours before. Are those 40 microstrain from the temperature change or have you yielded the test article? a two wire Wheatstone bridge hookup cannot tell you the difference.

If you are, for some reason, forced into using the two wire circuit (and we have been due to feed through connector pin limitations into vacuum environments) use high resistance strain gages (500Ω or 1000Ω) and extension cables of short length and very low resistance. This technique is minimization by division and will be discussed in a later chapter.

5.4.2 The Three Wire Circuit

Equal $\Delta R/R$s in adjacent bridge arms cause no bridge output as you saw when we developed the bridge equations. The problem in the two wire circuit is that the $\Delta R/R$s caused by lead wire resistance change with temperature appear in series with the legitimate $\Delta R/R$ from the strain to be measured.

Can you arrange the bridge so the ΔR_{L1} and ΔR_{L2} terms appear in adjacent bridge arms and cancel each other? Can you promote them "sideways" and out of the line of fire? Can you promote them out of the bridge? The fact that this section is titled "Three Wire Circuit" says yes.

This is done by adding a third wire and moving the C bridge corner from the bridge completion network in the signal conditioning out to the actual strain gage on the test article as shown in Figure 5.6. Notice the R_{L1} is now in bridge arm 1 and R_{L2} have moved to bridge

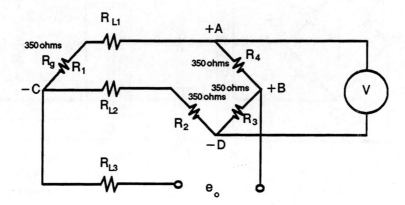

Figure 5.6 Three wire strain gage circuit

arm 2. Now, as the temperature changes, the $\Delta R_{L1}/R$ and $\Delta R_{L2}/R$ terms are approximately equal, in adjacent bridge arms, and tend to cancel each other. The remaining lead wire resistance, $R_{L3} = 1\Omega$ in this example, is now in series with the high input impedance (1MΩ in any case) of the measuring device reading e_o. Its effect is also now promoted sideways where it does no harm.

The three wire circuit is clearly the circuit of choice in the experimental mechanics measurements business. Single bonded foil strain gages should be wired with three wire circuits as a standard. The two wire circuit is never used for quarter bridge strain gages unless we are forced into its use.

5.5 SINGLE SHUNT CALIBRATIONS FOR WHEATSTONE BRIDGE TRANSDUCERS AND STRAIN GAGES

5.5.1 How and Where to Apply Single Shunt Calibrations

In Figure 5.7 you see a typical Wheatstone bridge transducer sitting at the far end of an extension cable from the signal conditioning. In the signal conditioning are a resistive "T" balance network for zero control, a precision regulated constant voltage power supply (DC, sine waves, or pulse trains), and a shunt calibration capability. This generic capability is inside 90% of the bridge signal conditioners and panel meters for bridge transducers in the world.

Note: The generic but inaccurate term *"shunt calibration"* is in use all over the world. The problem is the use of the word "calibration." The application of a bridge unbalance

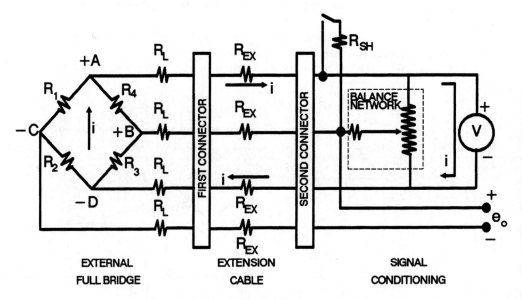

Figure 5.7   Single shunt calibration—improper method

using a resistor to shunt one or more bridge arms is in no way a *"calibration."* The transducer calibration must be previously available to understand the value of the shunted bridge unbalance. The purpose of the shunt is: (1) to verify the transducer is physically there and electrically operating; and (2) to create a known bridge unbalance associated with a similar unbalance determined at calibration time from real engineering unit inputs. The shunted bridge unbalance is used then to *span* the rest of the measurement system. It "calibrates" nothing. *Span verification* is a more technically rigorous term but is little used. In the aerospace industry, the measurement system user does not even have the right to use the term "calibration." That term is reserved with government agreement for an official metrology process with all the administrative record keeping and traceability that accrues to that function. As a user of measurement systems, we never *calibrate* anything. That term is reserved for our metrology organization. But we span verify everything! I will, however, use the term "shunt calibration" in the interest of readability and understanding.

In the figure, the shunt calibration resistor, R_{SH}, is applied across the leads going to the +A excitation and +B signal bridge corners. This would give a negative shunt calibration. The shunt also could be applied across +B and −D giving a positive shunt. This is a very convenient thing to do. Many vendors give the user access to these pins on the back of his instrument. Users also have breakout boxes laying all over the laboratory used to "break" into the signal path for signal injection and trouble shooting. These boxes often cause more trouble than they cure.

At least three basic rules of shunt calibration are violated in this example.

1. ALWAYS SHUNT THE EXTERNAL HALF OF THE BRIDGE.
   In this example, bridge arms 3 and 4 are shunted by the calibration. They are also shunted by the balance network. The voltage output from the shunt calibration will be a function of the balance position of the balance network. In other words, span is a function of zero, and that is the definition of a nonlinear system. Bridge arms 1 and 2 are unencumbered by the balance network and are the proper bridge arms for single shunt calibration.

2. ALWAYS SHUNT NONCURRENT CARRYING LEADS
   Note that one side of the shunt goes to the +A positive excitation lead. This lead is carrying the entire bridge current. Large calibration errors will result from the *IR* drop in this lead.

3. METROLOGY CALIBRATION VALUES ARE ONLY VALID WHEN THE METROLOGY LEAD CONFIGURATION HAS BEEN MATCHED (SOMETIMES DIFFICULT) OR A CIRCUIT USED THAT IS INSENSITIVE TO LEAD LENGTH.
   Note that the resistance of the extension cables, R_{EX}, is not accounted for in this setup. They probably were not even there in the metrology laboratory.

Figure 5.8 shows the correct way to approach this situation. Here, a seven wire (plus shield for a total of eight) system shows. A five wire system would be used for a quarter bridge strain gage application—three wires for the gage plus two wires for external shunt

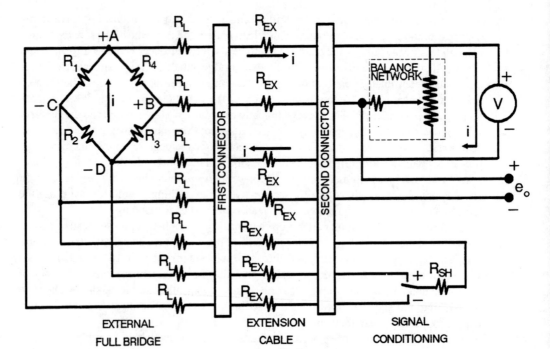

Figure 5.8   Single shunt calibration—proper method

calibration (plus shield). The calibration wires go all the way out to the transducer where they are connected to the appropriate bridge corners (plus and minus excitation and minus signal).

You can see that this wiring scheme allows only the shunting of the external bridge half, arms 1 and 2, satisfying rule #1. This wiring method precludes the shunting of current carrying leads, satisfying rule #2. It is, essentially, independent of extension cable length, satisfying rule #3. The extension cable resistance, R_{EX}, appears now in series with the shunt calibration resistor that has a minimum value of 25KΩ in our laboratories. These errors are negligible, since we have promoted them sideways and out of the way.

In reality, we run the shunt calibration wires to the transducer side of the first connector in most cases and make the connections to the wires from the bridge corners there. We can do this because the cabling from the installed strain gage bridge or commercial transducer to the first connector is very short.

I highly recommend this method of wiring for strain gages and Wheatstone bridge based transducers.

5.5.2 Metrology Calibrations and System Span Setting

Although the methods of a metrology organization will vary from company to university to national laboratory, you ought to be able to count on certain information being there in the calibration certificate. I'll use a TRW load cell calibration as a model for a fairly good calibration report. The calibration certificate shows in Figure 5.9. I've divided the certificate into six sections for clarity.

Section #1 defines the transducer type (Wheatstone bridge load cell), the manufacturer, model, capacity and engineering unit input, a unique property number, and the manufacturer's serial number. The property number is an administrative tool to account for the load cell as company property only. It has no other use.

The serial number, however, uniquely identifies that transducer from all others in the transducer universe. It is crucial that this serial number be affixed to all records pertaining to the cell's use on a test. The serial number also is the key to getting back into the vendor's data base for information he did not give you at procurement time.

Section #2 defines some of the electrical boundary conditions surrounding the six terminal transducer at calibration time. It states that the excitation was 10.000 VDC. It should state the source impedance of the power supply but does not. It gives the pin configuration for all necessary bridge corners and shunt calibration points.

Further, it states that the load cell was calibrated with a balance network composed of a 25KΩ potentiometer and a 10KΩ limit resistor as shown in Figures 5.7 and 5.8. These values match those in the signal conditioners used to operate these load cells in test. Here, we exactly recreate the boundary conditions of calibration, at test time. The balance network is in there so that the evil things balance networks can do to transducers[6] are, at least, recreated at test time. It does not tell us the input impedance of the reading digital multimeter, which it should. We know verbally that this impedance is greater than 10MΩ, so bridge output loading errors are negligible if this value, or larger, is used in test—which it is.

The actual tension calibration data for this load cell is in section #3. Millivolt outputs are given for increasing and decreasing loads in 10% increments. Column four shows the results of a second order polynomial curve fit for this calibration data. Compression data for the load cell are on a separate page.

Section #4 lists the diagnostic information regarding the calibration. On the left are the equation form and the coefficients for the second order fitted polynomial. We do not use this data at all. Setup on the floor assumes a linear model. It is highly likely that your use condition will too. Transducer nonlinearity of .012% FS (of full scale) and hysteresis of .041% FS are noted. Combining these two terms, as is sometimes done in combined linearity and hysteresis, is left to the student!

Output drift is a poorly named, but important number. Output drift tells you that the full scale millivolt value from the computer curve, 19.362 millivolts, has "drifted" .01% from the last calibration. It is, in fact, the change in transducer span from the last calibration. Zero balance is the open circuit output at zero load input for rated excitation measured with no balance network. This number should change very little over the years in a properly handled and used transducer. This value tells us if the transducer has been overloaded since its last calibration. Both the output drift and the zero balance are used to track the transducer's long-term health and performance.

Section #4 also tells you that the uncertainty at calibration time was ±.05% FS. The line right under that one, "NONLINEARITY, HYSTERESIS, AND DRIFT ARE CALI-

```
TRW
MEASUREMENT & COMPUTER RESOURCE CENTER        REPORT OF MEASUREMENT

DESCRIPTION:  LOAD CELL                        CAPACITY:   50 LBF
MANUFACTURER: INTERFACE                        I.D. NO: NP-100994         1
MODEL:        1410-AJ                          S/N:        19996
```

```
                        WIRING CONFIGURATION
           EXCITATION: 10.000 VDC         BALANCE WIPER (10 K OHM):B
           + INPUT : A , - INPUT : D      BALANCE POT(25 K OHM): A & D    2
           + OUTPUT: B , - OUTPUT: C      SHUNT CALIB: 6 & H
```

```
                        TENSION MEASUREMENT
   APPLIED        INCREASING      DECREASING         FITTED CURVE
  LOAD (LBF)    RESPONSE (mV)    RESPONSE (mV)        VALUE (mV)
 ----------    -------------    -------------        ------------
      0.0          0.000            .003               -0.000
      5.0          1.937           1.940                1.936
     10.0          3.871           3.877                3.872
     15.0          5.807           5.812                5.808
     20.0          7.742           7.750                7.744        3
     25.0          9.679           9.686                9.680
     30.0         11.617          11.622               11.616
     35.0         13.555          13.557               13.553
     40.0         15.489          15.491               15.489
     45.0         17.426          17.426               17.426
     50.0         19.361                                19.362
```

```
FOR: mV OUTPUT = A + B*(LBF  ) + C*(LBF  )^2  Nonlinearity =   .012 % FS
     A coefficient = -1.1888E-04             Hysteresis   =   .041 % FS
     B coefficient =  3.8715E-01             Output Drift =   .010 % FS
     C coefficient =  1.8648E-06             Zero Balance =   .150 mV      4
Standard Deviation = SQR(∑(d - d̄)^2/(n - m))              =  3.40E-3 mV
CSD Measurement Uncertainty: +/-   .050 % of applied load.
NONLINEARITY, HYSTERESIS AND DRIFT ARE CALIBRATED TO +/- .2  % FS
```

```
                        SHUNT MEASUREMENT
            SHUNT                RESPONSE              EQUIVALENT
       RESISTANCE (Ohms)           mV                    LBF
       ----------------          --------            ----------       5
          500  K                  1.748                4.515
          100  K                  8.722               22.526
           50  K                 17.420               44.986
           25  K                 34.721               89.644
```

EMC recommends the UUT only be used between 10 - 100% of Mfr rated capacity.

All Measurements are in accordance with MIL-STD-45662A.

This data was derived from TRW/Measurement and Computer Resource Center (MCRC) standards traceable to national standards, ratio type of calibration methods, or fundamental/natural physical constants. Supporting documents and data are on file for inspection at TRW Calibration Services Department (CSD), Redondo Beach, California.

```
Laboratory Ambient Conditions:    Measured by:  D. FRAZIER    Date:13 Jan 1992
   - Temperature  67.0°F
   - Humidity     45 % RH         Approved by:  C LULKA
```

Figure 5.9   Metrology calibration data for a load transducer

BRATED TO +/−.2%" fulfills part of our metrology's agreements with our government customers. If any combination of these three numbers exceeded .2% FS, our metrology organization would notify us that a threshold had been exceeded, and ask us to look at the transducer's uses since the last calibration and determine if any test objectives had been compromised.

Section #5 includes the determination of the shunt calibration bridge unbalances for four standard shunt resistances—25KΩ, 50KΩ, 100KΩ, and 500KΩ. We have a small disagreement with our metrology over this section. We prefer that these shunt equivalent loads be derived using the actual calibration data. Our metrology prefers to use the second order polynomial. We have not yet convinced them of the righteousness of our position. In any case, the differences are a few hundredths of a percent.

Section #6 includes information on applicable MIL-Specs, record keeping and calibration processes, and *the golden calf of traceability.*

Overall measurement system span is set using the shunt calibration values defined in section #6. What these numbers say is that, for any of the four shunt resistance values across the noted pins, the calculated "equivalent lbf" in column three would give the same unbalance as the corresponding shunt calibration.

The measurements engineer's relationship with the metrology folks in the organization is crucial to success on the floor. Both organizations have a difficult job to do. The smart measurement engineer takes care of the metrology folks by never surprising them and working closely when calibration requirements change.

NOTES

1. W. M. Murray and P. K. Stein, *Strain Gage Techniques;* University of California at Los Angeles, 1961; copyright by the authors; pp. 158–218.
2. Lawrence C. Shull, *Chapter 4—Basic Circuits;* Strain Gage User's Handbook, Elsevier Applied Science, New York, 1992, ISBN 1–85166–686–9.
3. Measurements Group Tech Note TN-139, *Errors Due to Wheatstone Bridge Nonlinearity*, Measurements Group, Raleigh NC, 1974.
4. See note 1.
5. Measurements Group Tech Note TN-502, *Optimizing Strain Gage Excitation Levels*, Raleigh, NC, 1979.
6. See note 1.

Chapter 6

Self-generating Transducers

6.1 SELF-GENERATING TRANSDUCERS

The other of only two types of transducers is the self-generating type. Self-generating transducers provide information and energy at their outputs directly upon imposition of the major input, with no requirement for a secondary, or minor, input. Examples include piezoelectric transducers for acceleration, moment, and pressure, and thermoelectric transducers (thermocouples) for temperature. This transducer class has distinct advantages and disadvantages in comparison to non-self-generating transducers. They are neither better nor worse—they are fundamentally *different*. Those differences will be discussed at length in this section.

Within the past year I have been disabused of a definition I've been using for years. The piezoelectric effect is, formally, defined only for crystalline materials such as quartz. Other organic and inorganic noncrystalline materials can exhibit "piezoelectric like" behavior, but they are not formally piezoelectric. Entire transducer businesses exist on this distinction. In this section I'll use the more liberal user's definition of the piezoelectric effect for clarity. If it looks like the piezoelectric effect, quacks like the piezoelectric effect and waddles like the piezoelectric effect—it's the piezoelectric effect!

There is much emotion on this self-generating versus non-self-generating transducer issue in the test measurement business. It's almost as if there was a contest going on among the adherents of each class. Most laboratories doing vibration, acoustics, shock, and modal survey testing use self-generating, piezoelectric transducers almost exclusively. Folks working in structural test laboratories use non-self-generating transducers equally exclusively. I've seen both sets of these folks on occasion go down in flames over a problem crying for a measurements solution in the other group's repertoire—but attempted with their hardware because that's all they understood. That is very unfortunate when it happens. To a craftsman whose only tool is a hammer, all problems look like nails. To a professional measurements engineer who understands both methodologies, the problem becomes one of using the right tool for the right job.

Editorial: In my experience, the most effective measurement organizations are those whose charter and skills allow them to cross testing discipline boundaries easily. They support the entire gamut of testing activities with an entire gamut of measurements tools at their disposal. When measurements groups are limited by artificial organizational boundaries—like the dynamics group makes only dynamics measurements, the structures group makes only structural measurements, the thermal group makes only thermal measurements, and the pressure group doesn't talk to anybody—the entire measurements function suffers. Organized in this manner, people tend to think measurements associated with their test discipline are somehow exclusively different from those in any other test discipline. In these organizations the measurements function is sort of bolted onto the *real* engineering function—which is controlling the test or producing the product. I've been at companies where from the transducer mounted on a turbine compressor blade to the analyzed data on an FFT plot, *five* organizational and *three* union boundaries are crossed. In this fractured system, who owns the measurement problem? When all the finger pointing is over, who stands up for the data? Save us all from this organizational myopia.

To a professional measurements engineer, a 100 KHz acceleration measurement on a shock test today is the same as a .01 Hz force measurement on a structural test next week. These folks daily cross artificial test discipline boundaries as if they were not there—which they shouldn't be. The technical approach, mathematics, and methods are consistent. Only the hardware and the frequency regime may be different.

6.2 THE PIEZOELECTRIC EFFECT

The Curie brothers in France in 1880 first discovered the piezoelectric effect. One of these brothers went on to marry Marie Curie and won the Nobel prize with her for another discovery. Again, we have a case of twenty-first-century problems solved with nineteenth-century solutions!

The explanation of the piezoelectric effect is best understood by considering pure quartz, which was the first material discovered to have these fascinating properties. Quartz is still in use today as a premier piezoelectric transducer material of choice. Pure quartz transducers are manufactured for the measurement of acceleration, force, pressure, and moment and torque. A similar, but more complicated, discussion can be made for man-made piezoelectric transducer materials such as ferroceramic lead zirconate titanate and bismuth titanate. However, these discussions would add little to your understanding as a user of the technology. If you need to know more, contact the transducer manufacturers who use them, such as Endevco,[1] Kistler,[2] and PCB Piezotronics.[3]

If a disc of monocrystalline quartz (silicon dioxide, SiO_2) cut normal to the crystallographic X-axis is loaded with a compression force, it will yield an open circuit electric charge of approximately 0.5 picocoulombs per pound. This charge occurs when the crystal deforms under load as shown in Figure 6.1. If the disc is cut normal to the crystallographic *Y*-axis, it will generate approximately 1.0 picocoulomb open circuit per pound of shear force in a certain direction as shown in Figure 6.2. The charge is, again, caused by the deformation of the crystal under shear load. This property of generating electrical charge when the crystal deforms is called piezoelectricity. The term comes from the Greek verb "Piezein"—to press.

Forces applied in other orthogonal directions will generate no charge. These crystals are highly directional, elastic, very stiff, show almost no hysteresis, and are thermally stable. Quartz is, in short, an almost perfect transducer material. Other man-made piezoelectric materials are not as good as quartz in these performance criteria, but show advantages in other areas that will be discussed below.

The charge generation effect can be understood as follows. As the crystal deforms due to the applied compression or shear force, the silicon atoms surrender four electrons in their outer shell to the two oxygen atoms in each SiO_2 molecule, whose outer shells are completed with two electrons each. The silicon atoms now have a quadruple positive charge and the oxygen atoms have a double negative charge. Electrodes covering the ends of the crystals collect this charge distribution. The examination of the physics of man-made piezoelectric materials is much more complicated, but the result is the same—charge is generated due to the deformation of the material in response to an applied force. In short, all piezoelectric transducers are actually force transducers masquerading as pressure, load, and acceleration transducers.

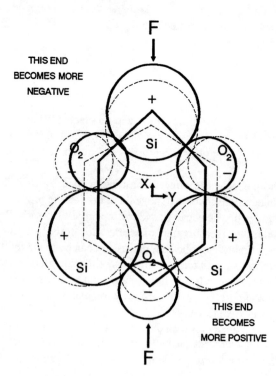

Figure 6.1 Piezoelectric effect of X-axis compression on quartz

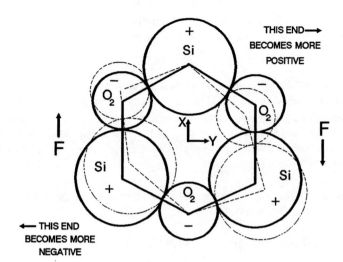

Figure 6.2 Piezoelectric effect of Y-axis shear on quartz

If building a force transducer interests you, let the force being measured act directly on the transducer material itself.

If acceleration is what you're after, use the relationship from physics that says

$$ACCELERATION = FORCE/MASS$$

There would be a proof mass that, by nature of its being accelerated, applies inertial force to the crystal.

If pressure measurement interests you, use the relationship

$$PRESSURE = FORCE/AREA$$

The pressure to be measured is collected by the area of a diaphragm. The resulting force is applied through a flexure system to the piezoelectric material.

6.2.1 Major Characteristics of Piezoelectric Transducers

Piezoelectric transducers have several characteristics that make them very suitable for certain applications and unusable for other applications.

Span to Threshold Ratio.   Properly designed piezoelectric transducers can exhibit ratios of full scale charge to threshold charge of 100,000,000 to one! They have an inherently huge dynamic range.

Directionality.   Piezoelectric materials, quartz in particular, have highly directional responses. They are very good in applications where orthogonal components of force or acceleration are to be measured separately, such as use in a multicomponent dynamometer.

Size and Weight.   Piezoelectric transducers can generally be made, all else being equal, in smaller and lighter packages than bridge type transducers for the same level of performance.

Stiffness.   The elastic modulus of the transducer material can be very high. This is true for natural quartz and the man-made materials. They are stiffer than the general metallic transducer materials. The result is piezoelectric transducers can be made with very high resonant frequencies. This, in turn, means they can be used over a wider frequency range. Some commercial transducers for the measurement of high frequency pressure or acceleration have first resonances of over 1000 KHz.

Temperature Range.   Depending on the actual materials used and the design, these transducers can have operating ranges from –450°F to over 700°F.

Hysteresis.   These transducers generally show very little hysteresis in their loading and unloading cycles.

Long-Term (static) Measurements.   Piezoelectric transducers are charge generators. A statically deformed crystal will generate a static charge for open circuit conditions. If you place, across the terminals of the transducer, a device with a noninfinite input impedance, this static charge will drain off through that impedance in an exponential fashion. Thus, piezoelectric transducer-based systems cannot be used for static measurements. Systems are available for stretching this low frequency time constant—but these systems still will not respond to a static level charge. This behavior can be used to the designer's benefit if the problem is measuring the dynamic components of some measureand in the presence of a DC level of no interest. An example would be the measurement of small dynamic pressure variations in the presence of a static chamber pressure in a rocket engine or chemical laser combustor, or vertical seismic accelerations in the presence of 1g.

Self-Generating Behavior.   Piezoelectric transducers are, of course, self-generating. The electrical analog of the input measureand is available to you directly with no minor or secondary excitation required. This generally makes for simpler signal conditioning. However, since no excitation is required, the power of using non-DC excitation waveforms for noise level identification is not available to you. Whatever comes out—comes out.

T-Insertion Capability.   Piezoelectric transducers, along with the circuitry to condition them, usually called charge amplifiers, can be operated in a very advantageous and unique mode. This is called the T-insertion mode and will be discussed in detail below. This mode allows in-situ: (1) independent span verification (see chapter on Measurement System Operations); (2) verification of the measurement system electrical transfer function and linearity; (3) verification of channel ranging; (4) detection of damaged or grounded transducers and certain cable problems. This capability is, unfortunately, seldom used in industry. It is *institutionalized* in our laboratories because of the spectacular benefits available from the methodology. Piezoelectric transducers with internal signal conditioning (unless designed for this capability) cannot be operated in this mode, which is a major drawback to this configuration.

6.2.2 Low Frequency Considerations

Piezoelectric transducers can be signal conditioned with either external or internal voltage or charge sensitive electronics. Manufacturers of internally conditioned transducers generally do not tell you which type is used. You have to test to find out.

When using charge sensing electronics, the input impedance of the signal conditioning is made colossally high—like 10^{12} Ω. The overall system low frequency limit is driven by the series voltage amplifier after the charge converter. The low frequency behavior is due to the low cutoff frequency of this series voltage amplifier and its terminal roll-off. The transducer and its cable capacitance have little to do with the governing low frequency behavior. In fact, charge converters can be procured that give a low frequency time constant of over 1000 seconds if the series voltage amplifier is DC coupled. This allows quasi-static calibrations.

In the voltage mode the situation is different. There are two characteristics governing low frequency behavior. The first is the RC time constant formed by the transducer and cable capacitances and the input resistance of the voltage amplifier. This system will roll-off at the low end with a slope of +1. This time constant is then followed by an AC coupled voltage amplifier with its own cutoff frequency and slope of at least +1 at the low end. The series sum of these two systems will roll-off at the low end with a slope of +2 or more. This slope will cause this system to undershoot to a step as shown in the chapter on frequency response. The point is that the low frequency response of any piezoelectric system will be a function of whether charge or voltage sensitive signal conditioning is used.

6.2.3 High Frequency Considerations

The system high frequency response will be governed by two characteristics that you have seen before. The signal conditioning electronics will have its own high frequency cutoff frequency and terminal roll-off that the manufacturer should specify. Be careful of internal low-pass filters that may lurk inside the signal conditioning case. This first limit has nothing to do with the transducer and its cabling.

The second characteristic is the (usually) highly underdamped, resonant response of the transducer to high frequency excitation. The overall system looks like a highly resonant component in series with a measurement system with a high frequency cutoff of its own. This is the case presented in the previous section on nonlinear system performance creating higher frequency components.

6.2.4 Sensitivity

Piezoelectric transducers have the unique situation in which their sensitivity to the major input, pressure, acceleration, or torque, can be specified in two ways. The first is to state that sensitivity in terms of charge generated per unit input, typically picocoulombs per PSI, G, pound or inch-pound. This value is highly insensitive to the capacitance (length) of the connecting coaxial cable. This is the major advantage of operating in the charge mode. As you increase the cable length, noise levels will come up, but the charge sensitivity essentially remains constant.

The second method is to specify the sensitivity in the voltage mode in terms of volts per input unit, typically millivolts per inch-pound, pound, PSI, G. This specification is valid only if the capacitance external to the transducer in its cable is stated. As this external capacitance increases with increasing cable length, the voltage sensitivity decreases.

There are a number of relationships that can be used to calculate and check sensitivities. Endevco has done the best explanatory job in their *Piezoelectric Accelerometer Instruction Manual.*[4] The discussion presented in that manual applies for pressure, load, and torque transducers as well.

There is a very interesting zinger lurking in this discussion. The Transfer function of certain piezoelectric transducers is a function of the type of circuit of which they are a part. The circuit in which the transducer resides actually affects its *internal transfer function.* I am not speaking of the system transfer function now—just the transducer's transfer function. No other transducer type exhibits this unique property to my knowledge.

How can this be? The sensitivity of piezoelectric transducers is a function of a number of parameters including the material parameters (piezoelectric coefficient, frequency constant, coupling coefficient, dielectric constant, Curie temperature, pyroelectric coefficient, elastic modulus) and the manner in which the transducer was designed. One of these material properties, the piezoelectric coefficient, d_{ij}, is the charge output over the force input, Q_i/F_j (for crystal cut in direction i, and force applied in direction j). This is similar to gage factor in a strain gage. For certain materials this coefficient is, in the charge mode, a function of frequency. This can be seen in Figure 6.3 for an Endevco 2225 shock accelerometer that uses a man-made ferroceramic element.

Note, in the frequency response plot at the page bottom, that the charge frequency response (solid line) is not flat. The charge sensitivity decreases at about −2.5%/decade until it rises for the inevitable first transducer resonance. The sensitivity in the voltage mode (dashed lines) is flat but rolls off on the low end as discussed before. This is an example of a an actual transducer characteristic being dependent on an outside agent.

For other piezoelectric materials this is not the case. Figure 6.4 shows the frequency response for a quartz transducer, an Endevco 2252 shock accelerometer. The frequency response at the page bottom is, for both voltage and charge, flat until it rises for the first resonance. For quartz, d_{ij}, is not a function of frequency.

The general batting order for piezoelectric transducer materials is: (1) naturally occurring, quartz, tourmaline and bismuth germainum oxide; and (2) man-made quartz, bismuth titanate, lead zirconate titanate (compression), and lead zirconate titantate (shear). The dielectric coefficients for these materials[5] vary from 2.2 picocoulombs per newton for quartz to 450 picocoulombs per newton for lead zirconate titantate in the shear mode! You can see why manufacturers went to the trouble of developing their own materials. Most piezoelectric transducers are, in fact, made of lead zirconate titanate, a ferroceramic. So far, it looks like you should be using man-made materials, doesn't it?

MODEL 2225

to 20,000 g
High Resonance Frequency

SHOCK ACCELEROMETER

The ENDEVCO® Model 2225 Accelerometer is specially designed to provide accurate data for high amplitude shock and vibration studies.

The accelerometer's high resonance frequency (80,000 Hz) results in minimum output distortion due to ringing when measuring impact transients with durations as low as 75 µsec.

The Model 2225 incorporates Piezite® Element Type P-8 in a shear design that is capable of measurements to 20,000 g and is not affected by transverse components of acceleration. Small size, light weight, rugged construction, and stable operation under high energy impacts all combine to make this an extremely useful accelerometer.

The Model 2225 is a self-generating piezoelectric transducer and requires no external power for operation.

Specifications for Model 2225 Accelerometer
(According to ANSI and ISA Standards)

DYNAMIC

CHARGE SENSITIVITY . . . . 0.7 pC/g, nominal
VOLTAGE SENSITIVITY[1] . . . . 0.65 mV/g, nominal
0.55 mV/g, minimum
FREQUENCY RESPONSE (±10%)[2] . . . 2 to 15,000 Hz. See curves below.
MOUNTED RESONANCE FREQUENCY[4] . 80,000 Hz, nominal
TRANSVERSE SENSITIVITY . . . . 5% maximum, 3% on special selection.
AMPLITUDE LINEARITY . . . . Sensitivity increases approximately 1% per 2000 g, 0 to 20,000 g.
TRANSDUCER CAPACITANCE . . . . 800 pF, nominal
RESISTANCE . . . . 20,000 MΩ minimum, at +25° C (+77° F)

NOTES

[1]With 300 pF external capacitance.
[2]In shock measurements minimum pulse duration for half-sine or triangular pulses should exceed 75 µsec to avoid excessive high frequency ringing. (See Endevco Piezoelectric Accelerometer Manual.)
[3]Use ENDEVCO® Model 2700 Series or Model 2640 Series Charge Amplifiers.
[4]The transducer has an additional resonance at approximately 38 kHz.

TYPICAL TEMPERATURE RESPONSE

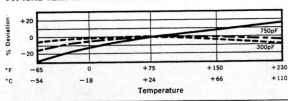

The solid line is the nominal charge-temperature response. The broken lines show the nominal voltage-temperature response with the indicated external capacitances.

TYPICAL FREQUENCY RESPONSE

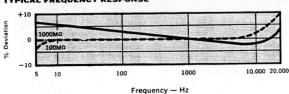

The solid line shows the typical charge-frequency response. The broken lines show the voltage-frequency response with the indicated loads and 300 pF of external capacitance. Estimated calibration errors:

5 to 1000 Hz: ±1.5%
1000 to 10,000 Hz: ±2.5%

Figure 6.3 Model 2225 shock accelerometer specifications (Printed through the courtesy of Endevco)

ENDEVCO PRODUCT DATA

MODEL 2252

Shock Measurements to 5000 g
Quartz Crystal

SHOCK
ACCELEROMETER

Reliable and accurate shock measurements up to 5000 g can be made with the ENDEVCO⁴ Model 2252 Accelerometer. The combination of a Piezite® Element Type P-2 with special construction features and single-ended compression design has resulted in an accelerometer with many unique characteristics.

The Model 2252 is insensitive to thermal transients. It can be subjected to shocks of 5000 g without zero shift, and its high resonant frequency (35 kHz) ensures linear operation from 0.1 Hz to 7000 Hz when used with the ENDEVCO® Model 2740B Charge Amplifier. The Piezite® Element Type P-2 also allows charge and voltage sensitivity to remain linear over a broad range of temperatures.

This accelerometer is a self-generating piezoelectric transducer and requires no external electric power for operation.

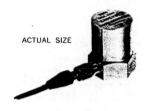

ACTUAL SIZE

Specifications for Model 2252 Accelerometer
(According to ANSI and ISA Standards.)

DYNAMIC

CHARGE SENSITIVITY . . . . 2 pC/g, nominal
1.5 pC/g, minimum

VOLTAGE SENSITIVITY¹ . . . . 5 mV/g, nominal

TRANSDUCER CAPACITANCE . . . . 100 pF, nominal

MOUNTED RESONANCE FREQUENCY . . 35 000 Hz, nominal

RESISTANCE . . . . 20 000 MΩ, min., at
75° F (24° C)

FREQUENCY RESPONSE (±5%)³ . . . . . 0.1 Hz to 7000 Hz³
2 Hz to 7000 Hz with
1000 MΩ load

TRANSVERSE SENSITIVITY . . . . 5% maximum

AMPLITUDE LINEARITY . . . . Sensitivity increases approximately 1% per 2500 g,
0 - 5000 g.

NOTES

¹With 300 pF external capacitance.

²In shock measurements the minimum pulse duration for half-sine or triangular pulses should exceed 0.15 milliseconds to avoid excessive high frequency ringing. (See Endevco Piezoelectric Accelerometer Instruction Manual.)

³With ENDEVCO® Model 2740B Charge Amplifier (±5% with pulse duration of 660 milliseconds).

TYPICAL TEMPERATURE RESPONSE

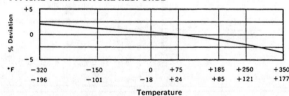

This curve shows the typical voltage and charge-temperature response of the accelerometer. Temperature response will remain unchanged for any length of cable.

TYPICAL FREQUENCY RESPONSE

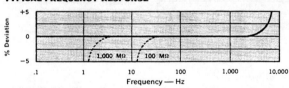

The solid line shows the charge-frequency response. The broken line shows the voltage-frequency response with the cable supplied and with loads shown.

Figure 6.4   Model 2252 shock accelerometer specifications (Printed through the courtesy of Endevco)

160

6.2.5 Temperature Considerations

The temperature performance of piezoelectric transducers is complex. It results from the temperature characteristics of the transducer material itself, called its pyroelectric behavior, and its interaction with its internal preload system and the transducer case.

All else being equal, quartz will show the best temperature performance in terms of minimized pyroelectric output (at zero frequency, measured with an electrometer) and sensitivity. This also is generally true for transient temperature inputs from whatever cause.

Charge is generated whenever a piezoelectric material's temperature changes. This is the transducer's pyroelectric output. If the transducer is sitting open circuit, this temperature induced charge is waiting for you when you connect the transducer cable, and may be significant. For certain accelerometers, the open circuit charge waiting for you varies from a minimum of the equivalent of .2 G/°F for quartz, to 24 G/°F for lead zirconate titanate! Thus, open circuit piezoelectric transducers make excellent thermometers. This is the major reason these measurement channels go momentarily off scale when connected.

Figures 6.5 and 6.6 show how the basic transducer sensitivity varies with temperature for several Endevco piezoelectric materials. The second curve in Figure 6.5 is for Endevco P-2 material, quartz. Notice that only a single curve is given for charge and voltage modes. This is characteristic and an advantage of quartz. The designer has to keep only one temperature performance curve in mind. For the other four Endevco materials (P-1, P-4, P-7, P-10), this is not the case, as the curves show. This is a disadvantage for these materials. You have to remember in which circuit you are going to run the transducer and account for that in your design.

What about temperature transients or gradients? The manufacturer will typically not specify this except in the sales literature. But in most cases, quartz gives better transient temperature performance than any other piezoelectric material. As you saw, however, its basic sensitivity is low. The measurements engineer is always sick—the best you can do is choose the sickness so it doesn't hurt you in today's measurement.

6.2.6 Transducer Mounting Considerations

One of the major advantages of piezoelectric transducers is their high stiffness and resonant frequency for a given mass. This means they can be used over a broad frequency range for measurands that are changing rapidly.

When making high frequency measurements, be aware that transducer mounting, as well as resonant frequency and mass, becomes critical. This shows in Figures 6.7 and 6.8 for a generic Bruel and Kjaer general purpose accelerometer with a 25KHz internal first resonance. The overall transducer resonance can be moved from the original 25KHz to below 1.5 KHz by varying the mounting method! Be careful and know what you are doing here. This is more and more important as the bandwidth for measurement goes up. It is of the utmost importance when going after shock-induced motions. The best literature for these types of measurements is in the proceedings of the Shock and Vibration Symposium and the Aerospace Testing Seminar of the Institute of Environmental Sciences.

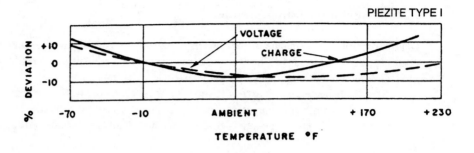

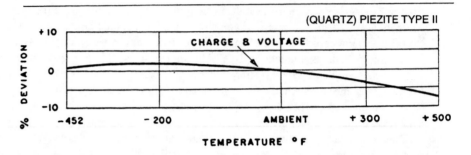

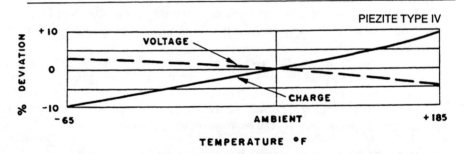

Figure 6.5 Sensitivity as a function of temperature for several commercial piezoelectric transducer materials (Printed through the courtesy of Endevco)

6.2.7 Other Environmental Effects

A second class of mounting troubles that can seriously degrade data validity are case strain effects. Here, the transducer case is strained during the measurement and that strain is transmitted to the piezoelectric material. When this occurs, strain-induced charge is generated having nothing to do with the measurand. But, it generally occurs in the same time frame and frequency range as the measurand, so its effects cannot be filtered out. It must be prevented by engineering the transducer mounting properly in the first place.

Piezoelectric measuring systems are extremely unforgiving in this regard. If the

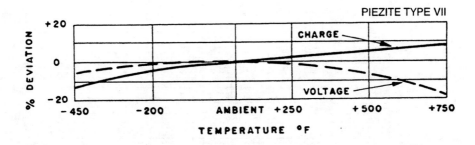

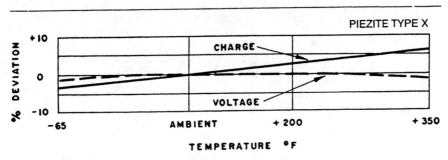

Figure 6.6 Sensitivity as a function of temperature for two commercial piezo-electric transducer materials (Printed through the courtesy of Endevco)

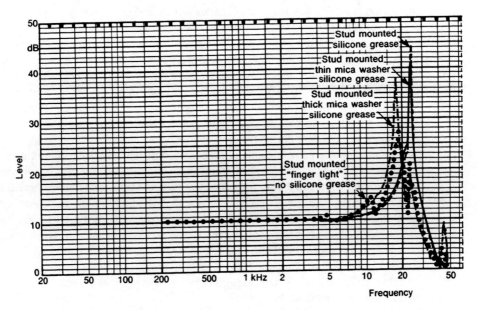

Figure 6.7 The frequency response of a stud-mounted general purpose acceler-ometer using slightly different mounting techniques (Printed through the cour-tesy of Brüel & Kjaer)

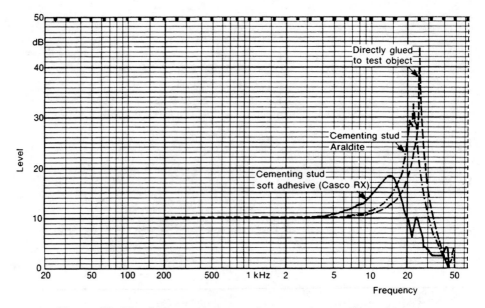

Figure 6.8 The frequency response of a general purpose accelerometer mounted using adhesives (Printed through the courtesy of Brüel & Kjaer)

measuring crystal is deformed, it generates charge and out comes a signal no matter what the cause. This is of particular importance when using flush diaphragm pressure transducers mounted in the walls of pressure vessels. As the internal pressure loads the vessel walls, the walls load the transducer causing outputs from case strain and from internal pressure. These cannot be distinguished from each other, cannot be filtered or otherwise separated. Case strain problems can only be eliminated by choice of transducer with low inherent case strain sensitivity and a mount that isolates the transducer from the structural strain in the first place. This design method also implies a foreknowledge of the dynamic strain field the transducer will undergo during test.

Knowing that environmentally caused noise levels will occur in test, some vendors build piezoelectric transducers that are internally compensated for certain noise levels. The acceleration compensated pressure transducer is an example. The thought here is that some pressure environments involve acceleration at the transducer mount. Acceleration is a noise level to a pressure transducer. There is, within the pressure transducer case, an additional piezoelectric accelerometer whose function is to compensate for acceleration caused charge generation occurring during the pressure measurement. Figure 6.9 shows data for a PCB 112A 3000 PSIG pressure transducer with an acceleration sensitivity of .002 PSI/G. PCB has designed a very small accelerometer into this pressure transducer, and you can see by the numbers that it works!

Figure 6.10 shows similar data from the Kistler AG of Switzerland catalog. These three models show acceleration sensitivities of from .03 to .05 PSI/G. These numbers are 15 to 25 times those of the PCB 112A noted above for roughly the same pressure range.

MINIATURE, QUARTZ, CHARGE-MODE
HI-SENSITIVITY PRESSURE TRANSDUCER

Series 112A

PCB PIEZOTRONICS

PRESSURE

use with charge amplifier

- offers acceleration-compensated, sensitive quartz element
- provides wide amplitude, frequency and temperature ranges
- responds statically for short term; calibrates statically
- involves very low strain and motion sensitivity
- includes floating clamp nut with American or metric thread
- combines flush-welded, flat diaphragm

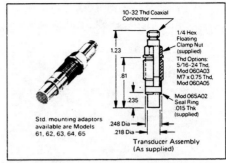

Series 112A is designed for compression, combustion, explosion, pulsation, actuation, cavitation, fluidic, blast, turbulence and sound pressure measurements, especially at higher operating temperatures.

Model 112A universal quartz minigage measures near static and dynamic pressures from full vacuum to 3000 psi (15000 psi optional). It measures transient or repetitive phenomena, relative to the initial or average pressure level, over wide amplitude and frequency ranges and under the most adverse environmental conditions. System voltage sensitivities range from 0.01 mV/psi to 1.0 volt/psi, depending upon the charge amplifier involved. The electrostatic charge signal from this conventional piezoelectric transducer is converted into voltage signal in a PCB or similar charge amplifier.

The structure of this sophisticated instrument contains a rigid multi-plate, high-sensitivity quartz element with an integral compensating accelerometer to minimize vibration sensitivity and partially suppress internal resonance effects.

Miniature quartz transducers are installed flush or recessed in existing or new minigage ports directly in the test object or in a variety of threaded mounting adaptors. A separate floating nut, with either a 5/16-24 or M7 x 0.75 thread, clamps the transducer in place, isolates against strain, and facilitates installation and removal of the sealed transducer assembly.

SPECIFICATIONS: Model No.		112A
Range	psi	3000
Maximum Pressure	psi	10000
Resolution	psi	0.004(1)
Sensitivity (nominal)	pC/psi	1.0
Resonant Frequency	kHz	250
Rise Time	μs	2
Linearity (zero based BSL)	%	1
Polarity		Negative
Insulation Resistance (room temp.)	ohm	1 x 10^{12}
Capacitance	pF	18
Acceleration Sensitivity	psi/g	0.002
Temperature Coefficient	%/°F	0.01
Temperature Range	°F	±400
Flash Temperature	°F	3000
Vibration; Shock	g's peak	2000; 20000
Case	SS	17-4PH
Diaphragm		invar
Weight	gm	5

Optional Ranges:

Low-Range	100 psi	112A02
High-Range	10000 psi	112A03

(1) Resolution dependent on range setting and cable length used in charge systems.

Add prefix to model number to specify options:

'H'	Hermetic Seal	e.g. H112A
'M'	Metric Thread (M7 x 0.75)	e.g. M112A
'HM'	Hermetic & Metric	e.g. HM112A

INSTALLATION: Mounting Ports

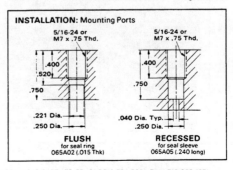

TYPICAL SYSTEM

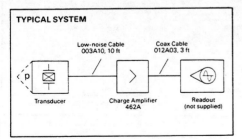

112583

PCB PIEZOTRONICS, INC. 3425 WALDEN AVENUE DEPEW, NEW YORK 14043-2495 TELEPHONE 716-684-0001 TWX. 710-263-1371

Figure 6.9 Model 112A piezoelectric pressure transducer specifications (Printed through the courtesy of PCB Piezotronics, Inc.)

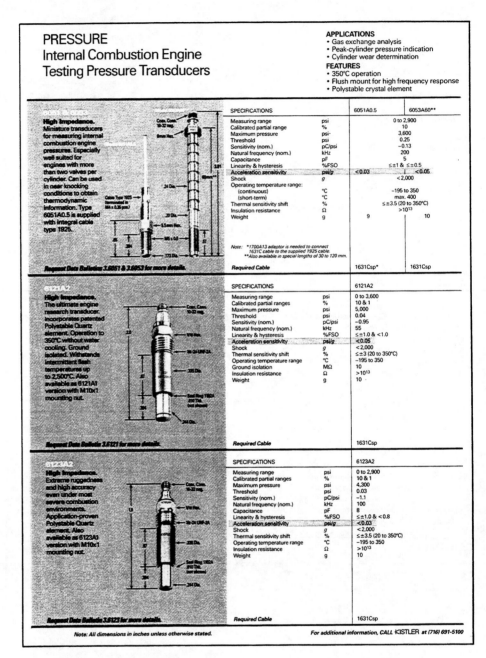

Figure 6.10 Piezoelectric pressure transducer specifications (Courtesy of Kistler)

Kistler chooses not to acceleration compensate these pressure transducers even though they are designed for internal conbustion engine applications with acceleration as a part of the environment. They do build acceleration compensated pressure transducers—but these are not them! They show here to illustrate the point.

6.2.8 Spatial Effects

Measurement system designers must be aware of two spatial effects when specifying piezo-electric transducers. These effects arise because of the extremely wide frequency range over which these transducers operate, their relatively small size, and the manner by which they are manufactured. The effects are transverse sensitivity and spatial averaging. These effects also occur in other transducer types, but the unique set of parameters mentioned here make piezoelectric transducer sensitivities to these effects more important.

Transverse sensitivity.   Transverse sensitivity is a problem in transducers de-signed to measure vector quantities such as acceleration, torque, and force. Here, you are interested in a measureand and its direction. I'll discuss this effect for acceleration trans-ducers.

Transverse sensitivity is defined as a maximum percentage of full scale charge or voltage sensitivity. For example, an acceleration transducer has a specified transverse sen-sitivity of 3%. This means that if a full scale acceleration is applied along any axis normal to the sensitive axis, out will come no more than 3% of the output you would get if the acceleration was applied along the sensitive axis. This is a blatant noise level. You get 3% of full scale out—for zero input along the sensitive axis.

This noise level occurs for a number of reasons. The primary reason is that although piezoelectric materials are highly directional when originally processed (very good), they cannot be installed in the transducer case to the extremely tight tolerances necessary to take advantage of this precise directionality (not so good!).

Some manufacturers tell you about it and plot it for you. Figure 6.11 from Bruel and Kjaer shows the transverse sensitivity plot for a B&K accelerometer. In this example, B&K's 100% corresponds to our 3%, and the transducer sensitive axis is the 90–270 degree axis on the plot. B&K even tells you the axis of minimum cross axis sensitivity on the transducer case with a small red dot. But all is not well. This manufacturer only runs this test at a single frequency. And guess what? Transverse sensitivity is a function of frequency with its own magnitude and phase. What about the frequencies you are interested in? No information is available.

What does this all mean to you as a system designer? Consider an example you might run into every day in normal testing. Consider a point on a structure that, at some frequency during a vibration test, experiences an acceleration 10Gs in the X direction and 1G in the Y direction. Your design job is to measure these accelerations as accurately as you can. Fur-ther assume that no other errors are made—everything else is perfect. Assume that you are using a standard accelerometer from a reputable vendor with the 3% cross axis sensitivity specification, and that these transducers are exactly aligned along the X and Y axes. The output relationship is simple: the output is the acceleration along the primary axis times the

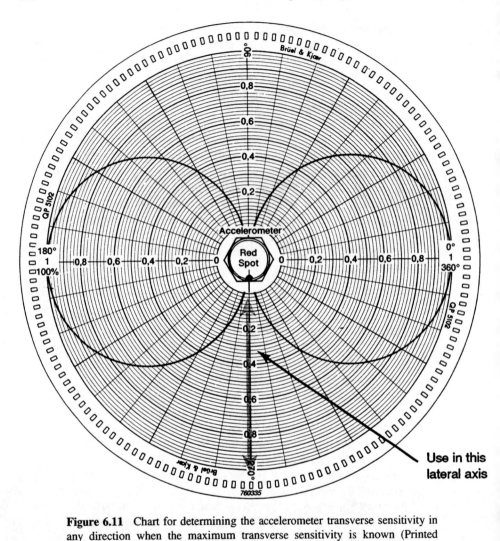

Figure 6.11 Chart for determining the accelerometer transverse sensitivity in any direction when the maximum transverse sensitivity is known (Printed through the courtesy of Brüel & Kjaer)

sensitivity in that axis plus the lateral acceleration times the cross axis sensitivity. For example:

$$X\text{-axis: output}_x = 10G(1.0) + (.03)(1G) = 10.03G, \text{ or a } .3\% \text{ error}$$

$$Y\text{-axis: output}_y = 1G(1.0) + (.03)(10G) = 1.30G, \text{ or } a\ 30\%\ error\ !$$

In the high acceleration axis you get a .3% error—almost negligible. But in the lateral axis where you're trying to measure 1G in the presence of 10Gs of noise—you get a 30% error, which is definitely not negligible!

And yet, cross axis acceleration data is published *without caveats* every day in almost every vibration laboratory in the world. And what do you think a fancy modal survey system would do with that lateral axis data corrupted from cross axis sensitivity? There isn't a model survey system in the world that takes this major error into account. Every one of them just grinds away trying to curve fit this corrupted data creating modes where none exist.

If you have triaxial data from a vibration test you can at least make some judgment about cross axis sensitivity errors. You'll see this later in the chapter on Knowledge-Based data acquisition systems. You might blank out portions of the response spectrum where cross axis sensitivity has caused known errors above a limit you set. With less than three orthogonal axes of data at a point, you cannot even make a judgment.

Spatial Averaging.   Spatial averaging occurs because transducers have some finite size. Since they do not have zero size, they tend to integrate what is happening at the interface to the phenomenon and report the result of that integration. This is a particular problem in high frequency measurements for which piezoelectric transducers are so well suited. An example is the measurement of dynamic pressure at the wall of a pressure line through which the pressurant medium is flowing. Figure 6.12 shows this effect.

Here, a flush diaphragm pressure transducer is installed in the wall of a duct in which the pressurant fluid flows. The fluid has pressure variations which are of interest. What is the character of these pressure variations in terms of time history and spectral content? This

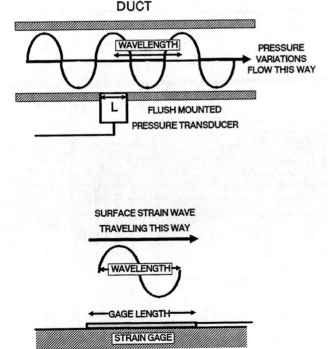

Figure 6.12 Examples of spatial averaging

type of problem occurs in detonation testing, hydraulics, combustion research, and chemical laser, gas turbine and rocket engine testing. It also occurs in acoustics testing if the direction of the incident pressure waves is not known. A strain gage undergoing a dynamic strain field shows at the bottom of Figure 6.12. If the wavelength of surface dynamic strain is the same as the gage length, you can see that the gage will integrate this to zero!

The problem arises because of the nonzero size of the sensing portion of the transducer. In the pressure measurement example noted above, the transducer has a diameter of L inches. If the diameter of the transducer is small compared to the wavelength, λ, of the lateral pressure, the effects of the integration over the diaphragm will be small. The transducer will follow the pressure faithfully and the data will be valid. If the wavelength, λ, is equal to the diameter, L, then one full wavelength exists over the sensing diaphragm and the transducer will *integrate this to zero*. Nothing will come out of the transducer for a very real pressure condition. The same thing will occur when there are integer numbers of wavelengths across the sensing diaphragm. The actual transfer function shows at the bottom of Figure 6.13. Between an $L\lambda$ ratio of one and two, the phase of the output is 180 degrees from the phase of the pressure. The transducer will report negative pressure for a positive dynamic pressure input! Between a ratio of two and three, the phase is zero again.

This new transfer function, caused by spatial averaging, acts in series with the transducer's own mechanical transfer function. If the pressure waves came at the diaphragm normal to its surface, the measurements problem does not exist. In this case, the transducer's own mechanical transfer function alone will govern its behavior.

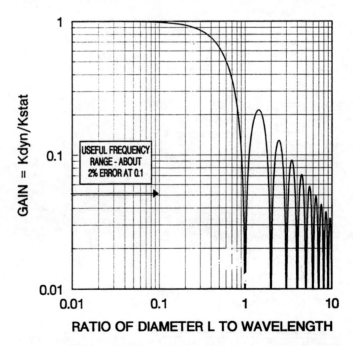

Figure 6.13   New serial transfer function caused by spatial averaging

6.2.9 "Charge" Amplifiers

The "charge" amplifier as it is known in the industry was patented by Walter Kistler of Switzerland in 1950.[6] The best description of the charge amplifier is in literature[7] from the company that still bears his name, Kistler AG of Winterthur, Switzerland. Figure 6.14 shows the basic configuration of the charge amplifier. The charge, Q, is generated in the piezoelectric transducer at the left. C_{ex} accounts for the capacitance of the transducer itself, the interconnecting cable, and the feedback range capacitor in the charge amplifier. The amplifier is an operational amplifier with a huge, negative open loop gain of $-V_i$. In this configuration it acts as a current integrator and not as a charge amplifier. The current into the amplifier is integrated to establish the transducer's generated charge, Q. The term charge amplifier is used in error, but is used popularly for this type of work and I use it in the popular sense here. The feedback capacitor, C_r, determines the range of the amplifier, with several values usually supplied for ranging purposes. R_m is the input impedance of the readout device and e_o the voltage read by that device.

The governing equation for the charge amplifier shows in the figure. Note that the system output voltage, e_o, is a function of five variables: the generated charge, Q; the feedback capacitance, C_r; the amplifier gain, $-V_i$; the external cable capacitance, C_c; and the transducer capacitance, C_t. The measurements engineer's job is to hold four of the variables constant by design for the duration of the setup and test and allow only Q to change. Since the value of V_i is so large, changes in $C_{ex} = C_t + C_c + C_r$ due to changes in cable length are almost negligible. This is one of the major advantages of this measurement methodology. System sensitivity, in terms of charge/engineering unit over a frequency range, holds closely for varying lengths of connecting cable, although system noise levels will increase with cable length.

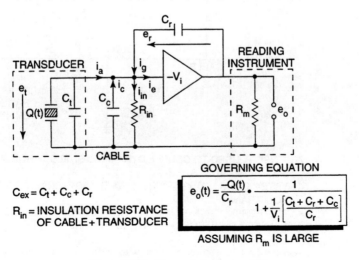

$$C_{ex} = C_t + C_c + C_r$$

$$R_{in} = \text{INSULATION RESISTANCE OF CABLE} + \text{TRANSDUCER}$$

GOVERNING EQUATION

$$e_o(t) = \frac{-Q(t)}{C_r} \cdot \frac{1}{1 + \frac{1}{V_i}\left[\dfrac{C_t + C_r + C_c}{C_r}\right]}$$

ASSUMING R_m IS LARGE

Figure 6.14 The real charge amplifier

6.2.10 The T-Insertion Technique

The use of piezoelectric transducers with charge amplifiers has an advantage over any other measurement method known to me. That advantage is the ability of the circuit to be T-inserted in-situ as shown in Figure 6.15. An insertion T has been installed in the circuit just before the charge amplifier. An insertion T is a six terminal device that passes the high side of the signal from the transducer directly through to the amplifier. The low side is broken by a low value resistor, usually of 100 ohms in commercial insertion Ts. Taps from either side of the 100 ohm resistor exit the insertion T as shown. Note that one side of the insertion T is tied to the low side ground on the readout device. In this configuration, you can see that a grounded transducer will short the 100 ohm resistor and prevent the insertion T from operating. In fact, this is a bonus because the insertion T method can be used in this manner to detect grounded transducers before the test is run and the data corrupted by powerline noise levels. A commercial insertion T, the Endevco 2944.1, shows in Figure 6.16.

During a sinusoidal transducer calibration, the RMS voltage and waveform across the resistor in the T is noted with a high impedance reading instrument (DMM, RMS meter) for certain values of engineering unit amplitudes and frequencies. These RMS voltage values are supplied to the users along with the primary calibration information which is the charge sensitivity, usually picocoulombs per engineering unit as a function of amplitude and frequency. At setup time, the reading device is replaced with a calibrated RMS voltage source.

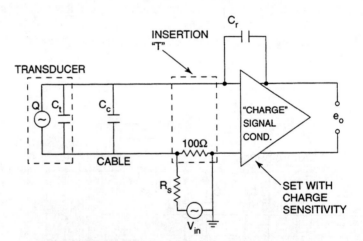

- T–INSERT USED TO COUPLE DYNAMIC SIGNALS THROUGH ENTIRE MEASUREMENT SYSTEM
- SYSTEM DOES NOT KNOW THAT INSERTED SIGNAL DID NOT COME FROM TRANSDUCER
- USED FOR INDEPENDENT SPAN VERIFICATION, AND IN-SITU SYSTEM TRANSFER FUNCTION CHECK (END-TO-END)

Figure 6.15 T-insertion method with piezoelectrically based measurement systems

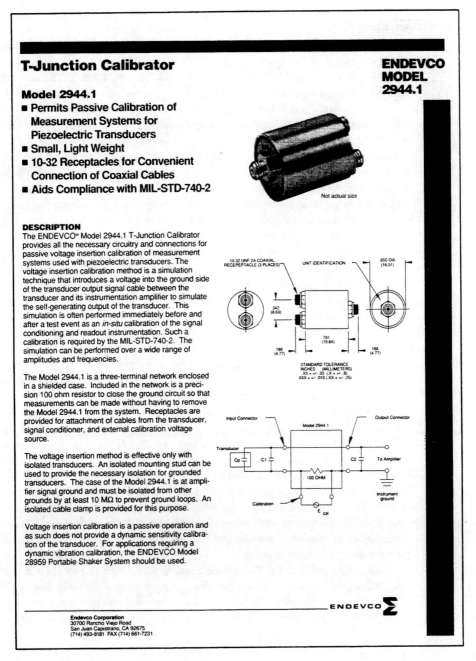

T-Junction Calibrator

Model 2944.1

■ Permits Passive Calibration of
  Measurement Systems for
  Piezoelectric Transducers
■ Small, Light Weight
■ 10-32 Receptacles for Convenient
  Connection of Coaxial Cables
■ Aids Compliance with MIL-STD-740-2

ENDEVCO MODEL 2944.1

Not actual size

DESCRIPTION
The ENDEVCO® Model 2944.1 T-Junction Calibrator
provides all the necessary circuitry and connections for
passive voltage insertion calibration of measurement
systems used with piezoelectric transducers. The
voltage insertion calibration method is a simulation
technique that introduces a voltage into the ground side
of the transducer output signal cable between the
transducer and its instrumentation amplifier to simulate
the self-generating output of the transducer. This
simulation is often performed immediately before and
after a test event as an *in-situ* calibration of the signal
conditioning and readout instrumentation. Such a
calibration is required by the MIL-STD-740-2. The
simulation can be performed over a wide range of
amplitudes and frequencies.

The Model 2944.1 is a three-terminal network enclosed
in a shielded case. Included in the network is a precision 100 ohm resistor to close the ground circuit so that
measurements can be made without having to remove
the Model 2944.1 from the system. Receptacles are
provided for attachment of cables from the transducer,
signal conditioner, and external calibration voltage
source.

The voltage insertion method is effective only with
isolated transducers. An isolated mounting stud can be
used to provide the necessary isolation for grounded
transducers. The case of the Model 2944.1 is at amplifier signal ground and must be isolated from other
grounds by at least 10 MΩ to prevent ground loops. An
isolated cable clamp is provided for this purpose.

Voltage insertion calibration is a passive operation and
as such does not provide a dynamic sensitivity calibration of the transducer. For applications requiring a
dynamic vibration calibration, the ENDEVCO Model
28959 Portable Shaker System should be used.

Endevco Corporation
30700 Rancho Viejo Road
San Juan Capistrano, CA 92675
(714) 493-8181 FAX (714) 661-7231

ENDEVCO Σ

Figure 6.16 Endevco T-junction calibrator (Printed through the courtesy of
Endevco)

The following statement can now be made: If the same RMS voltage is impressed across the resistor in the insertion T as was read at calibration time, the current conditions created in the circuit are the same as existed during the calibration. This condition is independent of the waveform of the inserted voltage and means that the circuit does not "know" the currents being generated are not coming from the transducer itself. The insertion T allows perfect simulation of an active transducer seeing engineering unit inputs.

Why is that condition so advantageous? The insertion T allows a full measurement system independent span verification (see chapter on Measurement System Operations) to be performed in-place after transducer installation. Using the insertion T method, you can absolutely verify system span before the test on the ranges you are going to run in the test. We T-insert all piezoelectric transducer data channels for span verification before use as a matter of our measurements soul (see chapter on Leadership and Management Issues). The T-insertion is the final, bulletproof verification of system span. We are extremely careful in setting up our measurement systems. Even being careful we find span errors consistently in about 5% of the data channels with this method. These are, of course, fixed before the test. If we find span errors at the 5% level with this methodology, what do you think the span error rate is on tests outside our laboratories where this verification method is not used? You can bet it is significantly higher than 5% and these span errors are not caught and corrected. This methodology is so powerful in this regard, it is amazing that other laboratories neither use nor understand it.

The T-insertion method has other advantages. The entire measurement system electrical transfer function can be checked end-to-end and in-place with this method using either wideband random insertion inputs or transient inputs. If the frequency content of the inserted signal has enough bandwidth, say 50KHz–100KHz, the frequency and damping of the installed transducer resonance can be measured and verified. This is how you detect damaged transducers while they are still installed. How do you know, on a shock test for instance, that the transducers or their mounts have not failed after the first firing? Some of them are buried inside the spacecraft structure and are neither visible nor inspectable. Even if you could see the installation you might not be able to detect a transducer mount debond. A wideband T-insertion check verifies this. System linearity, other than transducer linearity, can also be verified by inserting voltages of various amplitudes and monitoring system outputs. The T-insertion method also finds transducers that have open circuit or grounded conditions due to damage and certain cable problems. If the piezoelectric transducer is not connected or is grounded, the channel will not insert properly to the right voltage levels, giving a clear indication of trouble.

The price paid for the T-insertion capability to diagnose problems in-place is serial time. It does take time to perform this check. Our feeling is that it is absolutely worth the price to detect the problems before the test is run. We recently ran a major 170 channel major league system level sine vibration test. On a previous similar test it had taken a two-man crew about 12 hours to complete the T-insert function on 170 acceleration channels. In order to cut that cycle time, we built an automated T-insert system that performed the entire 170-channel check in 1 hour including problem solving and required a single operator! This represents a 12:1 cycle time decrease and a 24:1 productivity increase. That automated T-insert capability is now in use in all of our vibration laboratories.

You can see from the circuit diagram in Figure 6.15 that you need to place the 100-ohm insertion resistor between the transducer and any signal conditioning components. Most reputable piezoelectric transducer manufacturers now market internally signal conditioned models so that high level low source impedance signals are available directly at the transducer output. These models have some important advantages. One of their significant disavantages, however, is the user's inability to T-insert the channels because the correct access is not provided inside the transducer prior to the charge converter. Our preference is not to use these models for this reason.

6.3 THE THERMOELECTRIC EFFECT

The second self-generating transducer set I'll discuss is the thermocouple for the measurement of temperature. This is probably the most widely used transducer type in the world for experimental mechanics temperature measurements and certainly for process control. Dr. Robert Moffat of Stanford University is a preeminent expert in explanations of, and research into, the thermoelectric affect. Dr. Moffat's article, *Thermocouple Theory and Practice*, reprinted here, is a classic in the field. It has withstood the test of more than 20 years of use in the teaching of the principles of measurement engineering. The information in this article is about measurement and thermophysical principles. Therefore, it will be current and useful for decades to come. You can't make that statement about much in the engineering literature with technology changing as fast as it does.

His use of the graphical analysis method for complex thermocouple circuits is, in my opinion, a fundamental contribution to the literature on thermocouples and their use for valid temperature measurement. It allows both valid thermocouple responses and associated problems to be easily understood and communicated. This approach should be in every measurement engineer's toolkit. Since I could not write this section in a manner that approaches the effectiveness of Dr. Moffat's work, I've reprinted it here with his permission.

Please note that all italics are mine. They underscore points that are fundamentally important yet often misunderstood in the workplace, and misstated in the literature. Paragraph numbers are not included here as they were not in the original text. The figure numbers, however, have been changed for the sake of continuity.

Chapter 6

Thermocouple Theory and Practice

Dr. Robert J. Moffat
Stanford University
Stanford, California

INTRODUCTION

Thermocouples are perhaps the most commonly used instrument for temperature measurement, with the possible exception of the clinical thermometer. They are inexpensive, small in size, and remarkably accurate when used with an understanding of their peculiarities. The objective of these notes is to provide a quick introduction to the use of thermocouples for those totally unfamiliar, and to introduce some recent innovations in describing thermocouple behavior which may be of value to both novices and advanced thermocouple users.

The present paper presents the fundamental theory of thermocouple operation and a technique for the graphical analysis of circuits which facilitates the understanding of complex interconnections. Elementary and advanced circuit concepts are discussed.

THERMOCOUPLE MATERIALS

Any pair of thermoelectrically dissimilar wires can be used as a thermocouple. The wires need only be joined together at one end and connected to a voltage measuring instrument at

TABLE I[8]

100°C	500°C	900°C
antimony	Chromel	Chromel
Chromel	nichrome	nichrome
iron	copper	silver
nichrome	silver	gold
copper	gold	iron
silver	iron	$Pt_{90}Rh_{10}$
$Pt_{90}Rh_{10}$	$Pt_{90}Rh_{10}$	platinum
platinum	platinum	cobalt
palladium	cobalt	Alumel
cobalt	palladium	nickel
Alumel	Alumel	palladium
nickel	nickel	Constantan
Constantan	Constantan	
copel	copel	
bismuth		

the other end to form a usable system. Whenever one end of the loop so formed is at a different temperature than the other, a voltage will be developed which is related to the temperature difference between the two ends. Several metallic materials are listed in Table I, in order of thermoelectric polarity: each material is "positive" with respect to all beneath it in the listing. In an iron-palladium thermocouple, the cold end of the iron wire will be positive with respect to the cold end of the palladium.

There have been many instances in which the operating temperature of machinery elements have been measured using the machine structure as part of the thermoelectric circuit (cutting tool tip temperatures, cam shaft/rocker arm contact temperatures, etc.). In such cases it is necessary only to calibrate the material, measure the EMF (electromotive force, voltage), and interpret the signal.

The alloys usually used for thermoelectric temperature measurement are listed in Table II. These have been selected, over the years, for the linearity, stability, and reproducibility of their EMF versus temperature characteristics and for their high temperature capability.

Detailed information regarding the physical properties of these alloys can be found in the National Bureau of Standards Monograph 40, "Thermocouple Materials," by F. R. Caldwell. Calibration data are contained in NBS (now NIST) Circular 561 and are also available from instrument companies and thermocouple suppliers.

Thermocouple material can be purchased as bare wires, as flexible insulated pairs of wires, and as mineral insulated pairs encased in stainless steel tubes for high temperature service. Prices range from a few cents to several dollars per foot depending on the type of wire and the type of insulation. There are many suppliers: the telephone directory in San Francisco lists 12 in the yellow pages!

TABLE II

ISA Type	Typical Elements	Max. Temp.	Output mV/100°F
	60% rhodium + 40% iridium versus iridium	[1]3800F	0.3
S	90% platinum + 10% rhodium versus platinum	[2]2700F	0.5
K	Chromel-Alumel[3]	[2]2300F	2.3
E	Chromel[3]-Constantan	1600F	4.0
J	iron-Constantan	1400F	3.0
T	copper-Constantan	700F	3.0

1. Requires neutral atmosphere or vacuum.
2. Requires oxidizing or neutral atmosphere or vacuum.
3. Registered trademark of Hoskins Mfg. Co., Detroit, Michigan.

Any instrument capable of reading low DC voltages (on the order of millivolts) with a 5–10 microvolt resolution will suffice for temperature measurements with accuracy dependent on the voltmeter used. The signal from the thermocouple depends on the difference in temperature of the two ends of the loop; hence, the accuracy of the temperature measurement depends on the accuracy with which the "reference junction" temperature is known as well as the accuracy with which the electrical signal is measured.

Galvanometric measuring instruments can be used but, since they draw current, the voltage available at the terminals of the instrument depends not only on the voltage output of the thermocouple loop but also on the resistances of the instrument and the loop. Such instruments are normally marked to indicate the external resistance for which they have been calibrated. Potentiometric instruments, either manually balanced or automatically balanced, draw no current when in balance, hence can be used with thermocouple loops of any resistance without error. High input impedance electronic amplifiers draw only minute currents and, except for very high resistance circuits, pose no problems. When in doubt, check the input impedance against the circuit impedance to be sure.

The input stages of many instruments have one connection which is grounded. Ground loops can be caused by using a grounded junction thermocouple with such an instrument. If the ground potential at the place where the thermocouple is attached is different from the ground potential where the instrument is plugged in, then a current may flow through the thermocouple wire. The IR-drop in the wire due to the ground loop current will be mixed with the thermoelectric signal and may cause a large error.

A thermocouple loop produces a voltage in proportion to the difference between the temperatures of its two ends. To measure an unknown temperature, it is necessary to have one of the two ends at a known temperature (the "reference temperature"). Roughly speaking, *any error in the knowledge of the reference temperature is carried through as error in the measured temperature.* For this reason, considerable care should be given to the preparation of the reference junctions.

ANALYZING THERMOCOUPLE CIRCUITS

The material in this section is largely based on an earlier paper, *The Gradient Approach to Thermocouple Circuitry.*[9]

It will prove of great value to have a qualitative appreciation of the temperature-EMF calibrations of the more common materials. The data shown in Figure 6.17 was derived from NBS Circular 561 and other sources. It represents the output that would be derived from thermocouples made of material X used with platinum when the cold end is held at 32°F and the hot end is held at T. Note that those elements commonly used as "first names" thermocouple pairs [i. e., *Chromel* (-Alumel), *iron* (-Constantan), *copper* (-Constantan), etc.] have positive slopes in Figure 6.17.

It can be shown, from either free-electron theory of metals[10] or from thermodynamic arguments[11] alone, that the output of a thermocouple can rigorously be described in terms of the contribution from each of the lengths of material comprising the circuit: *the junctions are merely electrical connections between the wires.* Formally

$$E_{net} = \int_O^L \varepsilon_1 \frac{dT}{dX} dX + \int_L^O \varepsilon_2 \frac{dT}{dX} dX$$

where ε is the total thermoelectric power of the material, equal to the sum of the Thomson coefficient and the temperature derivative of the Peltier coefficient; T is the temperature X is the length along the wire; and L is the length of wire.

When the wire is exactly uniform in composition, so the ε is not a function of position, X, then

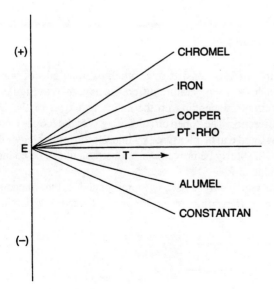

Figure 6.17   EMF vs. temperature calibrations for several materials

$$E_{net} = \int_{T_0}^{T_L} \varepsilon_1 dt + \int_{T_L}^{T_0} \varepsilon_2 dt$$

One usually assumes that the wire is homogeneous without worry but it is, after all, a special case of a more general problem. Used thermocouples may be far from uniform.

When only two wires are used in the circuit, then it is customary to further simplify the problem. Since both wires begin at one temperature (say T_0) and end at the other (T_L), the two integrals above can be collected

$$E_{net} = \int_{T_0}^{T_L} (\varepsilon_1 - \varepsilon_2) dt$$

Note that there are several simplifications built into this reduced equation.

1. ε is not a function of position; i. e., the wires are homogeneous
2. only two wires
3. each wire begins at T_0 and ends at T_L

These are the (only) conditions for which the EMF-temperature tables are intended.

This view of the thermoelectric EMF as originating in the wire rather than the junction does not negate or contradict any of the findings of Peltier, Thomson, or Seebeck but is entirely consistent with their results. Such a view offers a great operational advantage—it opens the way to graphical analysis of thermocouple circuits. Graphical analysis allows the behavior of the most complex (i. e., many materials) circuit to be unambiguously described. The procedure will be illustrated with circuits of increasing complexity.

CIRCUITS OF HOMOGENEOUS MATERIAL

The simplest thermoelectric circuits consists of two unbroken lengths of wire (A and B), joined at the ends and subject to a temperature difference as shown in Figure 6.18. A current will flow in such a circuit, depending on the nature and sizes of the wires and the amount of temperature difference. The "open circuit voltage" of the loop is a measure of the tendency of current to flow, and can be measured by interrupting either leg of the circuit, as shown in Figure 6.19. It will shortly be proven that an instrument can be connected to read E without affecting the value of E.

The "pattern circuit" which will be used to represent an ideal thermocouple circuit is shown in Figure 6.20. The difference between Figures 6.19 and 6.20 is simply that the

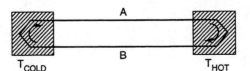

A

B

T_{COLD} T_{HOT}

Figure 6.18   The simplest thermoelectric circuit

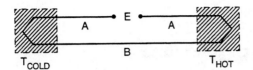

Figure 6.19 The "open circuit" thermoelectric circuit

interruption in the circuit is placed at the reference junction in Figure 6.20, rather than midway in one of the legs. Such a circuit is easiest to analyze by the methods to be presented here. More complex circuits (i.e., those containing switches, connectors, etc.) will be judged acceptable or not by whether or not they are thermoelectrically equivalent to the pattern circuit shown in Figure 6.20.

The simplest practical circuit consists of two wires joined together at one end and connected directly to a measuring instrument as shown in Figure 6.21. Such a circuit is thermoelectrically ideal (providing that the materials of the instrument do not affect the reading!) since it contains no switches, connectors, or leadwires. The output of this system will be graphically analyzed using the calibration data shown in Figure 6.17. The procedure is illustrated in Figure 6.21 and is as follows.

First, identify each "point of interest" by a number and assign some nominal temperature level to each point. The temperature need not be exact, simply "possible." Next, construct an EMF vs. temperature (E-T) coordinate framework, millivolts versus degrees. Now, locate point 1 on the E-T coordinates at zero millivolts and at the assigned temperature for point 1. Next, construct a line through point 1 having the slope of the calibration curve of the iron wire which connects point 1 and 2 (see Figure 6.17). Terminate this constructed line at the temperature corresponding to point 2 where the iron joins the Constantan. Beginning at point 2, construct a line parallel to the Constantan calibration curve extending back to the temperature at point 3. Both 1 and 3 are presumed to be at T_{AMB}. The difference in the ordinates of points 1 and 3 measures the net EMF generated in this circuit between points 1 and 3. Note that there is an implicit sign convention associated with thermoelectric materials such that if point 3 lies above point 1 in an E-T diagram, then point 3 is electrically negative with respect to point 1.

This simple circuit illustrates the general method of graphical analysis. The circuit analyzed in Figure 6.21 requires frequent measurement of the temperature at the instrument terminals and is seldom used where accuracy of better than ±2°F is required (chiefly because it is difficult to measure the temperatures of points 1 and 3 more accurately than ±2°F).

A reference zone of controlled temperature eliminates the need for frequent measurements of the ambient temperature. The reference zone may be an ice point, a triple point, or an electrically controlled high temperature reference zone box. Such a circuit is shown in Figure 6.22 assuming the reference temperature to be an ice point bath.

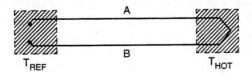

Figure 6.20 The "pattern circuit" for thermoelectric thermometry

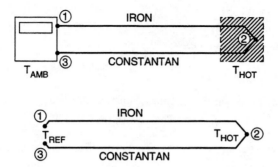

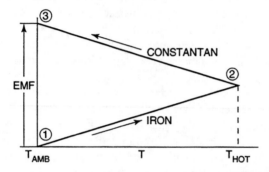

Figure 6.21 Temperature measurement using the ambient temperature as the reference

The analysis of this circuit follows the same pattern as in Figure 6.21. Note that the objective of the analysis is to determine whether or not the actual circuit and the pattern circuit have the same output EMFs. First, all points of interest in the actual circuit are assigned numbers and "possible" temperatures. Then, the *E-T* coordinates are constructed and the analysis begins. Point 1 is located physically at one terminal of the measuring instrument. It is presumed to be at ambient temperature and zero millivolts. Point 2 is at 32°F (0°C) and is connected to point 1 by copper wire. A line in the *E-T* diagram is constructed through point 1 having the slope of the copper calibration curve (from Figure 6.17) and extending from ambient temperature to the ice point. At point 2 the copper wire is connected to the iron wire which extends from 32°F to T_{HOT}. A line segment is drawn through point 2 and parallel to the calibration curve for iron. This line extends to $T = T_{HOT}$ which locates point 3: the end of the iron wire and the beginning of the Constantan. The circuit is completed in this fashion, with each material represented by a line segment of the proper slope.

The output of this circuit is the EMF between point 1 and 5 (those connected to the instrument terminals). The ideal circuit would have had the output given by EMF(2–4). The graphical construction shows that EMF(2–4) is equal to EMF(1–5) since the segments 1–2 and 4–5 each represent the same material (copper) over the same temperature interval (T_{AMB}-T_{REF}) and the wires are so connected as to cancel these EMFs. Thus, the actual circuit is thermoelectrically equivalent to the ideal circuit. Note that the copper lead wires (1–2 and 4–5) play no role in determining the output of the circuit provided that: (1) the calibration of the two pieces of copper are the same and (2) the temperature intervals across the copper

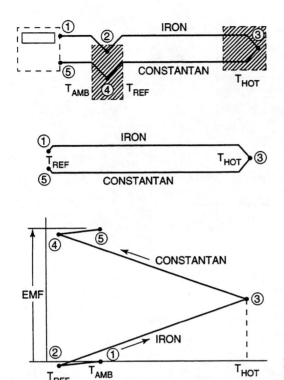

Figure 6.22 Temperature measurement using an ice bath as the reference

wires are the same. It should be clear from Figure 6.22 that if the temperature of point 4 were different from that of point 2, then the output of the circuit would be affected and the equivalence would be lost.

When several thermocouples are used in a single test, far from the measuring station, significant economies can sometimes be achieved by using a common zone box and substituting copper lead wires for the thermoelectric material. Such a circuit is analyzed in Figure 6.23. Again, the objective of the analysis is to determine the conditions under which the actual circuit is equivalent to the pattern circuit.

The function of the zone box is to provide a region of uniform temperature within which connections can be made. The temperature of the zone box need not be constant and it need not be known: it need only be uniform. Techniques for making zone boxes will be discussed in a later section.

The actual circuit consists of a set of thermocouples extending from the zone box to their individual sensing points, a reference bath, a selector switch, and a readout instrument connected together with copper wires. Providing the selector switch and the lead wires introduce no spurious EMFs, the behavior of any one thermocouple, read through this circuit should be the same as the pattern circuit shown. In the *E-T* diagram, the copper lead wires are shown "passing through" the selector switch with no acknowledgment of its existence: the switch is assumed uniform in temperature at the ambient temperature. a more complex analysis could be made by assigning "point of interest" inside the selector switch.

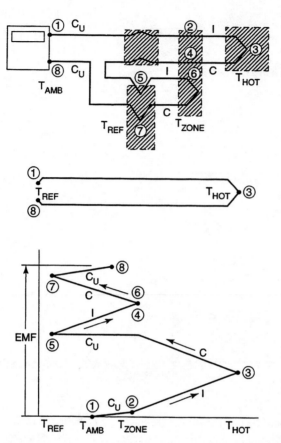

Figure 6.23 Multiple measurements using a zone box and selector switch

The present analysis shows that the copper wires do not interfere with the accuracy and that each individual circuit is equivalent to the pattern circuit provided that: (1) all the copper wire has the same calibration and (2) every point in the zone box is at the same temperature as every other point. The analysis can be made as detailed as desired by identifying more "points of interest" and assigning appropriate "possible" values of temperature to each identified point. In particular, note how this circuit adds the zone box temperature into the common output leads from the selector switch so that every individual thermocouple selected by the switch is properly referenced against the ice bath.

THE LAWS OF THERMOELECTRICITY

Various authors have attempted to summarize the behavior of thermocouples by means of sets of "laws" ranging from 3 to 6 in number. One of the more detailed sets is given by Doebelin.[12] Each law is easily proven by recourse to an EMF-temperature sketch as illustrated below, using the first three from Doebelin's list.

1. The thermal EMF of a thermocouple with junctions at T_{HOT} and T_{REF} is totally unaffected by temperature elsewhere in the circuit if the two metals used are each homogeneous.

In Figure 6.24 it is presumed that $T_{HOT} < T_{CANDLE}$. Providing the wire is uniform in calibration on both sides of the hot spot, the "hysteresis loop" closes and no net EMF is generated by the hot spot. The principle importance of this "law" is its comment on the "point-wise" nature of the thermocouple sensor. Providing the wires a and B are uniform in composition, then the thermocouple responds only to the temperature of the AB connection and not to the distribution of temperature along the wire.

2. If a third homogeneous material C is inserted into either A or B, as long as the two new thermojunctions are at like temperatures, the net EMF of the circuit is unchanged irrespective of the temperature of C away from the junctions.

Figure 6.25 shows a third material inserted into the A leg and then heated locally. It is presumed that the temperatures at the end point of C remain equal. As was found in the first law, if the material C is homogeneous, the EMF excursion induced from 2 to 3 is cancelled by that from 3 to 4 and no net signal is produced.

3. If metal C is inserted between A and B at one of the junctions, the temperature of C at any point away from the AC and BC junctions is immaterial. So long as the junctions AC and BC are at the temperature T_1, the net EMF is the same as if C were not there.

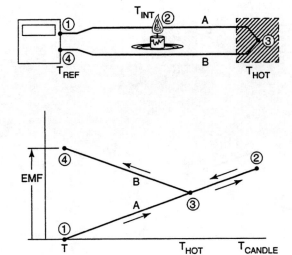

Figure 6.24   Illustration of the law of interior temperatures

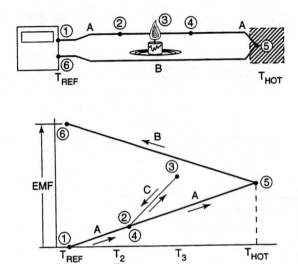

Figure 6.25 Illustration of the law of inserted materials

Figure 6.26 illustrates this case. An intermediate material, C, is inserted between A and B at the measuring junction. The diagram once more shows no net EMF if the inserted material does not undergo a net temperature change and is homogeneous.

The law of intermediate materials is of great practical importance because it speaks to the question of manufacturing techniques and how they affect thermocouple calibration. For example, the third material (material C) might be the soft solder, silver solder, or braze material used to connect the two materials A and B. The analysis shows that the output of the thermocouple is independent of that third material provided that it begins and ends at the same temperature. Practically speaking this is most usually accomplished by requiring

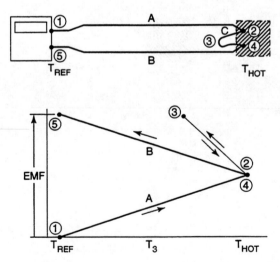

Figure 6.26 Illustration of the law of intermediate materials

that the third material be isothermal. If it is isothermal then certainly it makes no contribution to the output of the thermocouple. This proof should not be taken as a blanket license to connect thermocouple wires together without some care—some installations may result in temperature gradients near the junction and those do require a "two material" circuit. Many cases, however, provide a large isothermal zone around the junction and those cases are insensitive to the material used to join the thermoelements together.

WHERE IS THE EMF GENERATED?

The use of the graphical analysis technique focuses attention on the *regions of temperature gradient as those regions which determine the output of a thermocouple circuit. Material which is isothermal does not contribute to the net EMF of a circuit, regardless of the composition of the material.* (Note from C. P. Wright: This fundamental fact, that EMFs do not come from properly constructed thermocouple junctions but come *only* from temperature gradients in the thermocouple wires, is misunderstood more often than any other in discussions about thermocouples. I have seen this fact blatantly misstated in print three times in the last year—once in *Sensors* magazine in an article by an otherwise reputable manufacturer of PC data acquisition cards, and twice in literature from other manufacturers of PC data acquisition cards. Do you sense a common denominator here?) A typical thermocouple circuit may consist of 10 to 50 feet of wire which experiences only a 5°F or 10°F change in temperature along its length and a few inches of wire which enters a high temperature zone. It is the calibration of that few inches of wire in the temperature gradient which principally determines the output of the whole 50 feet.

When a thermocouple circuit misbehaves and produces strange values, there is no point in searching along the lead wires for the trouble (except for ground loops, pickup, galvanic EMF generation, or piezoelectric affects): *the signal is being generated where the temperature gradient is and that is where to look for the trouble.*

SWITCHES, CONNECTORS, ZONE BOXES, AND REFERENCE BATHS

The principal requirement of a switch or connector is that it does not affect the signal it is passing: that is, it should not produce EMFs which would contaminate the temperature signal from the thermocouple. Switches and connectors, by their very nature, involve several materials connected together and are susceptible to the generation of thermoelectric EMFs. In particular, the thermoelectric power of an alloy is a function not only of its chemical composition but also its mechanical state (i.e., cold work, etc.) so there is no assurance that a cast and machined component of a switch will have exactly the same calibration as a drawn and annealed wire of the same nominal composition. Metallic oxides may have thermoelectric power values measured in volts per degree instead of millivolts per degree. Thin films of metals may have different thermoelectric properties from bulk material of the same name. Thus, the probability is high that any switch or connector will be susceptible to generation of thermoelectric voltages if there are temperature gradients in the materials.

The principal defense against spurious EMFs is to ensure an isothermal region, not only "on the whole," but in detail. Switches and connectors should provide good conduction paths within their structures to equilibrate temperature inside the body. Outer surfaces should be insulators to impede heat transfer from the surroundings. Switch points, in particular, should provide good heat conduction paths. The mechanical energy dissipated as heat when the switch is moved appears first as a high temperature spot on the oxide films of the two contacts. It then relaxes through the structure of the switch raising the average temperature. If the switch points are not good conductors, substantial temperature gradients may persist for several seconds after a switch movement: just when the data are being taken!

Connectors frequently are used to join a thermocouple to lead wires, and this is often in a location very near the test apparatus. Radiant heating, free convection heat transfer from the heated body, and conduction along the stem of the thermocouple all contribute heat loads tending to cause temperature gradients within the connector. Again, the defense lies in insulating the outer shell, breaking the conduction path into the connector, and providing good conduction paths inside the connector to equilibrate the temperature.

Switches and connectors made of "thermocouple grade" alloys will minimize the troubles caused by poor thermal protection, but good thermal design is the ultimate defense.

The errors which can be introduced by using a nonisothermal connector are illustrated in Figure 6.27. An all-copper connector is presumed, with its connection points numbered 2, 3, 5, and 6 as shown. Two cases are examined: one in which the A and B wires both enter at one temperature and both leave at another ($T_2 = T_6$ and $T_3 = T_5$, but T_2 does not equal T_3) and one in which the A wire enters and leaves at one temperature and B enters and leaves at another temperature ($T_2 = T_3$ and $T_5 = T_6$ but T_2 does not equal T_5). The E-T diagram shows that the latter situation causes no error regardless of the type of material in the connector since there is never a temperature gradient along the connector material. In the first case, the resultant signal has "lost" the entire amount of the temperature interval across the connector.

Zone boxes are subject to the same constraint as switches and connectors: the temperature must be uniform inside the zone box. It need not be known and it need not be constant, but it must be uniform. It is, practically speaking, difficult to have a uniform temperature unless it is nearly constant as well. The transient characteristics would be different for different parts of the box structure, so one part would likely "lag" another in a transient giving rise to nonuniformities.

Zone boxes are usually made with relatively heavy copper or aluminum inner walls (1/8 to 1/4 inch thickness) and are covered with at least 1 inch of good insulation. Binding posts or terminal strips inside the box provide for connections between thermoelements and copper lead wires. Ports, through which the wires enter and leave, should be sealed to prevent air circulating inside the box. Care should be taken to prevent accidental heat loads from being imposed on the zone box (i.e., sunlight, radiant heat from hot components, free convection plumes from power supplies, direct contact with air conditioning air flows, etc.). A zone box made with reasonable care will maintain a uniform temperature to within 0.1°F under all service conditions. Tighter tolerances simply require more care and more insulation.

Reference baths may be purchased which provide stable and uniform zones for the termination of thermocouple systems. These generally consist of a heated zone box thermo-

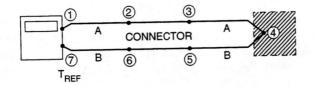

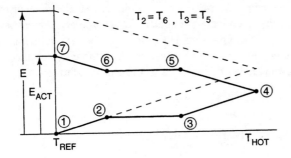

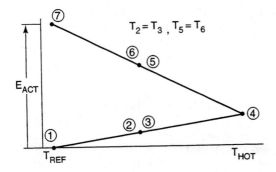

Figure 6.27 The effects of connectors

statically controlled to remain at some temperature higher than ambient. Laboratory users may well use an ice bath or triple point cell. For most engineering purposes a bath made with ice and water sufficiently pure for human consumption will be within a few hundredths of a degree Fahrenheit of 32°F if certain simple precautions are followed.

A Dewar flask or vacuum insulated bottle of at least one pint capacity should be completely filled with ice crushed to particles about 1/4 to 1/2 inch in diameter. The flask is then flooded with water to fill the interstices between the ice particles. A glass tube resembling a small diameter (1/4 inch) thin walled (1/32 to 1/16 inch) test tube should be inserted 3 or 4 inches deep into the ice pack and supported there by a cork or float. A small amount of silicone oil should be placed in the bottom of the tube to provide a region of good thermal contact for the junction. The reference thermocouple is then inserted until its junction is touching the bottom of the tube and secured to the top of the tube with a gas-tight seal. The object of the seal is to prevent atmospheric moisture from condensing inside the tube and causing corrosion of the thermocouples.

The assembly is shown in Figure 6.28 along with a proper connection diagram. Note that the relative polarity of the connection is different in Figure 6.28 from that in Figure 6.23. If a connection like Figure 6.23 is desired, then two glass tubes must be prepared, one each for the positive and negative elements. These tubes must be mounted farther apart than the size of the ice particles to assure adequate cooling. Into each tube is put one thermoelement and one copper lead wire.

The principal requirement of the reference bath is that its temperature be known accurately. Any region of known temperature can serve as a reference bath. Many instruments whose output is directly in units of temperature contain local reference regions and compensating circuits which augment the input signal to account for the local reference temperature. Such instruments can only be used with the type of thermocouple for which they are intended, since the compensating network presumes to know the calibration of the thermocouple being used.

OBTAINING HIGH ACCURACY WITH THERMOCOUPLES

One is sometimes faced with a problem of preparing a "matched set" of thermocouples to high precision, or of obtaining a thermocouple reading with an accuracy better than the manufacturer's tolerance. In such cases, the material should be tested for homogeneity and

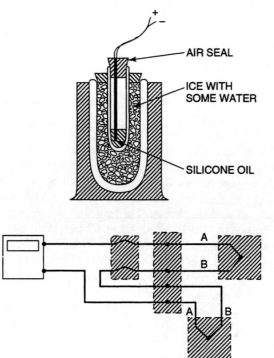

Figure 6.28   Construction of a reference ice bath

then the application should be designed to stretch the temperature gradient over as long a length of wire as possible to minimize the residual effects. This can be done by "bundling" the wires or enclosing them in plastic or metal tubes to prevent abrupt changes in temperature along the wires.

Further increases in accuracy can be obtained by taking advantage of the type of calibration relationship which thermocouples generally have. Most materials follow an equation which is given by:

$$EMF = A(T - T_{REF}) + B(T - T_{REF})^2$$

where the second coefficient is small. This being the case, the principal errors in calibration are the "slope" coefficient A. The absolute accuracy (i. e., the number is degrees of error in the result) can be improved by using a reference temperature close to the sensed temperature so that the difference to be measured is small. It would not, for example, be sensible to use a liquid nitrogen reference bath in measuring the temperature of a room when an ice bath would be equally well defined and would submit the thermocouple to smaller temperature difference.

THE EFFECTS OF INHOMOGENEITIES

The effects of inhomogeneities in the wires can be illustrated by the use of an E-T diagram. Consider Figure 6.24 as an example. Suppose that the calibration curve of the wire is not uniform. What affects the indicated temperature? The entire diagram in Figure 6.24! If the material is not entirely uniform, the loop between b and c might not exactly close: an error would have been introduced into the measurement. In general, one can say that the shorter the path on the E-T diagram, the lower the uncertainty of the reading. Large excursions of temperature along the wires leave the circuit prone to uncertainties and errors due to small deviations in the calibrations. The use of reference baths nearly at the sensed temperature plus the avoidance of unnecessary excursions both help to "clean up" a thermocouple circuit.

When thermocouples are used in unfavorable environments, or for very long times, the output voltage may drift with time due to the development of inhomogeneities in the wire in regions of appreciable temperature gradient. It will be difficult to identify this problem by subsequent recalibration, since the region of partially degraded material may be placed in a uniform temperature zone during recalibration and hence play no part in generating the signal under calibration conditions. No material or combination of materials which is isothermal can produce a thermoelectric signal. When one has a thermocouple suspected by being inhomogeneous, one has only one possible course of action: test for homogeneity. If the wire is homogeneous, then no calibration is required since the "used wire" is still the same as the unused portion, and that cannot have changed its calibration since it was not exposed to the environment. If the wire is not homogeneous, no recalibration can be of value because it will be impossible to place the temperature gradient in the same location for calibration as it was for service. This situation is described in Figures 6.29 through 6.31.

Assume that the thermocouple is exposed to the unfavorable environment only near the hot end, while in service, as shown in Figure 6.29. Assuming that both elements become less active as a result of the reaction with the environment, the output will drop as shown.

Recalibration places all of the affected material in a uniform temperature. The EMF is generated entirely by the material at the entrance to the furnace, which was never changed: the wires between 2, 3, 5, and 6 in Figure 6.30. Recalibration is misleading!

To test a thermocouple for inhomogeneity, clamp the junction at a constant temperature and heat the suspected points. A homogeneous material will generate no EMF, since both hysteresis loops will close as shown earlier in Figure 6.24. An inhomogeneous material will have different calibrations on the "upslope" and "downslope" sides of the temperature gradient, and will produce an error EMF. The process is illustrated in Figures 6.31.

The calibration of the damaged wire between points 2 and 3 and 3 and 4 is presumed different from the "as received" material between 1 and 2 and 4 and 5. On this account, the trajectories on the E-T diagram do not retrace the same paths and a net EMF is generated.

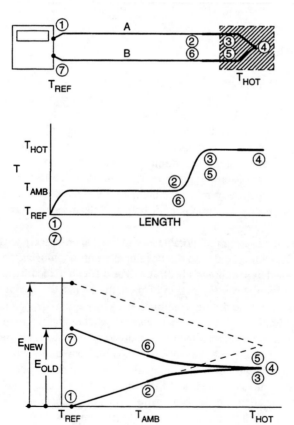

Figure 6.29 Temperature distribution along the thermocouple, in service, and the resulting output after deterioration of the wires

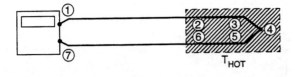

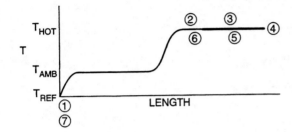

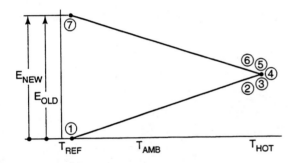

Figure 6.30 Temperature distribution along the thermocouple, during recalibration and the resulting output

In practice, it is frequently possible to execute this test using the fingers (or a pair of pliers) to grip the junction and applying heat either with a match flame or a soldering iron. Any signal derived is evidence of some trouble. Only "experience" will allow this quantitative test to become qualitative. The best resolution of a service-induced inhomogeneity is "side-by-side" running or comparison by replacement. Neither is very convenient but, on the other hand, there is no alternative. It is not within the state-of-the-art to be able to accurately interpret the readings from an inhomogeneous thermocouple in an arbitrary environment, *regardless of the effort expended.*

SOURCES OF SPURIOUS EMFs

Thermocouples produce DC signals in response to temperature differences. They can also pick up AC electrical noise and be sensitive to ground loops, all of which are tractable and can be handled as noise problems would be in any system. There are two problems peculiar to the thermocouple which bear mention: galvanic EMF generation and strain-induced EMFs.

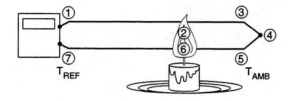

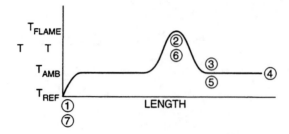

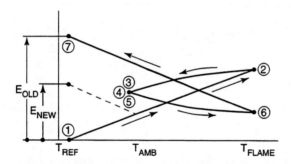

Figure 6.31 Homogeneity testing with a local hot spot and the resulting output

Thermocouples necessarily involve pairs of dissimilar materials and hence are liable to galvanic EMF generation in the presence of electrolytes. The iron-Constantan pair is the most vigorous, generating a signal of 250 millivolts when iron and Constantan are immersed in electrolyte.

Neither copper-Constantan nor type K material showed significant effects in tests conducted at Stanford.

Although type K materials do not display appreciable galvanic EMF, they are strain sensitive (i.e., cold work causes a change in calibration) and, during the act of straining, they generate appreciable EMFs. Temperature measurements made on pieces of equipment which vibrate may show apparent temperature fluctuations induced by the flexing of the thermocouple wires. Copper-Constantan and iron-Constantan are less active than type K, with the copper-Constantan being the least active. Spurious signals on the order of 100°F have been observed using type K materials on large pieces of apparatus (i.e., whole engine tests).

NOTES

1. Endevco, 30700 Rancho Viejo Road, San Juan Capistrano, CA 92675, 714/493–8181.
2. Kistler Instrument Corporation, 75 John Glenn Drive, Amherst, NY 14120–5091, 716/691–5100.
3. PCB Piezotronics, Inc., 3425 Walden Ave., Depew, NY, 14043–2495, 716/684–0001.
4. *Piezoelectric Accelerometer Instruction Manual,* Endevco, 30700 Rancho Viejo Road, San Juan Capistrano, CA 92675, Revised October 1979.
5. See note 1.
6. Walter P. Kistler, *Messverstarker zur Messung elektrishcer Ladung,* Swiss Patent 267431, Bern, Switzerland, 1950.
7. W. Mahr, *Measuring and Processing Charge Signals from Piezoelectric Transducers,* Kistler AG, Winterthur, Switzerland, 1982.
8. R. P. Benedict, *Fundamentals of Temperature, Pressure and Flow Measurements,* John Wiley and Sons, New York, 1969.
9. Robert J. Moffat, *The Gradient Approach to Thermocouple Circuitry,* Temperature and Its Measurement and Control in Science and Technology, Volume 3, Part 2; Reinhold, New York, 1962, p. 33.
10. F. Seitz, *The Modern Theory of Solids,* McGraw-Hill, New York, 1940, pp. 178–81.
11. A. Sommerfeld, *Thermodynamics and Statistical Mechanics,* Academic Press, New York, 1956, pp. 156–63 and p. 342.
12. Ernest O. Doebelin, *Measurement Systems: Applications and Design,* McGraw-Hill, New York, 1966.

Chapter 7

The Transducer Model
and the Problem
of Noise

A prerequisite for the measurement of valid experimental data is the identification and documentation of the noise levels associated with the measurement. If one were to take an absolute position on the issue, data presented without noise level assessment is useless since it cannot be validated. Until you know how much is noise, you do not know how much is data. *Squiggly lines on paper do not valid data make.*

The purpose of this chapter is to give you a method for noise level identification and documentation that will work with any non-self-generating transducer based measurement system. I've chosen to focus on non-self-generating transducers for this discussion because they present the most complete set of documentation tools and opportunities to the designer. Techniques for self-generating transducers are equally valid, but somewhat more cumbersome since the designer does not have control over a minor input. This, in fact, is the primary advantage of non-self-generating over self-generating transducers—control over the minor input.

This discussion focuses on that part of the measurement system most intimately in contact with the process—the transducer and its cabling system. The reason is that, generally, only the transducer and its cabling system have the opportunity to affect the process through energy transfer. For this reason, its performance in your test application is crucial to your success. You can certainly have problems with other components such as amplifiers, signal conditioners, or multiplexers. These components also act as six terminal devices. Problems here can generally be understood within the following transducer model. In some discussions on this subject matter, the authors discuss *all* measurement system components as transducers. Therefore, a discussion of the non-self-generating transducer can provide an overarching set of approaches to measurement system noise level problems.

7.1 THE GENERAL TRANSDUCER MODEL

As stated before in this book, all measurement systems respond to all imposed environments in a predictable fashion all the time, every time. This causal factor for noise levels makes them a subject, in this book at least, not to be obfuscated with statistics. Statistical descriptions tend to hide the fundamental causal reasons for a noise level's existence. Statistics is what you do if you do not understand your noise levels.

During any measurement, several environments are in effect—temperature, time, pressure, strain, motion, etc. The transducer responds to all portions of this complete environment. The noise level problem is characterized by a transducer that exhibits both non-self-generating and self-generating responses to all parts of the environment simultaneously, all the time. The voltages caused by this complex omelet of responses, only *one* of which you want, add to a first approximation within the transducer. The summed responses appear at the transducer output for your inspection. If a crime was committed on the data, it likely occurred before you ever got to see it. The summed output is then signal conditioned, zero suppressed, amplified, filtered, modulated, tape recorded, tape reproduced, demodulated, filtered again, anti-aliased, A/D converted, transfer functioned, FFTed and correlated! No wonder problems occur in this business.

I would venture to say that, at least in the aerospace business, 99% of the signal conditioning for non-self-generating transducers provides, in fact, only DC excitation. Transducers requiring non-DC excitations such as linear variable differential transformers or variable reluctance transducers are excerpted from that statistic. In general, if a transducer can be run with DC excitation, it, usually and perhaps unfortunately, is—right or wrong. With such DC excitation, the transducer model will show that there is no way at the transducer output to separate the effects of the nonself- and self-generating outputs. These effects irrevocably mix in the transducer. You see only the sum. You can't even use some of the fancy techniques of time and frequency correlation because in some cases the noise level that is killing your measurement is coming from what you want to measure in the first place. The "piezoelectric" strain gage mentioned in the section on response syndromes is an example of this.

The general model of this non-self-generating transducer shows in Figure 7.1. Only two environments show. The rules stated in the previous section on response syndromes are still in force. For this example, the desired part of the environment is strain. Any other part of the measurement environment could have been chosen as the measureand—this is a generic model. Each of the other parts of the measurement environment should have its own box for completeness. For this example, however, all other parts of the measurement environment exist in the second box labeled undesired environments.

In the middle of the modeled transducer are the two response types, self-generating and non-self-generating, that are always active in any transducer. The secondary energy input, the excitation, shows feeding the non-self-generating response. The outputs from the two responses add inside the transducer. This is an accurate description for a close first order approximation. The summed output becomes available for the succeeding system components. Between the desired and undesired environments and the transducer, four

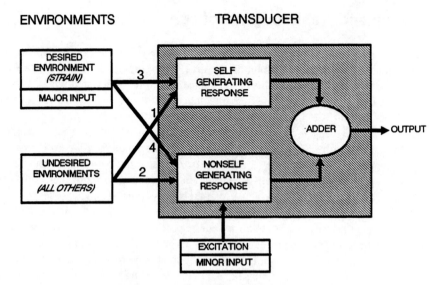

Figure 7.1 General transducer model

paths show. These paths represent the mechanisms by which the desired and undesired environments affect the two response mechanisms.

7.1.1 Path #1: Self-generating Response to the Undesired Environment

Thermoelectric voltages caused by temperature gradients on dissimilar metals within a strain measuring circuit are examples of this path. In a strain measuring circuit using copper wires and Constantan strain gages there are two thermocouples in every quarter bridge, and at least eight in every full bridge!

7.1.2 Path #2: Non-self-generating Response to the Undesired Environment

Examples of this mechanism include gage resistance changes due to temperature, pressure, time, corrosion, or magnetic fields.

7.1.3 Path #3: Self-generating Responses to the Desired Environment

Examples of this path include magnetic induction noise levels caused by the gage grid moving in a shock event and cutting magnetic lines of flux. Loops of wire moving in a magnetic field create current—noise. Constantan, a popular gage foil choice for certain applications, is piezoelectric under shock loadings. Charge will be generated in the foil having nothing to do with the changing strain field under the gage. Further, certain strain gage backings also show this effect. Noise levels generated along this path are particularly difficult to deal with since they are *generated by what you are trying to measure*. You cannot delete the desired environment when you run your test. You're paid to measure this one—it is not in your control as are some undesired environments. You must control noise levels occurring from this path by design.

7.1.4 Path #4: Non-self-generating Response to the Desired Environment

This, at last, is the data you want, and only the data you want. This is the legitimate strain-induced gage resistance change you desire in this example.

In the absence of the detailed knowledge provided by this model, the world of responses from your transducer set can look like that shown in Figure 7.2. Here, the responses from the environments look as if they are jumbled together within the transducer. If the inner workings of the transducer model are not understood, this confusing picture may be all you can get on your test. With this level of knowledge, it is impossible to separate the data wheat from the noise chaff.

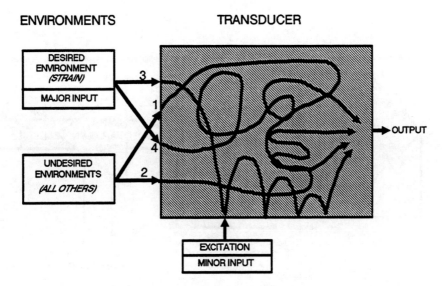

Figure 7.2   What transducers look like in the absence of the general transducer model

7.2 THE FIX FOR THIS COMPLEX SET OF RESPONSES

A fundamental problem presented to the measurement system designer and user is the control and identification of what is going on in the measuring system along each of the four paths as the test runs. This identification can easily be made if your system design provides for it. The necessary provisions show in Figure 7.3. This figure was included before in the chapter on Wheatstone bridge signal conditioning. Please note switches 3 and 4. They are the key to the control and identification of noise levels within self-generating transducers and the measuring system within which they operate. These switches already exist in some top line bridge signal conditioners. Others have one or the other. In any case, they should be mandatory. If they do not exist in your systems, you should have them installed.

A five-step process is needed to perform a conclusive noise level identification.

7.2.1 Step #1: Close Switch 3

Closing switch 3 presents a short circuit to the input of the following components of the measuring system. This is the definition of zero in a voltage sensitive system. Closing this switch identifies all noise levels, from whatever cause, occurring in the bridge output cabling, the signal conditioning and its amplifier, and all downstream components. Reduce the noted noise levels as required. Open the switch.

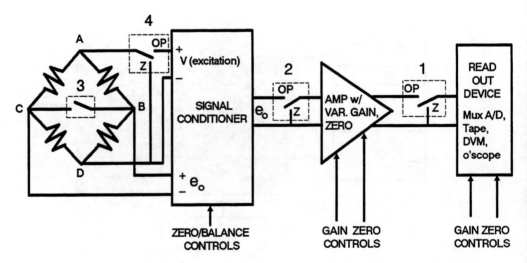

Figure 7.3 Wheatstone bridge signal conditioner with zero and noise level verification switches

7.2.2 Step #2: Create a Condition of Desired Environment (Strain) Off, and Excitation Off

This step isolates and identifies path #1 noise levels. Use switch 4 to cause the excitation to go to zero.

Note: switch 4 is shown here for the constant voltage mode of excitation. If your system used constant current excitation, the switch wiring would be arranged to open the excitation lines instead of short them. Zero is an open circuit—not a short circuit—for a current-sensitive system.

You have arranged for zero strain at this point, so path 3 is inoperative. Paths 2 and 4 are inoperative because the excitation is zero. Any noise levels noted at the system output are, therefore, due to path 1 only. Another example of this path would be electrostatic pickup in the transducer input cabling. Reduce path 1 noise levels as required.

7.2.3 Step #3: Create a Condition of Desired Environment Off, and Excitation On

Return switch 4 to the operate position which turns the excitation to the transducer ON. Maintain the condition of desired environment OFF as before. Apply as many of the undesired environments as possible. Paths 3 and 4 are inoperative because there is no desired environment. Path 1 was isolated, controlled, and documented in the previous step. Only path 2 noise levels remain. These are caused by self-generating responses to the undesired environments. Examples would be temperature changes in lead wires, drift due to transducer heating, and noise levels caused by ripple in the excitation voltage or current source. Reduce path 2 noise levels as required.

7.2.4 Step #4: Create a Condition of Desired Environment (Strain) On, Excitation Off

Turn off the excitation again. Apply the desired environment. Paths 1 and 2 noise levels have been previously controlled and documented. Path 4 is inoperative because the excitation is off. Now, path 3 noise levels are identified. These are undesirable noise levels caused by the desired environment. A previous example was strain-induced noise levels in shock testing. Reduce path 3 noise levels as required.

7.2.5 Step #5: Run the Test by Creating a Condition of Desired Environment (Strain) On, Excitation On

Paths 1, 2, and 3 noise levels were isolated, controlled, and documented in the previous four steps. The only remaining path, #4, is the data you want—the non-self-generating response to the desired environment. If you've done your job in the previous steps, what remains must be valid, noise free experimental data. The following table summarizes the noise level documentation procedure for you.

TABLE 7.1 NOISE LEVEL DOCUMENTATION PROCEDURE FOR
NON-SELF-GENERATING TRANSDUCERS

Path	Major Input	Minor Input	Response Type	Output Source
1	Off	Off	Self-generating	Undesired environment
2	Off	On	Non-self-generating	Undesired environment
			Self-generating	Minor input
3	On	Off	Self-generating	Desired environment
4	On	On	Non-self-generating	Desired environment valid, noise free data!

7.3 SUMMARY

The transducer model presented in this chapter, and the noise level isolation and documentation procedure that follows from it, are very powerful ideas having several advantages. First, this transducer model admits the perversity of the measurements universe. In this very real universe, all measurement systems react to all parts of the environment all the time, every time. Other models that do not take this into account are not nearly as useful for understanding the nefarious noise level omelet. In particular, it includes the simultaneous effects of the self-generating and non-self-generating responses. Further, this model clearly shows the relationships between the response types and the total measurement environment, only one component of which you want to measure today. Last, a simple step-by-step procedure is given for this noisel level documentation. In some advanced systems, these documentation procedures occur automatically before every test run. Even if certain noise levels are not under control, this method assures their identification. Such conclusive identification assures that they will never be reported or misinterpreted as valid data.

The method is applicable to any measurement made with a non-self-generating transducer with any excitation waveform: constant voltage or current, DC levels, pulse trains, or sine waves. The method can determine sources of noise levels directly by locating those points where noise is contaminating valid information. Once this identification has been made, the reductions of the noise levels can proceed in a straightforward manner.

Last, the method can prove conclusively the validity of the data *while the test is in progress*. After all identified noise levels are under control and documented, the remaining information must, by definition, be valid, noise free data.

Chapter 8

Noise Level Reduction Techniques

In the previous chapter on the transducer model and noise level documentation procedures, I discussed a foolproof method for the detection, isolation, and documentation of noise levels in non-self-generating measurement systems. That is only half the battle. Now that you have identified the noise level culprit, what do you do about it if it is clearly too large?

We define three general methods for noise level reduction. There are more complicated models in this area, but I do not find the complication necessarily useful. As far as I know, all noise level control methods used in experimental mechanics measurements can be discussed within the following model. These three methods are:

1. Minimization by Division
2. Mutual Compensation
3. Information Conversion

This chapter will discuss the first two methods. The third method, information conversion, will be given its own chapter later in the book. Examples from your probable experience will be used to illustrate the methods. Be aware that what is important is the general method—not the individual example given.

This subject is complicated. To support ease of understanding, the section appears in outline form so the underlying structure is clear.

1. Minimization by Division (8.1)
   - Reduce the environmental parameter causing the problem (8.1.1)
     - Grounding and shielding
     - Filtering environmental inputs
     - Isolating the system from the environment
   - Reduce the transducer's sensitivity to the environmental parameter (8.1.2)
     - Self-compensation
     - Put the noise level "in jail"
2. Mutual Compensation (8.2)
   - Basic technique (8.2.1)
   - Subtracting circuits in general (8.2.2)
     - Common mode rejection performance
     - Systems that measure differential quantities directly
3. Information Conversion (discussed in a later chapter)

8.1 MINIMIZATION BY DIVISION

Minimization by division is the incremental reduction of the effect of offending noise levels in the measurement system by one of two methods. It is useful to think of this system/environment coupling as the product of the level of the environment and the system's sensitivity to the environment. You can work on either part of the product to reduce the total effect. The first method is to reduce the environmental parameter itself at the measurement location until its amplitude no longer matters. The second method is to leave the environmental parameter alone, but reduce the measurements system's sensitivity to that parameter. This will be more easily understood via the following summary outline for this method.

8.1.1 Reduce the Environmental Parameter Causing the Problem

Grounding and shielding.   Grounding and shielding of measurement systems refers to the reduction of noise levels typically caused by electrostatic, magnetic, and radiation environments by electrical methods. Discussed in detail, this subject is clearly outside the scope of this book. Entire books exist on this subject matter alone.[1,2,3]

We refer to them when in deep trouble on this issue. Then, they are invaluable refererence materials. A measurement system designer ought to stay out of deep trouble by proper system design in the first place.

However, there are some simple cookbook procedures for the elimination of noise levels at power line frequency and its harmonics[4] that represent most problems you might face with your systems. This procedure may seem a little "down home" to you but I assure you that it rests on solid physical and engineering principles. It seems simple because it is written for university junior year engineering students who have never yet faced a real engineering problem in their lives. I know because I was once such a junior taking this Measurement Engineering course.

From Stein:

1. All line-operated instruments should be connected to a single outlet. (This author expands this to include the same circuit breaker service.)

2. All line-operated instruments should have a separate ground connection in the form of a third prong on the power connector. Check all connectors to see if this third prong has been cut off by the previous mis-user. If it has, then the case of the instrument may need to be separately grounded by procedures noted below.

An instrument that is grounded through the power cord to the multiple outlet junction box should not also be grounded to another instrument in the measurement system.

3. Shield all components, cables, breadboards, etc. (by wrapping in aluminum foil if necessary). Do not yet ground any shield.

An electrostatic shield (author's comment: braided, woven or foil wrapped cable shield, metal box, etc.) has no purpose until it is grounded. If there is a sin worse than not grounding a shield, it is providing more than one ground connection.

Not every shield needs to be grounded for every type of setup. There may be an optimum ground point that needs to be discovered. The sections below will systematize this ground determination.

(Author's note: We find that on occasion it makes no difference to system noise levels whether a particular component shield is grounded or not. In this case, ground it anyway. Most test facilities I have ever visited have notoriously poor power conditions. These conditions are largely not under the control of the test organization. Changes outside the facility, that are blind to the test organization, can profoundly change the grounding/shielding requirement within the test facility. I once spent $60,000 of hard earned capital money learning this lesson. Your approach should always be to protect your measurement system from the evils of outside power condition changes that are not under your control by use of line isolation and power conditioning equipment.)

4. Start at the readout device with the following procedure:

4a. Touch the case. If the noise level changes at the system output (either gets larger or smaller), the case must be grounded.

4b. Wrap your hands around the cable connecting the readout device to the preceding link in the measurement chain. If the noise level changes at all, the cable shield must be grounded. The increased capacitive coupling of your body to the signal cable will increase the coupled electrostatic noise levels if the cable needs shielding and grounding.

Repeat these two steps for every component and cable in the setup. Two phenomena to watch out for are the instrument case that you *think* is grounded and is not, and the cable shield inadvertently connected to the plus or minus signal leads.

A single ground point should be selected and all grounds connected to this point. Then, the final ground may be moved to find its optimum position. This procedure is, however, seldom necessary.

5. Another technique, to be used with the preceding one, is the short-circuiting of system component inputs from the readout device sequentially toward the transducer. (Author's note: You've seen this procedure before in the chapter on the Wheatstone bridge and its proper setup procedures.)

Short circuit the readout device to find its intrinsic noise level. Below this noise level you may not operate. No amount of proper grounding and shielding can give you a noise level less than this. It is the absolute design floor. Remove the short circuit. (Author's note: Stein here is referring to a voltage sensitive portion of the measurement system. If the portion you are working on is charge sensitive, replace the short circuit with an open circuit. If you were working at the input to a charge amplifier, you would apply an open circuit for noise level checking.)

Disconnect the component input cable and apply a short circuit at the forward end as shown in the figure. Note the noise level at the readout instrument output.

In this manner one may locate an instrument with high intrinsic noise level outputs. These may then be treated by replacing the instrument or by common mode rejection techniques.

These short circuit techniques are illustrated in Figure 8.1 for two links in the (measurement system) chain and carry through the entire system.

These techniques will handle most electrostatic noise level problems. They will do nothing about magnetically induced noise levels since the copper, aluminum, or steel shields around your cables and instruments are, essentially, transparent to magnetic fields. Magnetically absorbing materials (such as Mu-metal, Armco Steel, Middletown, OH) are needed to provide shielding for this environmental parameter. It also should be remembered that magnetic fields are vector quantities having a magnitude and direction. Knowing this directionality, you can sometimes optimize your setup by rotating components to minimize magnetically induced noise levels.

Filtering undesirable environmental inputs. This method presumes that what you want to measure today occupies a certain range of frequencies, and that the noise level causing parameter occupies a different range of frequencies. Three cases show in Figure 8.2. In the first case, the data and noise frequencies are widely and clearly sepa-

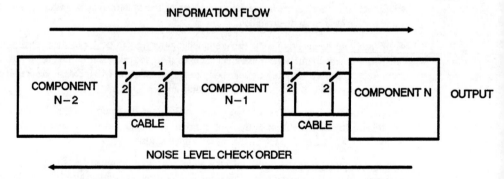

Figure 8.1   Method for checking noise levels in components and cables

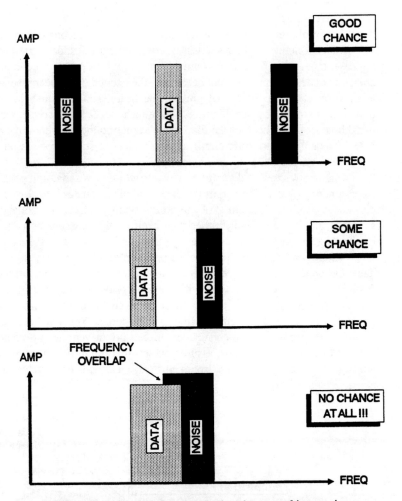

Figure 8.2   Filtering the undesired environmental input noise

rated—a good case for frequency selective filtering. In case two, the data and noise frequencies are closer together causing a problem for the filtering approach. Adequate filtering of noise levels here may force compromise with the valid data frequencies. You may or may not be able to filter successfully here. The third case shows overlapping data and noise frequencies. Here, filtering is clearly not an option.

For instance, looking back to the chapter on resonant systems, consider a piezoelectric accelerometer used for making shock acceleration measurements. You might want to measure shock valid shock acclerations out to 10 KHz with a transducer highly resonant at 40 KHz. The 40 KHz resonance excited by the shock event may drive the measurement system into a highly nonlinear region of the output/input characteristic with all kinds of attendant problems.

What to do?

This problem may be handled in two ways. Approach one handles the problem in the mechanical domain—right at the desired environment/transducer interface. In this interesting case, a portion of the desired environment, high frequency apparent acceleration, becomes an undesired environment because of the excited transducer resonance. It is possible to make mechanical filters that prevent the troublesome frequencies from entering the transducer in the first place. Figure 8.3 shows a Bruel & Kjaer condenser microphone with bleed hole from the back of the diaphragm vented to the ambient atmosphere. This bleed hole forms a high-pass filter eliminating the effect of static pressure on the diaphragm. It has filtered out the DC pressure component.

Figure 8.4 shows a diagram of a tuned Helmholz resonator in parallel with a tap into a pressure transducer. The purpose of the Helmholz resonator is to absorb pressure frequencies at the resonant frequency of the port/pressure transducer resonance. This energy is absorbed into the resonator, rather than exciting the pressure port into resonance, thus broadening the measurement frequency response. This is a legitimate form of filtering.

The second approach handles the problem in the frequency domain after the transducer has done its job and been affected, to whatever extent, by the offending environment frequencies. This approach uses an in-line filter prior to the first stage of signal conditioning. Figure 8.5 shows such a filter from Endevco for use in piezoelectric measurement systems. You have to be careful using this technique. You must assure yourself that the offending environment's frequency content has not driven the system nonlinear; if so, all bets are off. Further, you must assure yourself that the in-line filter has not been affected adversely by the undesired environment itself and has been matched to the transducer's characteristics if required.

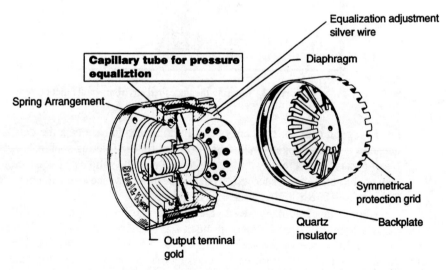

Figure 8.3   Sectional view of a 1″ Brüel & Kjaer microphone cartridge (Printed through the courtesy of Brüel & Kjaer)

DYNAMIC PRESSURE ENVIRONMENT
TO BE MEASURED

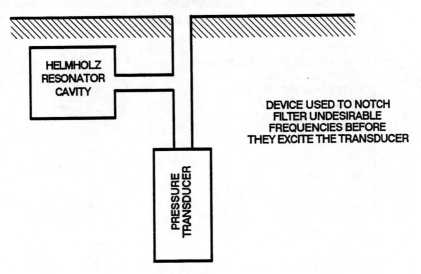

Figure 8.4 A Helmholz resonator is used to "filter" undesirable pressure frequencies

Remember, when filtering out the unwanted frequency effects of the undesired environment, the rules of frequency content and waveshape reproduction still apply. You still have to worry about the total measurement system transfer function. In practice, this means that the data frequencies and the offending frequencies must be separated enough on the frequency axis to allow you to drop in the needed additonal filter, while maintaining the integrity of the data. In the absence of this frequency separation, filtering is not a viable candidate and will certainly do more harm than good.

Isolate the system from the environment.   This method relies on the isolation of the measurement system component, typically the transducer, from the offending, noise producing environment. Examples of this type of noise level control that are in general use follow.

This method is a powerful one in that it does not have to last forever. It only has to last until the measurement is over. After this, no one cares. As an example, strain gage installations typically have extensive protective coatings applied after the gages have been installed and wired. Their purpose is to prevent damage to the installation and prevent the intrusion of moisture. Moisture intrusion will eventually reach the gage foil, corrode it, and cause a positive zero shift that looks exactly like strain change. Eventually, the invasion of moisture into the gage installation will occur—this is inevitable. The protective coatings delay this intrusion and its effects until the test is over.

Airborne Low Pass Input Filter

ENDEVCO
MODEL
2690

Model 2690

- For Use With PE Transducers and
  Charge Amplifiers
- Small Size, Light Weight
- Rugged Construction
- Two Pole Low Pass Filter
- Passive, No External Power
  Required

Actual size

DESCRIPTION

The ENDEVCO® Model 2690-XXX-YYY Series of
Low Pass Input Filters are designed to change the
frequency response of measurement systems
utilizing a piezoelectric transducer and a charge
amplifier. The input filter connects between the
transducer and charge amplifier and the filter should
be located near the charge amplifier.

The frequency response of the input filter is a function
of external source capacitance consisting of transducer and cable capacitances and therefore each
filter is designed for a specific value of source capacitance. With the specified value of source capacitance
the typical frequency response will be a two pole
Butterworth response.

The -XXX describes the upper cutoff frequency
(-5% point) per Table 1. For example, a -101 is a filter
which is flat to 100 Hz, a -501 is a filter which is flat
to 500 Hz.

The -YYY describes the source capacitance, (C_s) per
Table 2. For example, a -101 is a filter designed for
100 pF source capacitance, a -502 is a filter designed
for 5000 pF source capacitance.

-XXX	-5% [Hz] Upper Cutoff	-3dB [Hz] Upper Cutoff
101	100	200
201	200	400
501	500	1000
102	1000	2000
202	2000	4000
502	5000	10 000
103	10 000	20 000
251	250	500

TALE 1: FREQUENCY RESPONSE

-YYY	Source Capacitance (Cs)	Maximum Internal Capacitance
101	100 pF	10 200 pF
201	1000 pF	9300 pF
252	2500 pF	7800 pF
502	5000 pF	5300 pF
752	7500 pF	2800 pF
103	10 000 pF	150 pF

TABLE 2: SOURCE CAPCITANCE

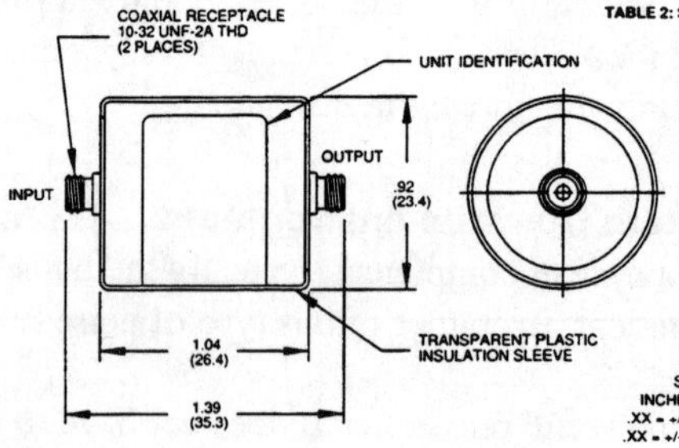

COAXIAL RECEPTACLE
10-32 UNF-2A THD
(2 PLACES)

UNIT IDENTIFICATION

OUTPUT

INPUT

.92
(23.4)

1.04
(26.4)

1.39
(35.3)

TRANSPARENT PLASTIC
INSULATION SLEEVE

STANDARD TOLERANCE
INCHES   [MILLIMETERS]
.XX = +/- .02   [.X = +/- .5]
.XX = +/- .010   [.XX = +/- .25]

ENDEVCO Σ

Endevco Corporation
30700 Rancho Viejo Road
San Juan Capistrano, CA 92675
(714) 493-8181 FAX (714) 661-7231

Figure 8.5 Lowpass filter for piezoelectric transducers (Printed through the courtesy of Endevco)

We run many deployment tests where the measureands are torque about an axis of rotation of a joint, and the deployment angle of the joint. These tests typically run at ambient as well as high and low temperatures to simulate the appendage operating temperature extremes. We isolate the transducers from the thermal environment by thermal shielding and use of materials with low thermal conductivities. We find this isolation easier in these tests than correcting the resultant data for the effects of the thermal environment.

In our precision mass properties load measurements, great pains are taken to prevent the application of noise causing extraneous loads from entering the load transducers. Such noise producing loads include bending, shear, torsion and eccentric loads. The load cells are carefully isolated from these noise causing parts of the undesired environment by very careful load fixture design and an optimum test procedure.

Another example is the routing of extension cables away from potential sources of noise such as power panels, motors, and generators. Here, you are isolating the measurement system by putting physical space between the system and the noise sources. We have also discovered that, in spite of the best grounding and shielding practices, certain fiber optic oscillographs should not be operated in close proximity to large electrodynamic vibration exciters. We use them successfully in vibration laboratories by putting space between them and the shakers.

In modal survey applications the measurement of low level accelerations is critical. These accelerations typically take place at frequencies below 100 Hz, which means that the deflections are large. Large deflections can cause high strains on the structure. The modal accelerometers are isolated from the surface strains by their mounting techniques, which might include using strain isolating phenolic isolating blocks. In this application, you allow the noise level producing environment to occur in a "dead" part of the measurement system—the phenolic mounting block where it has no effect on the data.

The preceeding sections have described methods for reducing the actual noise level causing environment at the measurement location. There will be times when this is not practical. Perhaps the problem environments are not really under your control or, worse, are caused by what you want to measure. In cases where you cannot reduce the level of the problem causing environments, the next best bet is to reduce the measurement system's sensitivity to those environments. Either approach will work. Here, you allow the environments to occur, but use a measurement system designed to be appropriately insensitive to those environments.

8.1.2 Reduce the Transducer's Response to the Noise Causing Environment

Transducer self-compensation.   In transducer self-compensation, the manufacturer (which could be you) builds into the transducer the necessary insensitivity to the problem causing undesired environments while retaining sufficient sensitivity to the desired environment. Here the transducer, exclusive of any circuit in which it operates, compensates internally for the noise level. The case of using the external electrical circuit as a compensator is discussed in sections on subtracting circuits in general.

Transducer manufacturers go to exquisite lengths to provide self-compensation to "standard" noise level sources in their transducers. The purpose of this self-compensation is so that you, as a user, need not worry about it. Well, guess what? You have to worry about it anyway. The manufacturer must make assumptions about the way the universe will look to their transducers at test time. The manufacturer assumes that his transducer remains a rigid body under shock loading, when in fact the transducer is made of jello at shock frequencies having numerous internal resonances. They assume isothermal test conditions, when your test necessarily creates thermal gradients in time and space. They assume no magnetic fields, when your test is being run to assess the effects of magnetic fields on your superconducting magnet support structure. They assume no electromagnetic coupling among your LVDTs (linear variable differential transformers) for deflection measurements, while on your test several are mounted close enough together for their carrier excitation frequencies to interact causing undetectable calibration errors. The point is that we, as measurement system users and designers, violate the manufacturer's overt, and unfortunately covert, design assumptions daily. Further, it does not matter if we know we are violating an assumption or not—the transducer knows! Mere wisdom here does not protect you. Your system design and procedures must protect you.

Transducer self-compensation is primarily affected by proper mechanical and electrical design, and excruciatingly careful control over internal flexure design and sensor mechanical, electrical, and chemical properties. This is the real expertise you are buying from a reputable manufacturer.

There is room for an entire book on this subject matter alone. To narrow the field significantly and present a discussion of reasonable length, I'll focus on examples using, again, bonded electrical resistance strain gages in this section. Further, I'll use only examples of characteristics from the strain gage world as a model for the overall field. Strain gages have been chosen for several reasons. Examples of transducer self-compensation are easily understood for this transducer. The concepts are directly transferable to other transducers. Last, understanding of these issues for strain gages is directly applicable to an entire class of strain-gage-based transducers for experimental mechanics measurements.

Strain Gage Factor as a Function of Strain Level. The gage factor of a strain gage is its internal measure of the gage's sensitivity to strain. It is:

$$\text{Gage Factor} = (\Delta R/R)/(\Delta L/L)$$

where R is the original, unstrained gage resistance and L is the original, unstrained gage length. The Δs refer to the changes in these quantities. As a user, you would not want this "constant" gage factor to be a function of strain. If it were a function of strain, then the gage output would be a nonlinear with its measureand. Figure 8.6 shows a plot at the top of $\Delta R/R$ function of $\Delta L/L$ for a Constantan strain gage alloy as received from the foil or wire manufacturer. The gage factor of the alloy is the slope of this curve. There is an elastic range of $\Delta L/L$ in the strain gage. Within this range, if the strain is removed from the gage it will return to its unstrained length. Note the steep slope in this range. As the strain increases, it moves into the gage's plastic strain range and the gage factor changes to approximately two. What happens is that a basic material property, Poisson's ratio (ratio of lateral contrac-

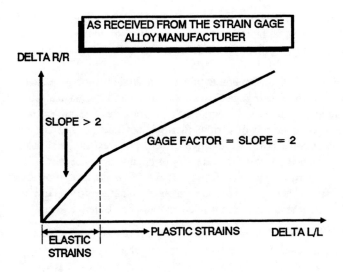

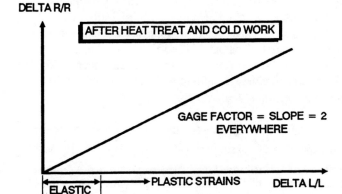

Figure 8.6 Gage factor as a function of strain

tion to axial extension), changes abruptly when the strain moves into the plastic region. In this form, the gage is highly nonlinear.

The strain gage manufacturer, knowing that we want linear strain gages, modifies the properties of the alloy so that its gage factor is also two in the elastic range. In this manner, the gage factor is always about two, and the gage behaves linearly with strain. The manufacturer performs this bit of apparent magic through judicious heat treatment and cold work of the gage 100 microinch thick foil or 50 AWG, .5 mil wire. How they do this is their equivalent of the formula for Coca Cola—very private. They have self-compensated the gage for strain and caused its properties to be linear over a very wide range of strains. Indeed, bonded resistance strain gages work very well cycling into and out of plastic strains in both tension and compression.

Gage Factor as a Function of Temperature. If measuring strain interested you, it would be advantageous if the gage factor was not a function of temperature. The manufacturer also handles this one for you by minimizing the gage factor's sensitivity to temperature.

Figure 8.7 shows this—this is actual manufacturer's data. It shows the gage factor sensitivity with temperature for two of their gage alloys, Constantan and Karma. This minimization is, again, done with heat treat and cold work on the foil or wire. The curves show for each alloy for installations on materials of various thermal coefficients of expansion. You can see by the slopes that for a fairly wide range of temperatures the gage factor changes less than 5%. For some test purposes this amount of change is negligible. If not negligible, corrections to the experimental data can be made for this noise level providing the temperature at the gage location is known. For the transducer manufacturer looking for 0.01% changes in sensitivity, other techniques are needed and will be discussed later.

Thermal Output as a Function of Temperature. If you had strain gages installed on a single piece of test material, and you hung that material in an oven and changed its temperature slowly and evenly, you probably would want the strain gages to read zero output over the temperature range. The reason for this is that, here, there is no thermal stress applied to the test piece and that case is, generally, not of interest to a stress analyst. Any output under these conditions, free thermal expansion on a single material undergoing slow, uniform temperature change, is the thermal output. It used to be called apparent strain, though this term is leaving the technical lexicon due to international agreement. Would the gages, in fact, give you zero output under these conditions? No, they would not. The thermal output from a strain gage is a function of three variables:

- Thermal coefficient of resistance of the strain gage alloy (microohms/ohm/degree)
- Thermal coefficient of expansion of the strain sensitive alloy (microinches/inch/degree)
- Thermal coefficient of expansion of the material on which the gage is installed (microinches/inch/degree)

All three are nonlinear functions of temperature, so the sum of the three effects is, not surprisingly, nonlinear with temperature. This shows in Figures 8.8 and 8.9. The manufacturer minimizes the thermal output over a temperature range based on the thermal expansion coefficient of the material to which the gage is bonded.

How does the manufacturer do this? They control the two gage alloy properties they can control—its thermal coefficient of expansion and its thermal coefficient of resistance. They do this, again, by heat treat and cold work of the gage alloys. The details of this lie quietly in the same safe as the secret for linearizing the gage factor.

The user has several other forms of self-compensation available through careful engineering of the gage installation. As you discovered earlier, if Constantan gages are used the typical copper/Constantan thermocouples in the circuit/gage leadwire junctions have a basic sensitivity of 20 microvolts/°F. A quarter bridge with 10 VDC excitation has a strain sensitivity of 5 microvolts/microstrain for a gage factor of 2.0. For every degree difference

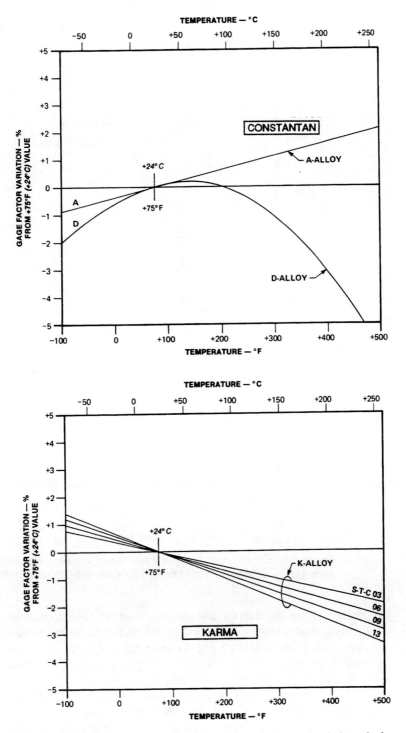

Figure 8.7 Gage factor as a function of temperature (Printed through the courtesy of Measurements Group)

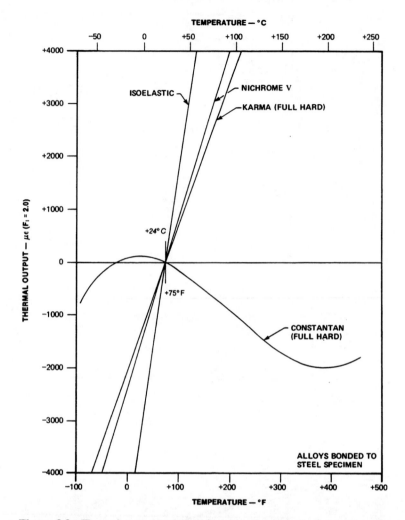

Figure 8.8 Thermal output (μe) as a function of temperature as received by strain gage manufacturer (Printed through the courtesy of Measurements Group)

in temperature at the thermocouple junctions within the bridge, a possible error of 4 microstrain exists (20 microvolts per degree/5 microvolts per microstrain). If this was your problem, you might consider using Karma gages that have a thermoelectric power with copper of 2 microvolts/°F—not the 20 of Constantan. Karma gages are simply much better self-compensated for this effect within the bridge.

The choice of gage alloy also can allow improved self-compensation to magnetic fields, hydrostatic pressure, corrosion, and a number of other environments.

The reputable transducer manufacturer goes to much trouble to self-compensate their transducers whether they are strain gages, strain-gage-based transducers, or any other trans-

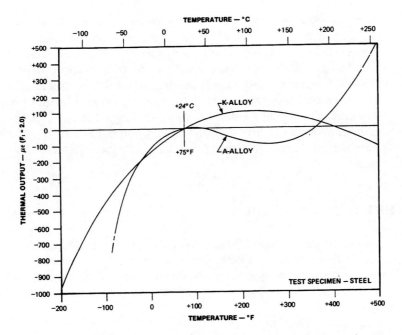

Figure 8.9   Thermal output (μe) as a function of temperature after strain gage manufacturer has adjusted performance (as sold condition) (Printed through the courtesy of Measurements Group)

duction mechanism. This is the place to put the noise level control mechanisms—right in the transducer where it will do the most good.

Put the noise level "in jail".   Put the noise level in jail?

This means you can, by design, arrange it so that the noise level that's bothering you is "in jail" where it will do your measurement system and your data no harm.

Say you had a rocket engine test stand out in the desert somewhere with the block house and your measurement system 1000 feet away. The 1000-foot cables get hot and cold with the weather. They get down to 20°F in the winter and up to 150°F in the summer. You need to run Wheatstone bridge transducers over the data lines. Further, in an economy move, your boss said to leave out the external sense leads in the cables—they cost money! The result is that the constant voltage excitation to your bridge transducers varies with the temperature during the day to a degree that is unacceptable to you. With constant voltage excitation you are a dead duck under this set of conditions. Here, the noise level, excitation changes due to temperature-induced lead wire resistance change, is in a part of the bridge where it kills you. It is not in jail.

How do you put the excitation "in jail"? Change to constant current excitation. Here, the excitation current to the bridge is constant no matter what the leadwire resistance is. By changing how the system works you've put the noise level in jail where it does you no harm. The three wire strain gage circuit shown in the chapter on the Wheatstone bridge is another example of this technique. You put the resistance changes that upset the two wire circuit in jail—in adjacent bridge arms where they do minimal damage to the data.

We use, in a vacuum environment, a 256 channel bridge signal conditioner/multi-plexer/A/D converter, called the SpaceMux, for resistance temperature transducers during thermal vacuum tests. This signal conditioner system is designed to spin with a rotating spacecraft. The sliprings on this test were very noisy. The SpaceMux operating inside the chamber allows us to drive only high level digital signals over these noisy sliprings. The noisy sliprings have been promoted to a point in the measurement system where they do no harm.

8.2 MUTUAL COMPENSATION

The second major method for the elimination of noise in measurement systems is mutual compensation. In this method two or more transducers are allowed to experience the noise causing environment(s) and their outputs due to those environments are appropriately sub-tracted from the data minimizing the effect of the noise.

8.2.1 Basic Technique

Mutual compensation is a particularly useful technique when using bridge-based transduc-ers with their inherent ability to subtract. The name of the mutual compensation game is to arrange things so that the outputs from the measureand you want add, while the outputs from the noise causing sources subtract. There are two advantages to this approach. First, the noise levels are minimized. Second, the data levels double, significantly increasing the signal-to-noise ratio.

You have already seen an example of this subtractive form of compensation in the acceleration compensated pressure transducers noted before. These manufacturers design so that inside the transducer are both a pressure and an acceleration sensor whose outputs due to pressure add, but whose outputs due to acceleration (the noise level) subtract.

The basic technique of mutual compensation is to: (1) provide two transducers that are both affected by the same noise causing environment; (2) simultaneously; which (3) provide the same outputs due to that environment; and (4) whose outputs are then sub-tracted perfectly.

Those are stringent conditions when you take them one at a time. The people who say, "Well, just put it in a bridge to handle that," have never been through this logic. Let's take those conditions one by one to see how close you can really come to perfect compen-sation. Again, I'll use strain gages to explain these conditions.

Condition 1: Two transducers in the same environment. If the environ-ment causing your measurement noise levels is time (your problem might be long-term drift) you have a chance! You can certainly get two transducers to exist at the same time.

What about other noise causing environments? Transducers have finite size and mass. Some noise causing environments are spatial, or vector, in nature. They have a direction and magnitude and maybe some frequency content. Magnetic fields are a perfect example of this. The magnetic fields drop as one over the cube of the distance from the field source. If your two sensors are not aligned alike, at exactly the same distance from the field source, with a duplicate materials distribution between them and the source, they cannot be expected to compensate for each other.

How about temperature transients and gradients? If you had a problem where the temperature of the test piece was causing you grief in the strain measurements, the chances of having that piece at absolutely constant and uniform temperature are minimal. After all, you do have to at least heat it up and cool it down to run the test (or vice versa).

Since transducers do have size, mass, and direction, it is unlikely that you can really get two transducers in the same environment. The best you can do is an approximation of this condition. You should not expect any more.

Condition 2: At the same time.   This condition goes with the first one. If time is your culprit, you have it made.

If there are any changes in the noise causing environment with either time or direction, then the chances of having each transducer experience the same environment at the same time is, again, minimal. Let me give an example from the literature. Figure 8.10 shows test hardware that had been strain gaged to measure axial load in the member. The tension member had been carefully designed and gaged using the best design and gaging practices. The test measureand of interest was transient load in the test article caused by

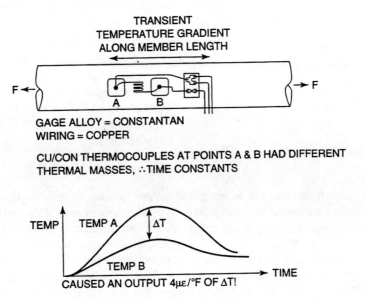

Figure 8.10   Violation of condition #2 "two transducers in same environment at same time"

thermal transients. The test is run and the measured load data from the carefully calibrated strain gage bridges does not make sense. How come?

In this case, despite the best practices, the input test environment caused axial temperature gradients along the instrumented member. Each copper/Constantan solder junction on the gage acts, as noted before, as a thermocouple. Under the conditions of an axial temperature gradient, there is no chance whatsoever that each thermocouple junction is at the same temperature at the same time, given the gage and terminal tab layout of this installation. Therefore, there was a net thermoelectric emf generated inside the bridges that invalidated the load data. This is an example of transducers, and the copper/Constantan junctions are the transducers, not being in the same environment at the same time.

Condition 3: Provide the same outputs to the noise causing environment.   Assume that you have passed the previous two difficult hurdles. You have created a situation where your two transducers are in the same environment at the same time. The next hurdle is that the two transducers must react to this environment in the same manner.

All transducers, indeed all measurement system components, are individuals if you look closely enough. Sometimes, you do not have to look too closely to discover this. There is no average member of a transducer family, just as there is no average engineer. They all act differently. That is why manufacturers give them serial numbers—a tacit admission of their individuality.

Transducers have a history just as you and I do. They are an absolute product of that manufacturing and use history. They were manufactured, after all, not produced by divine intervention. That integrated transducer history has several facets. An easy way to remember these facets is to use COMETMAN.[5] COMETMAN is an acronym named for the eight forms of energy recognized in the United States.

TYPES OF TRANSDUCER HISTORY
C = Chemical
O = Optical
M = Mechanical
E = Electrical
T = Thermal
M = Magnetic
A = Acoustic/fluid
N = Nuclear

Each transducer has an integrated history for each of these eight energy types. Further, every transducer remembers its individual history in each of the eight areas. If you exceed its previous history in any energy area, its performance will surely change.

Let me give you some examples. Strain gages have some performance parameters adjusted by the manufacturer's use of heat treat and cold work—thermal and mechanical energy history. Change the history, change the performance. Their properties are also surely functions of their metallurgy—chemical history. If you overheat a transducer, its properties will change—thermal history. If you irradiate a transducer by use in a nuclear environment near a reactor, its properties will change, perhaps severely—nuclear history.

Piezoelectric transducers are polarized by application of an electric field and temperature—a combined electrical and thermal history. Each transducer has its performance modified by the manufacturer by making sometimes minute changes in the COMETMAN history.

Why would you expect two transducers with different accumulated COMETMAN histories, to which they are extremely sensitive, to provide the same output to any noise causing environment? Indeed, you should expect the reverse—that two closely matched transducers in the same environment will never give the same output. This, again, is why serial numbers were invented.

Figure 8.11 illustrates the sum of the first three conditions. This figure shows the range of thermal outputs for ten Constantan strain gages on a piece of test material for which they were temperature compensated. These installations were identical as closely as we can control conditions: same gage type, gage lot, surface preparation, adhesive lot, cure and clamping procedure, solder and flux, solder cleaning procedure, leadwires, protective coatings, all installed at the same time by the same technician. This specimen was then subjected "slow and uniform" temperature change in an unstrained condition.

We are trying here to force, as closely as we can, the first two conditions: same environment at same time. Temperatures measured on the specimen varied no more than 3 degrees F, which is about the experimental tolerance expected for copper/Constantan thermocouples that have not been individually calibrated. The solid line defines the average thermal output for the ten gages later used in a computer data reduction to suppress this noise level. The dotted lines define the peak envelope of the maximum and minimum thermal outputs from the ten-gage set. These data closely reproduce that given by the manufacturer for this test condition.

Note the divergence of the envelopes as the temperatures extend away from room temperature in both directions. It is clear from these data that transducers simply cannot provide the same output for the same environment under the best of test conditions. What

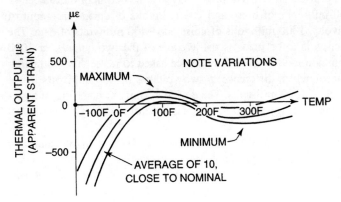

SAME: PIECE OF FOIL, GAGE LOT, SURFACE PREP, ADHESIVE MIX, INSTALLATION METHOD, CLAMPING METHOD AND PRESSURE, CURE SCHEDULE, PROTECTIVE COATINGS, AND INSTALLING TECHNICIAN

Figure 8.11   Apparent strain data for 10 strain gages

you are seeing here is the envelope of the performances of the ten basic transducers, made with the very best of today's technology, when subjected to the first three conditions. These data are typical for strain gages and other individual transducers. This is the type of thing you should expect.

Condition 4: Noise induced outputs are subtracted perfectly.   Now, presume that you've created the first three conditions: same environment, at the same time, and two transducers giving the same output. The last condition you need to perform mutual compensation is a perfect subtractor to perform the subtraction of the noise levels. The Wheatstone bridge performs this task very well. One also can think about any device that has a differential, or "difference making," input such as a differential oscilloscope or differential amplifer.

The degree to which a device, a subtractor, can successfully perform this subtraction is defined by its common mode rejection performance. The scenario is as follows: (1) we have arranged for two transducers to give the same output of the same polarity, (2) for the same noise causing environment, (3) at the same time, and (4) outputs of opposite polarity for the desired measureand. When the subtraction occurs, the noise levels ostensibly cancel, and the data outputs add. The simplifed equations are:

	Due to Noise	*Due to Data*	*Total*
Transducer #1:	N	D	N + D
Transducer #2:	N	–D	N – D

$$\text{System Output} = \text{Transducer \#1} - \text{Transducer \#2} = (N+D) - (N-D) = 2D$$

This, very simply, is how mutual compensation works. It can also work if the design provides noise of opposite signs with data of the same polarity. Be aware that no system will make this subttraction perfectly. There will always be error involved.

8.2.2 Subtracting Circuits in General

Common mode rejection performance.   Say that the output of transducer #1 was 1100 millivolts: 1000 millivolts of noise and 100 millivolts of data. The output of transducer #2 is 900 millivolts: 1000 millivolts of noise and –100 millivolts of data. The level at which the subtraction is to be made is the average of the two outputs, or 1000 millivolts. This is the "common mode" level. Any device asked to make this subtraction must be able to discern accurately a difference in two voltages of 200 millivolts in the presence of a common mode of 1000 millivolts. The device's ability to make that subtraction is defined by its common mode rejection ratio. This is a measure of its ability to reject the common mode level.

The device must reject (not react to) a common mode of 1000 millivolts and then make a subtraction of two much lower level signals and report the result as data. The common mode rejection ratio is a measure of the error created in the answer when this subtraction occurs. The common mode rejection is usually expressed in dB since the ratios are large. It is defined as the ratio of the common mode voltage to the error that results from the subtraction.

If the common mode rejection ratio were infinite, and the same voltage applied to both terminals of a differential device, the answer from the perfect subtraction would be zero. Unfortunately, this never occurs due to finite common mode rejection ratios.

In a typical measurement case, the subtractor is a differential amplifier fed from a Wheatstone bridge transducer. Half the bridge excitation voltage appears across each bridge arm in a balanced bridge. For a bridge excitation of 10 volts, the common mode voltage is 5 volts. Here, you need to measure the $\Delta R/R$ induced voltages at the bridge output corners in the presence of a 5-volt common mode. If you wanted to be sure that the common mode errors were less than 1 microvolt (considered negligible), what common mode rejection ratio would you need?

$$\text{CMRR} = \text{common mode voltage/allowable error} = 5 \text{ V}/1 \text{ microvolt}$$
$$\text{CMRR} = 5,000,000/1 = 134 \text{ dB}$$

A common mode rejection ratio of 134 dB is required at whatever frequency the $\Delta R/R$ occurs to perform this subtraction. There isn't an instrumentation or signal conditioning amplifier my firm owns that will do this job! That is probably the case at your laboratory too. Our standard laboratory digital multimeters (Datron 1062s) will do the job with 6 dB to spare—but only at DC.

Complicating the issue, the common mode rejection ratio also has a magnitude and phase that are functions of frequency. In fact, most differential measurement components are optimized for common mode rejection performance at 60 Hz and its harmonics. Most manufacturers will give you their CMRR magnitudes as a function of frequency. I've never seen a specification on CMRR phase as a function of frequency.

Systems that measure differential quantities directly.   Certain transducers, such as differential pressure transducers and thermocouples, are designed to measure these differential quantities directly. With these transducers, the average pressure at the two inlet ports or the average temperature of the two junctions is a noise level. You do not want these transducers to respond to this level. You want the transducer to measure the differential line pressure or differential temperature despite the common mode level.

How can you assess the performance of these systems? You can do it in the same manner as explained above with certain modifications. Figure 8.12 shows a generic model for any differential transducer. Let's use pressure for this example. Q_1 and Q_2 are the two pressures, whose difference, $Q_1 - Q_2$, we want to know. The model for the generic differential transducer includes two internal transducers, one for Q_1 and one for Q_2, their outputs subtracted, and the result of that subtraction shown to you as output. Each transducer has its own gain, T_1 and T_2.

If everything were perfect, the two internal transducer gains, T_1 and T_2, would be equal and feeding a perfect subtractor—that is never the case. The output from the unrealizable, perfect transducer would be

$$\text{Output} = (Q_1 * T_1 - Q_2 * T_2) = T(Q_1 - Q_2)$$

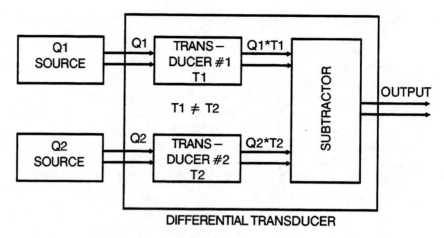

Figure 8.12   Model of a differential transducer

However, in real measurements life, T_1 and T_2 are never the same—maybe close, but never the same. The gains will have an average, a common mode gain if you will.

$$T_C = (T_1 + T_2) / 2$$

and a difference value, T_D, where

$$T_D = T_1 - T_2$$

The real output of a real differential transducer is, therefore

$$\text{Output} = Q_D * T_C + Q_C * T_D$$

$$\text{(data)} + \text{(noise)}$$

By inspecting the equations governing common mode behavior, it can be noted that the common mode rejection ratio can also be stated as the ratio of the two transfer ratios, or gains, T_C and T_D.

$$CMRR = T_C/T_D$$

In this manner, the final output equation can be stated as

$$\text{Output} = T_C(Q_D + Q_C/CMRR)$$

This equation states that the output of the device is the average of the two internal gains, T_C, times the sum of the differential quantity Q_D (this is the data we want) plus an error term equal to the average of the two quantities Q_C divided by the common mode rejection ratio. This is the most useful form of the governing equation. For the unrealizable, perfect subtractor, *CMRR* is infinite, and the error term goes to zero.

     Assume a differential pressure measurement. The problem is to measure a differential pressure of 20 psi at a line pressure of 1000 psi. You're planning to make the measurement with a transducer with a common mode rejection ratio of 60 dB or 1000, and you want to measure to an experimental uncertainty of 1%. Will this transducer do the job? Sounds easy, doesn't it?

$$\text{Output} = T_c(Q_D + Q_D/CMRR) = T_c(20 + 1000/1000) = T_c(20 + 1)$$

Inspecting the equation shows an error term of 1 psi will be created in the subtraction. That 1 psi error is 5% of the total differential pressure—not the required 1%. Therefore, *this transducer lacks any chance at all of making this measurement to a 1% experimental uncertainty.* Your only choice is to change to a transducer providing a higher common mode rejection ratio.

This test should be run in metrology every time a differential pressure transducer is calibrated. Both transducer pressure ports should be manifolded to the same, common pressure. That pressure should then be run up the maximum rated common mode pressure. The noted output, which will not be zero, defines the common mode performance of the transducer—at least at zero frequency.

As a second example, assume a differential temperature measurement using differential thermocouples. We use this technique in thermal vacuum testing to define temperature gradients across spacecraft mounting fixtures, although at lower temperatures than this example. The problem is to measure the temperature difference between one junction at 990°F and the second junction at 1010°F. Use the common mode equation to check the resulting experimental uncertainties in this measurement.

The thermocouple wire will have a nominal sensitivity of 20 microvolts per degree with variations of only +/− .12 microvolts per degree along its length. The manufacturer tells you this and you verify it with metrology calibrations. This represents a small error away from nominal sensitivity, about 0.6%. In terms of the common mode equation then

$$Q_C = (990 + 1010)\,/2 = 1000°F$$

$$Q_D = 20°F, \text{ the real temperature difference}$$

$$T_C = 20 \text{ microvolts/F}$$

$$T_D = .12 \text{ microvolts/F}$$

$$\text{Output} = Q_D{}^*T_C + Q_C{}^*T_D = (20)(20) + (.12)(1000)$$

$$\text{Output} = \underbrace{400 \text{ microvolts}}_{\text{data}} + \underbrace{120 \text{ microvolts}}_{\text{noise!}}$$

Your design will make a *31% measurement error* in the differential temperature before you ever get out of the transducer! This would clearly not meet anyone's specifications.

The lesson is this: differential transducers need very high values of common mode rejection ratios to make valid measurements at high common mode levels. Such performance costs money. There appears to be no way around that.

8.3 EDITORIAL

Noise level isolation, identification, and control is one of the cornerstone skills of the professional measurements engineer. It sets this engineer apart from amateurs in the field. It is fundamentally important to the production of demonstrably valid data.

When watching successful measurements engineers operate, I consistently notice that their thought processes are with the noise levels they expect or have seen on the test, as much as with the valid data they are seeking. You can watch their eyes as they scan the data being produced. They are continually asking, "Is this data, or noise, and how do I know?" They may not be expressing that audibly, but they are expressing it nonetheless.

In order to validate any experimental data, an assessment of the noise levels present in the data is mandatory. Your customer has the right to ask you for this assessment as a form of proof that your data is valid. Most customers in my experience never ask, and I have often wondered why. Is it because they make a covert and dangerous assumption about the data—that it is perfectly valid with zero noise levels? I've seen this assumption made a thousand times. Or do they think the people providing the data would, somehow, be offended?

All customers have the right to ask for this proof. Smart customers do ask for it. Their attitude is, and should be, "Show me." A customer who understands enough about experimental measurement processes to ask for this proof is a customer who understands your problems. They are almost always easier to work with than those who do not understand your problems. The measurements engineer's attitude should be, "You bet! Here are noise levels we're concerned about."

Smart measurements engineers provide this documentation quickly and effectively because they have thought about its provision in their measurement system design and operational procedures. Really good measurements engineers know their system's basic sensitivities to expected noise levels and have determined simple yet perceptive tests to demonstrate these susceptibilities.

Noise level documentation should be cherished and published—or at least reviewed with the paying customer. Spectacularly low noise levels should be celebrated!

NOTES

1. R. Morrison, *Grounding and Shielding Techniques in Instrumentation;* John Wiley and Sons, NY, 1989.
2. H. Ott, *Noise Reduction Techniques in Electronic Systems,* John Wiley and Sons, New York, 1976.
3. *Guideline on Electrical Power for ADP Installations,* FIPS PUB 94, U.S. Department of Commerce/NTIS, Springfield, VA, 1983.
4. P. K. Stein, *Measurement Engineering,* Stein Engineering Services, Phoenix, AZ; 1968, pp. 188–89.
5. P. K. Stein, *Sensors/Transducers/Detectors: The Basic Measuring System Components;* Proc. Joint Measurement Conference, Boulder, CO: Instrument Society of America, Pittsburgh, PA, June 1972, pp. 63–91.

Chapter 9

Noise Level Control
by Information
Conversion

In the last chapter we discussed the first two general methods for noise level control in measurement systems: minimization by division and mutual compensation. Information conversion is the third general method. Information conversion uses the unique ability of non-self-generating measurement systems to move information around the frequency domain by the judicious use of non-DC excitation waveforms. This frequency shifting can be used to separate self-generating noise levels from non-self-generated data in cases where the two previous methods fail miserably.

Assume that you have the following scenario. You have, in your measurement system, noise and data frequencies that are too close together to use frequency selective filtering. If you use this technique, you'll throw the valid data baby out with the noise wash! You might have the condition where you're interested essentially in DC data, but thermoelectric noise levels in a bridge circuit and system zero shifts exist at very low frequencies—like right on top of your data. Or, you might have structural data frequencies from 20–500 Hz that are of interest, with self-generated power line noise at 60 Hz and its first six multiples—120, 180, 240, 300, and 360 Hz. You might have shock data frequencies from 500–30,000 Hz measured with strain gages, with self-generated shock-induced noise levels all over the spectrum.

You've tried everything in your measurements bag of tricks provided by the first two methods—and you're still in trouble. Last, you've got a non-self-generating measurement system, but you are getting killed by self-generated noise levels. This is the unifying situation in each of the examples noted above. Self-generated noise levels can include coupled electromagnetic/electrostatic noise levels (other than those carried into the measurement system as noise on the excitation waveform itself), system zero shift with time, thermoelectric EMFs, piezoelectric effects, or triboelectric effects in cabling. This list is in no way exhaustive.

This scenario, shown generically in Figure 9.1, may seem restrictive to you. After all, I've put at least five restrictions on this scenario:

1. You can't filter
2. You can't minimize enough
3. You can't mutually compensate enough
4. Noise levels are caused by self-generating mechanisms
5. Desired data is caused by non-self-generating mechanisms

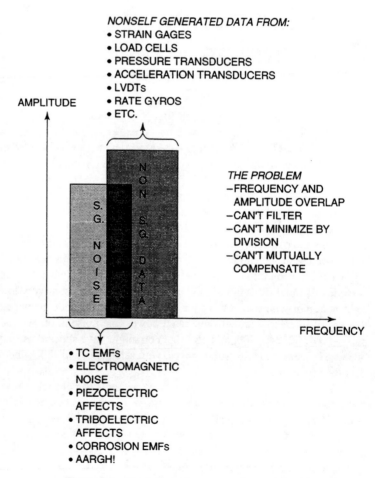

Figure 9.1   The information conversion opportunity

In fact, this represents a rather large family of problems encountered in most mechanical engineering test facilities all over the world. Unfortunately, most laboratories neither are aware of, understand, nor solve these noise level problems. This family of noise problems is large enough that a discussion of the techniques to solve them is warranted as a separate chapter. It is warranted because minimization by division and mutual compensation require a certain foreknowledge of the problems to be encountered and experience in their solutions. You may have neither the foreknowledge, nor the experiential knowledge, nor the time/budget to acquire them. When you find yourself in this situation, some bulletproof methodology is needed. Information conversion is such bulletproof methodology for this situation.

9.1 MULTIPLICATION OF WAVEFORMS

The key to solving the scenario's problem is the ability to "convert," or move, data from its original place in the frequency domain to another part of the frequency domain—namely away from the self-generated noise levels that are the problem.[1] This ability rests on a useful property of certain non-self-generating measurement systems, the property that allows the multiplication of waveforms within the system.

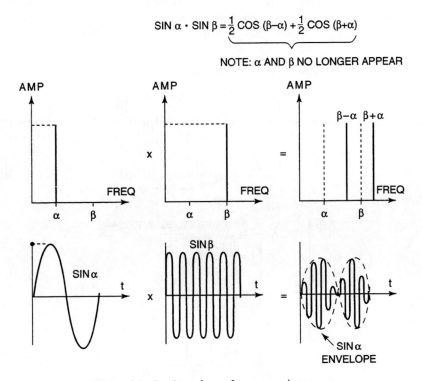

$$\text{SIN } \alpha \cdot \text{SIN } \beta = \frac{1}{2} \text{COS } (\beta-\alpha) + \frac{1}{2} \text{COS } (\beta+\alpha)$$

NOTE: α AND β NO LONGER APPEAR

Figure 9.2   Product of waveforms: two sine waves

Waveform multiplication shows graphically in Figure 9.2. The figure shows, perhaps, the most simple case, the multiplication of two sine waves at frequencies α and β. The equation for the product of two sine waves of unit amplitude is:

$$SIN(\alpha) \times SIN(\beta) = COS(\beta-\alpha)/2 + COS(\beta+\alpha)/2$$

When you put two frequencies into the multiplication process, α and β, the output occurs at two new frequencies, $\beta-\alpha$ and $\beta+\alpha$, the sum and difference frequencies. The original frequencies no longer exist, and no information is available at those frequencies. Whatever information was there at the original frequencies is moved (converted) to entirely new frequencies. You can see this in Figure 9.2 below the equation. $SIN(\alpha)$ shows furthest left as a single spectral line at frequency α. $SIN(\beta)$ shows in the middle as the second spectral line at frequency β. Their products show furthest right at frequencies $\beta-\alpha$ and $\beta+\alpha$. The multiplication in the time domain shows at the bottom of Figure 9.2. The product of the multiplication is a fully modulated waveform whose *envelope* is $SIN(\alpha)$. However, the frequency analysis of this waveform will show no information at that frequency. This result is a very interesting one for the measurement system designer.

There is a complication with this methodology. While $SIN(\alpha)$ is positive in its first half cycle, the product has both positive and negative values. A determination of the sign of $SIN(\alpha)$ cannot be made directly from inspection of the waveform of the product for this reason. Other techniques must be used to make this determination and these will be discussed later. Further note that there are times when the product has zero value when the input $SIN(\alpha)$ has finite value. At these zero product instants, there is no data available at all.

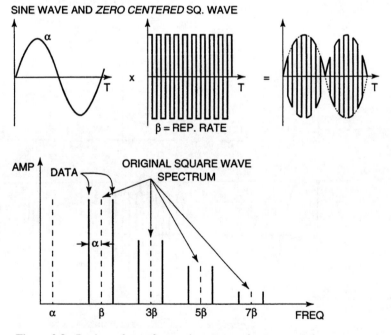

Figure 9.3   Product of waveforms: sine wave and zero centered square wave

Figure 9.3 shows the result of the multiplication of a SIN(α) with a zero *centered* pulse train wave of repetition rate β. The specific pulse train shown has a duty cycle of 0.5 making it a square wave. There is no limitation, however, on the duty cycle of the multiplying waveform. In some cases, discussed later, very small duty cycles, like .01 or less, are called for. The multiplication shows first in the time domain at the top, then in the frequency domain at the bottom. The time domain output is very similar to the first example. It is a fully modulated waveform, but with square corners! Note, in the frequency domain, that the original frequencies α and β and its harmonics no longer exist in the output. The SIN(α) information is now stacked as sum and difference frequencies around β and its harmonics. All original frequencies disappear in the multiplication, having been converted to new data frequencies. The ambiguities noted above for sine wave multiplication also apply to this multiplication by a zero centered pulse train.

The third case, shown in Figure 9.4, is for a sine wave multiplied by a zero *based* pulse train. The results are different from those with the zero centered pulse train. The multiplying pulse train has no negative information in it, having always positive or zero amplitude. As a result, the multiplied waveform in Figure 9.4 retains the phase sense of the original SIN(α) waveform. When SIN(α) is positive in its first half cycle, the output is either positive or zero. When the input is negative in its second half cycle, the output is either negative or zero. The phase ambiguity has disappeared in this case, but a new one has arisen! Half of the time, for a 0.5 duty cycle at least, the output is zero when the input is nonzero. You are losing one half of the information. How do you recover this apparently lost information?

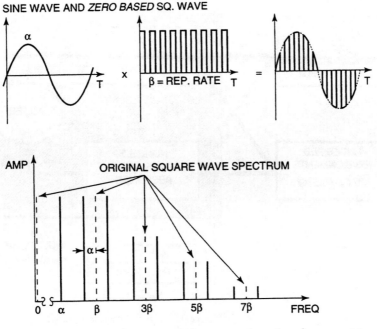

Figure 9.4   Product of waveforms: sine wave and zero based square wave

9.2 THE WHEATSTONE BRIDGE ALLOWS THE NECESSARY WAVEFORM MULTIPLICATION

There is a critical and common feature among these three examples of waveform multiplication. It is that such multiplication allows you to move (convert) non-self-generated information along the frequency spectrum under the designer's control. Where should it be moved? Obviously, the data should be moved far enough away from the frequencies of the self-generated noise to allow the noise levels to be effectively removed by filtering.

Look at the transducer model in Figure 9.5, reproduced from earlier in the book. In the language of the transducer model, you are working on a problem where path 1 and 3 self-generated noise levels are invalidating path 4 non-self-generated data. You can see here the built-in multiplier in a non-self-generating transducer. It is the excitation waveform.

Here, again, is the general equation for the output from a Wheatstone bridge:

$$\text{Output} = KV(\Delta R1/R1 - \Delta R2/R2 + \Delta R3/R3 - \Delta R4/R4)\,/4$$

Restating this:

$$\text{Output} = (\text{constant})\,(\text{Excitation})\,(\Delta R/R \text{ terms from measureand})$$

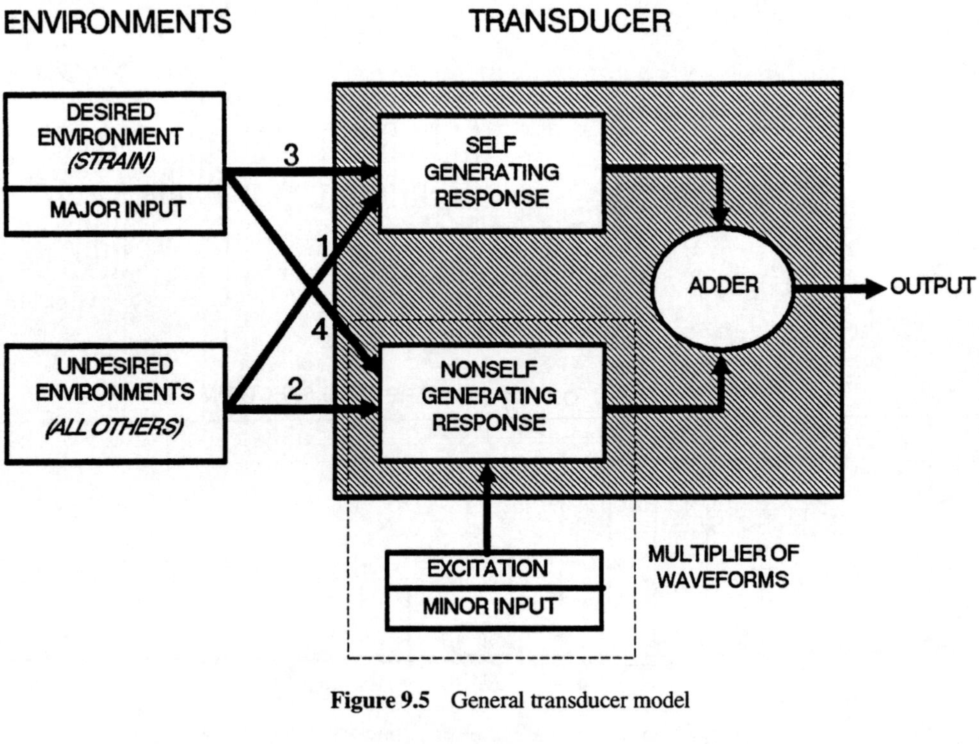

Figure 9.5   General transducer model

It is clear from this formulation that the needed multiplication is the excitation times the $\Delta R/R$ terms. The solution to the problem shows in Figure 9.6. Information conversion involves the moving of the data frequencies away from the noise frequencies by the judicious use of non-DC excitation waveforms, followed by successful noise filtering, followed by data reconstruction at its original frequencies. The generic steps are simple:

1. Apply a non-DC excitation waveform that will move the non-self-generated information along the frequency spectrum away from the self-generated noise.

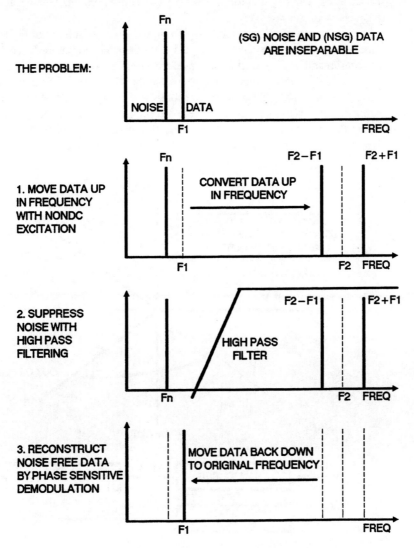

Figure 9.6 Information conversion using sine wave excitation

2. The self-generated noise will stay put at its original frequencies. You have, by definition, no way to move it anyway.

3. The data and noise frequencies have now been sufficiently separated to allow filtering. Filter out the noise frequencies.

4. Reconstruct the now noise free valid data by moving it back to its original frequencies (phase sensitive demodulation).

What in the world is phase sensitive demodulation? This function is carried out in a box shown, along with its output/input linearity curve, in Figure 9.7. Phase sensitive demodulators work with sine wave excitation.

As you can see in the block diagram in the lower left corner of Figure 9.7, the phase sensitive demodulator receives, as inputs, both the sinusoidal transducer output caused by the measureand and the excitation waveform itself. Its output is shown in the figure. The output is a DC-coupled voltage proportional to the RMS level of the transducer's sinusoidal carrier output signal. The demodulator looks at the relative phase between the transducer output and the excitation input. If these are in-phase, the demodulator assigns a positive sign to the DC-coupled level at the output. If they are out of phase, a negative sign is assigned to the DC-coupled level. This operation is the one that reconstructs the data at its

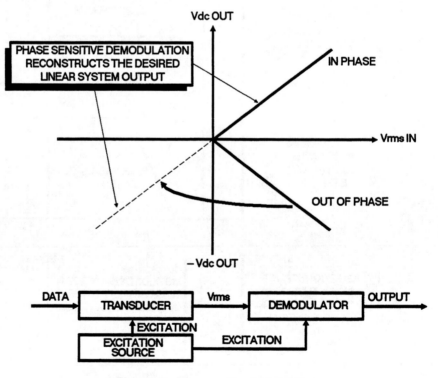

Figure 9.7   Phase sensitive demodulation

original frequencies with the proper sign and magnitude—with the self-generated noise removed. A good phase sensitive demodulator will operate over a frequency range up to about 25% of the carrier frequency.

9.3 EXAMPLES OF THE USE OF INFORMATION CONVERSION

9.3.1 Example 1: The Measurements Group P350 Strain Indicator

The P350 strain indicator shown in Figure 9.8 is the latest in a distinquished line of null balance strain indicators going back about 35 years and issuing from several vendors. The P350 uses a zero centered pulse train excitation system. The waveform is a square wave with a 1.5 KHz repetition rate at 1.5 volts rms. Square waves were chosen to diminish the system's sensitivity to capacitive unbalance problems caused in the cables between the active strain gage bridge components and the instrument.

If one had to operate forever more in an experimental stress analysis or structural test laboratory, and were allowed only *one* strain indicator, the venerable P350 would be the one I choose. The reason is that since the P350 is a carrier system and, therefore, it can separate self-generated noise from the non-self-generated strain data. No DC excited strain indicator can do this, regardless of price.

We had a graphic example of this several years ago. We were working with some 10,000 pounds full scale, 4340 steel instrumented bolts from a reputable vendor. These were internally strain gaged in a quarter bridge configuration. Initially, these bolts were calibrated in a test machine using DC powered strain indicators at our laboratories. The bolts showed extremely unstable zero performance due to handling alone! Zero would

Figure 9.8   P350 strain indicator (zero centered pulse train excitation) (Printed through the courtesy of Measurements Group)

slowly shift up to an equivalent 300–500 pounds just due to the handling required to install them in the test machine.

This was clearly unacceptable performance since the application was a tight tolerance load in a spacecraft separation system. Such a system is a single point failure in a spacecraft and demands the utmost in measurement reliability and certainty. Please note here that this *most important* measurement we make is made with a simple strain indicator—not with a complex $1,000,000 digital data acquisition system. If this measurement is invalid, we will certainly damage, perhaps destroy, a spacecraft and its *entire mission*. I would not like to pay the bill for that mistake.

We found that the noise levels were caused by self-generated thermoelectric (thermocouple) EMFs due to the unsymmetrical wiring of the bridge corners inside the bolt hole. The vendor had created an extra copper Constantan thermocouple junction on one side of the bolt. Some of the internal wiring would experience a thermal gradient just due to the heat of the operator's hand! We found that the vendor had calibrated the bolts using a P350 carrier strain indicator not understanding the instrument's unique insensitivity to self-generated noise levels. They obviously had never seen this problem at the factory and could not recreate it. On discovering this, we went AHA!, and changed to the carrier strain indicator. The problem instantly disappeared.

There isn't a DC-powered instrument in our stock, and we have 560 channels, that would have solved that problem. We solved it using the right methodology—using a *35-year-old instrument design*!

Before I leave this subject, remember the problem with this type of excitation—zero centered pulse trains. By inspection of the modulated output alone, the instrument cannot tell if the original bridge unbalance is positive or negative. Inside the P350 is a rather simple detection circuit that accomplishes this task. Further, since the information input to the strain indicator is a low level pulse train, non-DC coupled amplifiers can be used. The manufacturer does not have to build in DC stability in this stage of the instrument along with its inherent problems. These are two of the reasons why the P350 can still be made small, light, and rugged. It is perfect for general use in and out of the laboratory, and has been for 35 years.

9.3.2 Example 2: Schaevitz LVDT Signal Conditioning

Our standard signal conditioner for use with linear variable differential transformer (LVDT) displacement transducers is the Schaevitz Instruments Co. ATA-101. This device is a 2.4KHz, 3 Vrms excitation, sine wave carrier signal conditioning system and shows in Figure 9.9. This excitation frequency is chosen because it is near the zero phase shift frequency for most of Schaevitz's line of LVDT transducers.

This signal conditioning is a carrier system for an additional reason. The transduction mechanism within the LVDT is magnetic induction, which only responds to changing voltages and currents. DC excitation cannot even be used with this transducer type. The system does, however, make full use of the carrier system's ability to separate self-generated noise from non-self-generated data.

Figure 9.9   ATA-101 linear variable differential transformer (LVDT) carrier signal conditioner (Printed through the courtesy of Lucas Schaevitz)

9.3.3 Example 3: Zero Based Pulse Excitation in a Measurement System

Several years ago we deployed a system[2] designed to make very precise quasi-static strain measurements on a series of 6″ and 9″ diameter graphite epoxy tubes. The strain data, from a number of 45 degree rosettes installed on the composite material, were used to calculate stresses on the tube surface. The stresses were then cast into response bending moments and torques due to loading conditions.

The key design features were the abilities to: (1) match the power dissipation into the tubes from the rosette gage installations with that during earlier apparent strain testing; and (2) separation of legitimate low level, non-self-generated $\Delta R/R$ in the strain gages from self-generated noise levels caused by the thermoelectric EMFs in the bridge circuits, overall system zero shift with time (called drift), and 60 Hz noise levels. The key issue was that once the system was zeroed mechanically and electrically, it could not be rezeroed again for 18 months. How do you track system drift (a self-generated noise level) under those conditions?

We solved this difficult design problem by using a zero based pulsed excitation system. Figure 9.10 shows the main operating principle of this system. At the top you can see the pulsed excitation waveform. Its amplitude and duty cycle were chosen to exactly match the power dissipation into the composite materials seen during apparent strain runs done with a DC excited system. In this manner, the strain gage installations on the composite tubes run at the same temperature in both tests, this being critical for the apparent strain corrections to be valid.

EXCITATION WAVEFORM

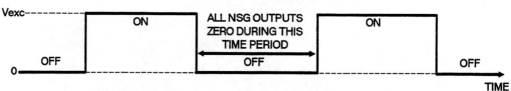

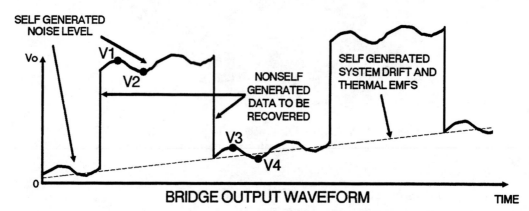

Figure 9.10   Data reconstruction after sampling with zero based pulse excitation

   The lower portion of the figure shows the low level, bridge output signal caused by the sum of the several self-generated noise sources (thermoelectric EMFs, system zero drift with time, and 60 Hz noise) and the non-self-generated strain data. The data were read at points V_1 and V_2 with the power on. These two points were sampled one-half cycle of the power line frequency apart. You can use this technique with *any* periodic self-generated waveform. These two readings were then averaged (AVG1), cancelling the effect of the power line frequency, 60 Hz noise level. The data were then read again, at points V_3 and V_4, with the excitation off. These points were sampled one-half cycle apart, canceling 60 Hz self-generated noise in these readings. Since the excitation is off during this time period, there by definition can be no non-self-generated signals from the bridge. Zero non-self-generated data is documented here. Readings 3 and 4 were then averaged as AVG2. The simple relationship:

$$\text{Non-self-generated data} = \text{AVG1} - \text{AVG2}$$

calculated only the non-self-generated data from the strain induced $\Delta R/R$, and compensated for the *three self-generated noise levels*. With this method, you don't care where the system's zero drifts to (as long as it remains within the system full scale), since you are correcting for zero drift every cycle of excitation. This method kills these three noise level birds with one information conversion stone.

9.3.4 Example 4: Zero Centered Pulse Excitation in a Measurement System

This measurement problem[3] was similar to that immediately above, except several aspects were worse. In this application we couldn't check the measurement system zero for 24 months! So, we had to maintain absolute control over this noise level and document it over that time frame. Further, the actual strains to be measured were lower than in the previous example.

The key to this design was the ability to keep maximum control of the temporal and spatial temperature gradients in the gaged parts caused by the strain gage bridge self-heating and laboratory temperature changes of +/− 5 degrees F. The approach chosen was to combine the information conversion capabilities of pulse train excitation, with the absolutely constant power dissipation provided by keeping the gages powered at all times, not for half the duty cycle as before. Whatever those thermal gradients were that would cause self-generated thermoelectric noise levels, we wanted them to stay put and not change. They make an easier target for noise level correction if the problem is stationary.

The key here was to use a zero centered pulse train for excitation. The excitation voltage is either plus or minus, but it's always on. Power dissipation is always constant, which holds the thermal gradients as constant as we can possibly make them.

The same self-generated noise level cancellation as before is made with a slightly different method. This shows in Figure 9.11. The difference is that in this zero centered

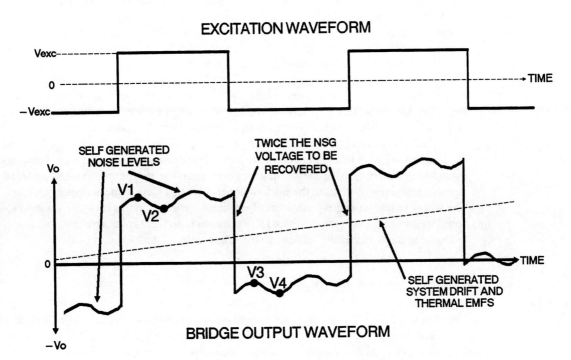

Figure 9.11   Data reconstruction after sampling with zero centered pulse excitation

pulse train case, the difference between AVG1 and AVG2 is caused by twice the level of the desired non-self-generated strain data. Here, the proper algorithm is

$$\text{Non-self-generated data} = (AVG1 - AVG2) / 2$$

The method neatly cancels the three self-generated noise levels with the added bonuses that the installation has improved stability and twice the sensitivity of the case mentioned above.

9.4 A HISTORICAL NOTE

In the 1950s there were very few decent DC excitation measurement systems on the market. The reason was that no one could yet make the necessary, high stability DC coupled amplifiers at a competitive price. Now, we take this technology for granted. But we've paid a price for this. Few engineers who have ever looked at PC based measurement "systems" have ever discovered non-DC excitation for non-self-generating measurement systems. The manufacturers of these systems haven't either. Most of them think DC excitation is all there is. It's a little like Hollywood—they're designing for the lowest common denominator of knowledge.

Thirty years ago almost all systems were carrier based because carrier system amplifiers did not have to be DC coupled and were, therefore, less expensive and bulky. After the sinusoidal or pulse train bridge output was amplified a little using low cost AC coupled stages, a low gain DC amplifier was used to drive the final system output voltage. The primary reason for their use in the 1940s, 1950s, and 1960s was not their unique self-generated noise level killing potential, but their low cost.

As progress was made in the development of cost competitive, stable, high gain, DC coupled instrumentation amplifiers, carrier systems fell out of favor and remain so to this day. They are, after all, a little more complicated to use since they have one extra control and you have to worry about those funny sine wave or pulse things for excitation.

The ability of carrier measurement systems to give you information conversion, and by doing so separate self-generated noise levels from non-self-generated data, is unsurpassed—regardless of what the marketplace tells you. The most expensive DC powered signal conditioning systems in the world will not solve this class of measurement problem. The shame is that an entire generation of engineers getting into this business via the PC-based instrumentation "system" think DC excitation is all there is because that's all that is available to them. Hopefully, this book will help a little.

9.5 A WARNING TO THE READER

Refer back to Figure 7.1 in a previous chapter—the unified transducer model. In this entire chapter, I've been describing situations where self-generated paths 1 and 3 noise levels were corrupting path 4 non-self-generated data. What about path 2 non-self-generated noise levels? These are non-self-generated noise levels from the undesired environment.

An example would be a temperature induced sensitivity shift on a piezoresistive accelerometer while making acceleration measurements on something getting hot as it runs.

Information conversion cannot help you solve the problem of the separation of undesired path 2 and desired 4 outputs. This separation must be made by other methods. Refer to the previous chapter on noise level control.

NOTES

1. Peter K. Stein, *Information Conversion as a "Noise"—Suppression Method;* Stein Engineering Services, Phoenix, AZ, 1975.
2. Charles P. Wright, *Large Graphite Epoxy Tubes as Moment and Torque Transducers,* Proc. of the Western Regional Strain Gage Committee, Society for Experimental Mechanics, Phoenix, AZ, March 1985.
3. Gregg Graham and Charles P. Wright, *High Stability Structural Strain Measurements Using Information Conversion,* Proc. of the Western Regional Strain Gage Committee, Society for Experimental Mechanics, Seattle, WA, August 1993.

Chapter 10

Frequency Analysis

Frequency, or signature, analysis is probably the most ubiquitous analysis function performed by the measurement engineer. We publish 14,000 to 20,000 of these analyses per year in our test organization alone. These analysis methods attempt to identify, at least, the distribution of frequencies and amplitudes coming from the transduced process information. Additional important aspects add to the recipe when the measurement engineer desires phase information among the frequency components and perhaps to an external reference. Further information may be available to the experimenter when two or more frequency analyses are used to determine common properties in the signals. This opens the door to the complex mathematics of transfer function analyses, coherence, and cross spectral properties. These analyses can be performed with purely analog components or can be synthesized in digital hardware via software.

Almost all frequency analyzers sold today operate with sampling and digital filters. This is both good and bad. It is good since modern frequency analyzers are faster, frankly more accurate, and infinitely more flexible than older analog systems. We use these almost exclusively for reasons of survival! However, the shift to digital technology in this area is not without its price. The price the measurements engineer pays in moving directly into digital frequency analysis techniques is that he may have no *feel* for the analysis processes going on. Modern frequency analyzers, whether two channels or 256, tend to foster use without knowledge. Their use in some laboratories can become a *button pushing exercise* (such exercises are anathema to a professional measurements engineer). You hear statements like, "How do I know it's right? It came out of the analyzer like that didn't it?" You could not get away with that with analog systems. You *had* to understand what was going on to be successful. Readers should consider themselves warned.

This chapter cannot begin to provide a thorough treatment of this subject matter. Two books this size couldn't cover this subject thoroughly. I will cover here a minimum set of subjects that must be second nature to measurement systems designers and users when working with dynamics-related measurements.

10.1 INFORMATION CARRYING TECHNIQUES

Frequency analysis fits within a general structure of information carrying methods in measurement systems. Information can be carried in a measurement system in several ways. Information about the process of interest must be carried on some pattern of properties of some waveshape of a physical or chemical quantity. The following table lists the presently identified methods.[1,2]

INFORMATION CARRYING METHODS

	Common Name
I. A CONTINUOUSLY VARIABLE CHARACTERISTIC	
1. amplitude of a level	*analog*
2. amplitude of a sine wave	amplitude modulation (AM)
3. frequency of a sine wave	frequency modulation (FM)
4. phase of a sine wave	phase modulation
5. repetition rate of a pulse	pulse modulation
6. amplitude of a pulse	pulse amplitude modulation
7. position of a pulse	pulse position modulation
8. width of a pulse	pulse width modulation
9. duration of a pulse	pulse duration modulation
II. PRESENCE OR ABSENCE OF A CHARACTERISTIC	
10. amplitude of a pulse	pulse code modulation (*digital*)
11. frequency of a sine wave	temperature sensitive paints
III. PATTERNS IN THE:	
12. amplitude-time domain	wave shape analysis
13. amplitude-frequency domain	signature (*frequency*) analysis

The two most commonly mentioned methods of carrying information, analog and digital, are only two of *thirteen* presently identified and general methods. Information carried on a quantity's amplitude patterns in the time domain is analog. The interpretation process for this data in the time domain is *waveshape analysis*. If the information exists in the pattern of sine wave frequencies of the analog data, the process that exposes the pattern is *frequency analysis*. The interpretation process for this frequency domain information is *signature analysis*.[3]

Signature analysis, based on frequency analysis data, is common to all fields of science and is, for that reason, generic. It is used in physics, astrophysics, chemistry, biology,

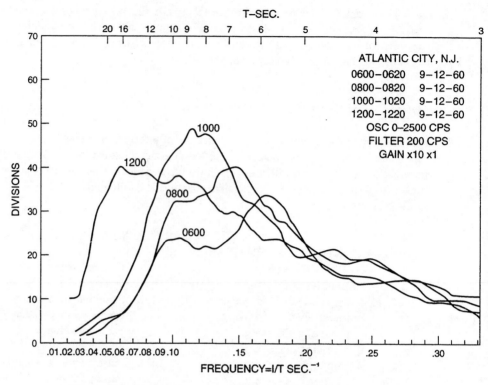

Figure 10.1 Signatures of wave heights during Hurricane Donna (9–12–59)

medicine, physiology, psychology, geology, economics, and engineering. It is a fundamental analysis method and must be well understood by the measurement system designer and user. The measurement engineer must be as comfortable with the complex frequency representation of a waveshape, and the method for obtaining it, as with the waveshape itself.

Frequency analysis is a useful tool across, perhaps, an incredible 50 orders of magnitude in the frequency continuum from the $.68 \times 10^{-16}$ Hz of a potentially oscillating universe with a minimum 28 billion year period (14 billion year presently expanding universe age times two) to the multigigahertz of the subatomic physicist. I don't know of another analysis method that is viable over such a broad range of frequencies.

Mechanical engineers are interested macroscopically in a fairly narrow range of frequencies in the middle of that spectrum. The data in Figure 10.1[4] represent a very low frequency portion of that range. These data are the frequency analyses of wave heights measured during the devastating hurricane Donna on September 12, 1959. Spectra show for four times in the morning. You can see from the spectra that the physics of the wave interactions with the shoreline changes significantly over the six-hour time frame represented here. You are very interested in this data if you are an engineer doing a dynamic loads analysis of a pier structure.

10.2 FREQUENCY SELECTIVE COMPONENTS

Frequency analysis as a technique relies on the use of measurement system components that are *highly* frequency selective. They are generally called "filters" although the term in this context is not nearly specific enough. An understanding of how these processes work is independent of how the manufacturer implements the "filters." For years, the only choice was analog filters. Now, many manufacturers offer high performance sampled frequency analyzers using a myriad of digital filter implementations. The difference is only one of implementation.

Figure 10.2 shows two general types of "filters" used for frequency analysis. The filter at the top looks, in fact, like a filter. It can be one of two types. The first type is constant bandwidth filter where the −3 dB bandwidth is constant despite the filter's center frequency. Almost all filters in digital frequency analyzers are of this type. The second type is the constant percentage bandwidth filter where the −3 dB bandwidth is a percentage of the filter center frequency—the filter gets wider as the center frequency increases. Octave and one-third octave filters for acoustics work are examples of this type. The shape factor, typically defined as the ratio of the −40 dB bandwidth divided by the −3 dB bandwidth, controls the shape and sharpness of the filter in either case. Typical analog fixed frequency filters for swept sine testing have shape factors of two to four, with two being very sharp.

NOMENCLATURE:

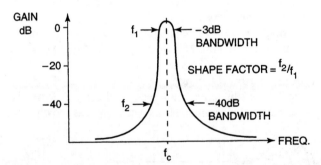

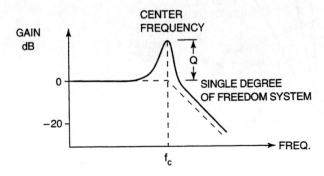

Figure 10.2   Frequency selective components

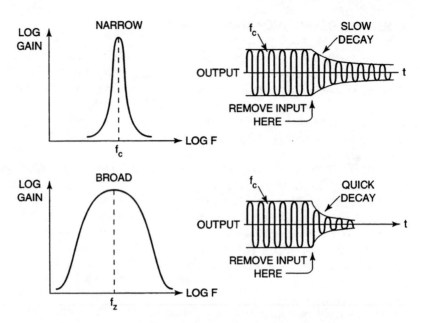

Figure 10.3   Bandwidth determines how a filter reacts in the time domain

Certain of our digital frequency analyzers have narrow band swept filters for sine testing that mimic, by design, these analog filters very closely.

The other type of filter is the single degree of freedom damped system with variable center frequency shown in Figure 10.2 at the bottom. This filter type is used in the determination of the shock response spectrum.

Figure 10.3 shows how wide and narrow filters react in the time domain. This figure shows at the top a narrow bandwidth filter (the terms being relative at best). Assume that the filter center frequency aligns with a constant amplitude sine wave input as shown to the

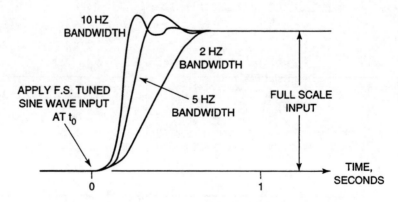

Figure 10.4   Comparison of constant bandwidth filter "loading" rates

right. If the sine wave input was removed, the filter would decay very slowly. It can dissipate energy only slowly due to its narrow bandwidth. Conversely, since it can decay only slowly, it can accept energy only slowly. This means that if the input sine wave was initially at zero amplitude and then instantly went to unity amplitude, the filter would react slowly—at the same rate it decays. This is called filter loading. It takes a finite amount of time for the filter output to reach steady state response. The somewhat wider filter at the bottom in Figure 10.3 can dissipate energy much faster. Due to its increased bandwidth, it also can take on steady state energy faster as shown by the much quicker decay upon input removal. The comparative loading rates for 2, 5, and 10 Hz constant bandwidth filters with shape factors of four show in Figure 10.4. The overshoot for the 2 and 5 Hz filters is a function of the post filter RMS detector in the filter set used for this demonstration, not of the filter itself.

10.3 SWEEP RATE EFFECTS

There are several circumstances in measurements engineering where a narrow filter, shown in Figure 10.2 top, is used to sweep over a data spectrum coming from fixed structural properties to extract information at the filter's center frequency. This case occurs most often in swept sine vibration testing. Often, the requirement is to measure the structural response only at the sweeping frequency of the sinusoidal excitation, while rejecting responses at any other frequency.

   The outputs from these filters are subject to effects caused by the nonzero sweep rate of the excitation.[5,6] Figure 10.5 illustrates this. Here, the response that would occur from a sine wave excitation with a vanishingly small sweep rate is shown by the solid line. This system response will approach the steady state solution to the mechanical system's differ-

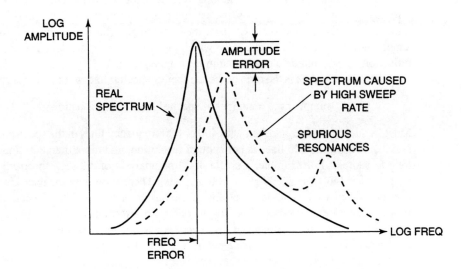

Figure 10.5   Sweep rate effects on narrow band components

ential equations. Any reasonable sweep rate will cause the response shown by the dotted line. This effect increases with decreasing system resonance bandwidth and filter bandwidth, and increasing sweep rate. It is a function of the instantaneous velocity of the excitation frequency, *dF/dt*. The measured response will be lower in amplitude than the steady state response and will occur at a higher frequency for a positive sweep rate. A negative sweep rate (higher to lower frequency) will cause the resonance to appear at a lower frequency with lower amplitude. If a positive sweep rate is fast enough, spurious resonances will occur at higher frequencies having nothing at all to do with the structure's properties at these frequencies. In summary, sweep rates that are too high for the structure's properties will cause lower amplitudes at measured resonances, higher or lower resonant frequencies depending on sweep direction, and spurious after resonances. The inverse corollary for the measurement engineer is that sweeping a fixed bandwidth filter over a data spectrum too fast will cause the same errors in the resulting data—resonance amplitudes too low, frequencies too high or low depending on the sweep direction, and spurious after resonances that do not exist. The analogy is exact.

10.4 SAMPLE TIMES

To produce a frequency analysis, the input waveform must be inspected over time. I use the term "inspected" carefully. The statement applies despite the analysis methodology. In the digital sense, it must be sampled over time. In the analog sense, it must be averaged over time. A frequency analysis cannot be produced in zero time, even if the customer implies that it can with his reduction request!

This discussion of sample time will, however, be limited to digitally sampled systems. I think analog systems will no longer be available for these analyses in the near future. In digitally sampled systems, the sample time is a function of the number of spectral lines in the analysis and the base band analysis range. The base band analysis range is simply the bandwidth, from zero, over which the analysis must be valid. It is the frequency reflection of the valid data bandwidth in the time domain.

The sample time is the ratio of the number of spectral lines to the analysis bandwidth.

sample time = number of spectral lines/analysis bandwidth

Most frequency analyses are run with 400 or 800 frequency lines in the spectrum. Analysts feel that fewer than 400 lines is not enough resolution, and more than 800 lines takes too long to sample! For example, a single 400-line analysis of a 2KHz spectrum takes 200 milliseconds (400 lines/2000 Hz = .200 seconds). There is nothing the measurement engineer nor the customer can do about this. If the customer desired 32 independent averages for this analysis, a period of $32 \times .200$ seconds = 6.4 seconds of data would be required. You will find that your customers will not understand this relationship. They will invariably ask you for large numbers of averages from a very short time record. There is only one solution to this request anomaly. That is education of your customer followed by an intelligent test plan allowing the sampling times needed. Figure 10.6 shows the relationships

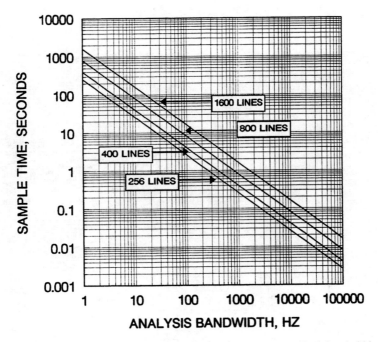

Figure 10.6 Relationships among number of spectral lines, analysis bandwidth, and sample time

between base band analysis bandwidth and sample time for a single 256, 400, 800, or 1600 line spectra.

10.5 AVERAGING OF FREQUENCY ANALYSES (SPECTRA)

Certain engineering processes have random frequency content in the data. Examples of such processes include acoustic, aero- and hydrodynamic, combustion and random vibration testing. A structure's response to random excitation must be random. It may, however, contain data frequencies driven by the structure's modes of vibration reacting to the random excitation. These modal responses are usually of interest to the customer for the data. Measurements engineers use spectral averaging techniques to highlight the modal responses while suppressing the random responses from the structure.

These techniques use the following concept for a given frequency line in the spectra and for multiple spectra. For a given frequency line, amplitudes caused by random processes will eventually average to near zero due to the necessary randomness of the amplitude and phase (the average of X and −X is zero) assuming enough averaging time. Amplitudes caused by deterministic structural responses, however, will be similar in all spectra and will average to their true value (the average of X and X is X). Figures 10.7 through 10.9 illustrate this.

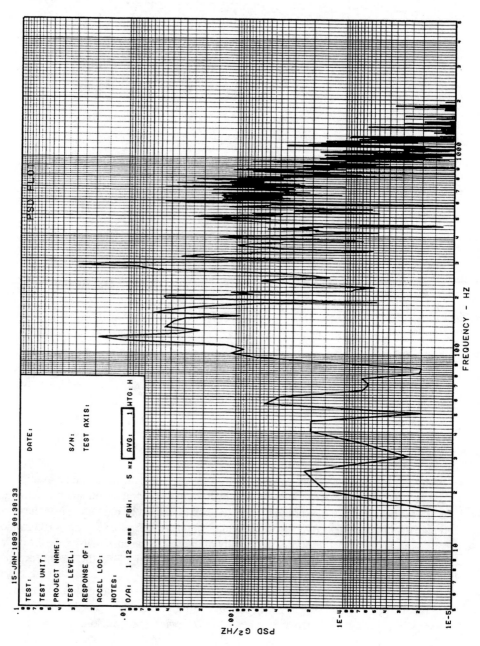

Figure 10.7 Autospectral density function (PSD) of acoustics data (1 average)

250

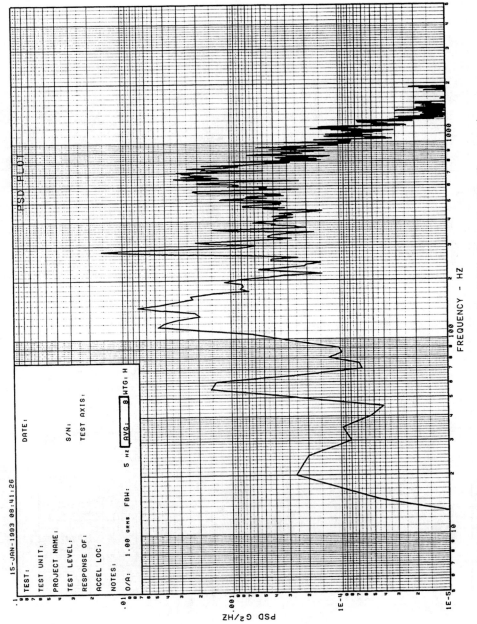

Figure 10.8 Autospectral density function (PSD) of acoustics data (8 averages)

251

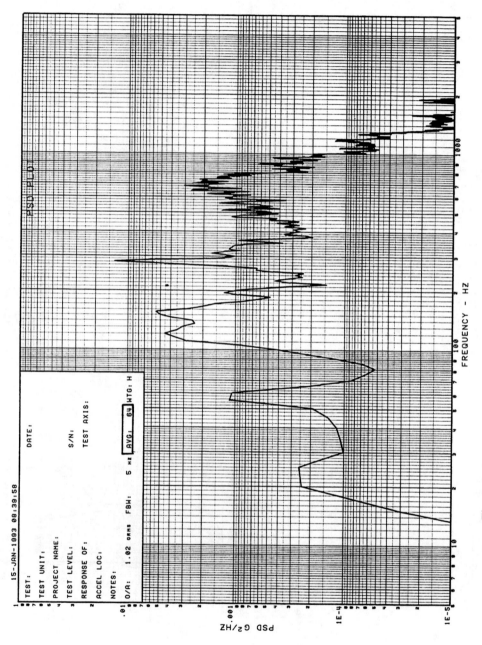

Figure 10.9 Autospectral density function (PSD) of acoustics data (64 averages)

252

Figure 10.7 is an autospectral density frequency analysis from a single average of random acceleration from an acoustics test. This is the raw autospectrum from a single pass at the data. Figure 10.8 shows the results of 8 averages of the same stationary data. Note how the random responses are suppressed, particularly at the higher frequencies, while the structural resonances continue to stand out. Figure 10.9 shows 64 averages of the same stationary data. Here, the random frequency content has been effectively suppressed with averaging, leaving the structural resonances evident.

Modern frequency analyzers provide this averaging function in at least three methods. Linear averaging gives all spectra an equal weight of one as shown in Figure 10.10 for N spectra. In this method, N spectra are created in sequence. The N spectral amplitudes at each frequency line are averaged and the averaged results plotted at the frequency lines. The averaged spectrum is not valid until all N spectra have been captured and averaged. This is the typical averaging method used in the random vibration world for stationary random data.

Exponential averaging occurs as shown in Figure 10.11. Here, the Jth through the Nth spectra have exponential weighting, with the latest spectrum, the Nth, having a weight of one. This averaging method emphasizes more recent spectra over older spectra. The averaging at each frequency line also is done with these exponential weightings. When the Jth through the Nth spectra have been averaged, the resulting spectrum is displayed. Then the Nth + 1 spectrum is taken, the Jth spectrum discarded, and another average calculated and displayed. The normal use of exponential averaging is in the detection of changes in the averaged spectrum with time as something in the process changes (nonstationary process).

The third averaging method is peak averaging and is, in fact, not averaging at all. It is simply misnamed but has nevertheless firmly come into the test lexicon. In peak averaging, shown in Figure 10.12, N spectra are taken, weighted equally, and inspected. The largest peak value in any of the spectra at any spectral line is reported at that spectral line. This display is the envelope of the worst case amplitudes at any frequency over the averaging time. No averaging occurs at all.

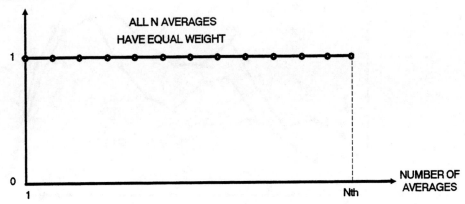

Figure 10.10   Weighting for linear averaging

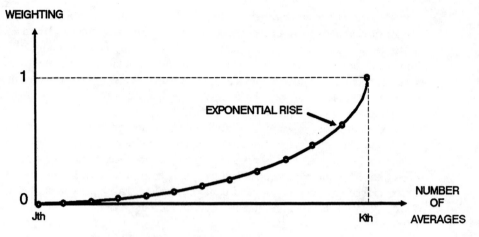

Figure 10.11   Weighting for exponential averaging

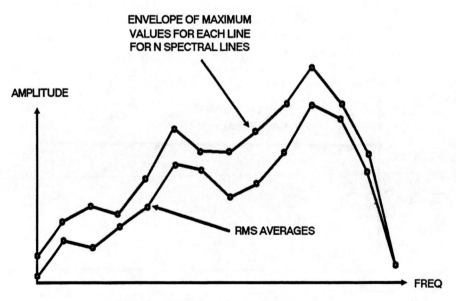

Figure 10.12   Weighting for peak "averaging"

10.6 A REAL WORLD SAMPLING AND AVERAGING EXAMPLE

Here is an example of how these sampling and averaging relationships must drive the test planning. We have, over the years, had to measure dynamic pressure in chemical laser combustion chambers. These complicated devices feature large steam generators to create a vacuum into which the laser combustion products exhaust. The lasers are fed with certain gaseous fuels and diluents that come into the device in a preset sequence. High pressure water flow cools the device. The device lases in the presence of a complicated group of vibration producing environments—steam generators, cooling water flow, and fuel and diluent gas flows. The requirement is to separate combustion-caused dynamic chamber pressures in the frequency domain (the data) from complex dynamic pressures caused by laser vibration accelerations (the noise).

This process occurs by carefully noting frequency analyses as the test process sequentially builds before the final combustion begins. Four to six conditions might be required, the last one being actual combustion. The reduction process is then to subtract the various spectra of the noise sources from the spectra of the combustion pressure plus the noise sources. What remains is the desired spectra of the combustion alone.

These types of frequency analyses require sampling and averaging of the data. Both processes take finite time. The test time line must reflect these minimum process times for the creation of the necessary frequency analyses, or the data cannot be provided despite the customer's insistence. In the absence of enough time at each condition, the measurement engineer has only two options, both of which compromise the result. The number of averages must be reduced or the base band analysis range must be increased (decreased resolution). There are no other options.

In summary, the measurement engineer must insert himself into the test planning process early enough to influence the experiment time lines. If the measurement process is taken for granted (as it all too often is), and the measurements people are brought into experiment planning too late, the planning crime may have already occurred and the data be severely compromised before the experiment is ever run.

10.7 POWER OR AUTOSPECTRAL DENSITY ANALYSIS

FFT (Fast Fourier Transforms) analysis determines the amplitudes of the distribution of sine wave frequencies in the subject wave shape. The result is a plot of RMS sine wave amplitude versus frequency. Power, or more commonly auto-, spectral density analyses are done on random data and provide spectra in terms of amplitude squared per Hertz. Miles and Thomson[7] give the best theoretical explanation of power spectral density analysis. I prefer a much more practical and historical derivation from a practicing measurement engineer's perspective.

In the 1950s and 1960s, this analysis was performed by recording the random data on magnetic tape. The operator then created a tape loop from a stationary portion of the data and played this tape loop continuously into a constant bandwidth swept sine analyzer. This forces the data to become quasi-random because it now has a fundamental repetitive pe-

ONCE UPON A TIME IN THE OLDEN DAYS.......

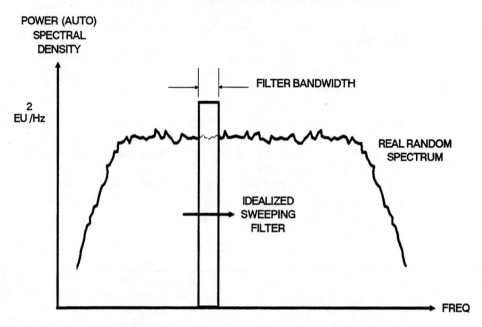

Figure 10.13   Power (auto) spectral density analysis

riod—the period of the tape loop. This was a necessary approximation. The background random data spectrum and the sweeping filter show in Figure 10.13. Analyses like these were being done in laboratories all over this country on wildly different sets of measurements hardware with conflicting procedures and methods. The government noticed that similar analyses in different companies and laboratories looked nothing like each other. In fact, different laboratories used different bandwidth sweeping filters. The following methodology evolved as an attempt to solve this intractable problem.

Presume the same random data reduction was being performed in three laboratories, A, B, and C. The intent is to get the same answer for the same data in all laboratories though different hardware is used in each. The laboratories use filters of 160, 40, and 10 Hz respectively.

Laboratory	Filter Bandwidth	Measured Output	Squared Output	Divide by Bandwidth
Lab A	160 Hz	4 EU RMS	16 EU RMS^2	.1 EU RMS^2/Hz
Lab B	40 Hz	2 EU RMS	4 EU RMS^2	.1 EU RMS^2/Hz
Lab C	10 Hz	1 EU RMS	1 EU RMS^2	.1 EU RMS^2/Hz

For a random input, the filter's RMS output is a function of the square root of the filter bandwidth. That's the nature of the problem—everybody gets a different answer to the same problem. Solution #1 is to force everyone to use the same bandwidth filters. This was rejected for reasons of capital investment. The chosen solution was to normalize the measured readings to account for the different filter bandwidths.

The simple solution is to square the measured output and divide by the filter's bandwidth as shown in the two columns to the right in the table. Using this methodology, all laboratories get the same answer in terms of power spectral density (RMS units^2 per Hz). Autospectral density is a better title for this analysis and this name is beginning to be seen in the literature. These analyses are done by the hundreds of thousands every year in vibration laboratories all over the world.

The other side of the coin is that test laboratories misapply this analysis more often than any other. The power spectral density analysis is designed for *random data only*. It makes no sense whatsoever for transient data or data with periodic content. The analysis of choice is the FFT analysis, not the PSD analysis, for data having transient or periodic content. You will find that some new dynamics analysts will request PSD data reduction for waveforms with transient or significant periodic content. This is because they have never seen any other analysis in their education experiences. The designing measurement engineer has the responsibility to educate the customer on this point.

Analysts commit a further crime when asking for PSD data reduction of dynamic strain, stress, pressure, or deflection data. What in the world does microstrain^2/Hz or Psi^2/Hz have to do with any physical process? No structure ever failed because of too many microstrain^2/Hz! This is the unit of the myopically educated engineer. Structures fail because of overload caused by too much strain, stress, deflection, or pressure—not because of microstrain^2/Hz.

10.8 OCTAVE-BASED ANALYSES

Acoustics measurements, usually of dynamic pressure (sound) generally use constant percentage bandwidth filters based on octave frequency ratios. These filters have the approximate frequency response shown in Figure 10.14. Octave-based analyzers have filters whose −3 dB bandwidths are determined by the relationship

$$F_{upper}/F_{center} = 2^N$$
$$F_{center}/F_{lower} = 2^{-N}$$

where $N = 1/2$ for octave, 1/6th for third octave, and 1/12th for sixth octave filters. Each filter has one half its logarithmic bandwidth stacked between F_{lower} and F_{center}, and one half between F_{center} and F_{upper}. Randall[8] presents the best explanation of these filters and their proper use.

Commercial analyzers providing octave and third octave analyses for acoustics work generally have the filters implemented in parallel in the analyzer. This shows in Figure 10.15 for a typical 33 parallel filter third octave set with the filter center frequencies spaced one third octave apart. All filters see the input waveshape simultaneously. Filter RMS outputs are plotted versus the center frequencies to document the analysis. A typical one third

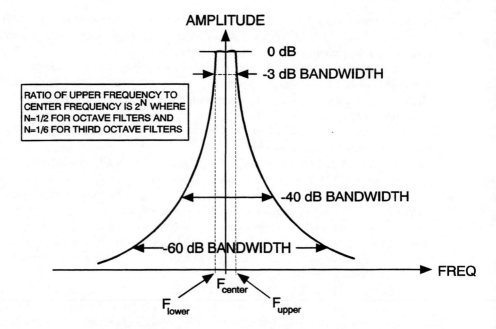

Figure 10.14   Octave and one-third octave filter transfer functions

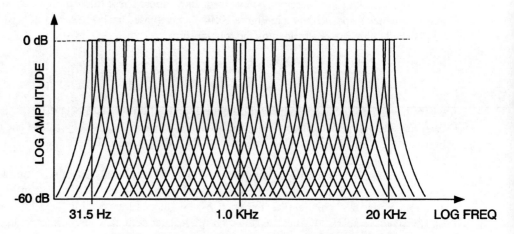

Figure 10.15   One-third octave filter set with 33 parallel filters

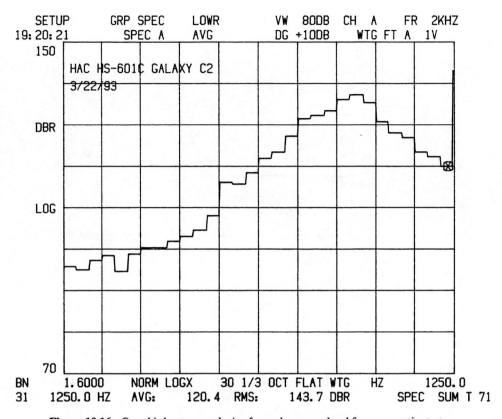

Figure 10.16   One-third octave analysis of sound pressure level for an acoustics test

octave analysis of an acoustic spectrum shows in Figure 10.16. An equivalent octave filter results if each three successive one third octave filters are grouped and their outputs measured together.

10.9 SHOCK RESPONSE ANALYSIS

Your customer is designing a device that will be subject to shock inputs. Perhaps it's a black box on a spacecraft subject to pyrotechnic release devices, a sight mounted in a tank turret subject to gunfire shocks, or a new valve installed on a jackhammer. They know the approximate resonant frequency, damping, and shape for a number of vibration modes of the device because they've analyzed them in the design process. The question is: Will the device be damaged by the shock inputs it will see in service? The shock response spectrum is a particular form of frequency analysis and is the tool for answering that design question.

The shock response spectrum (SRS) rests on the behavior of the single degree of freedom, damped spring mass system shown in Figure 10.17. The single degree of freedom

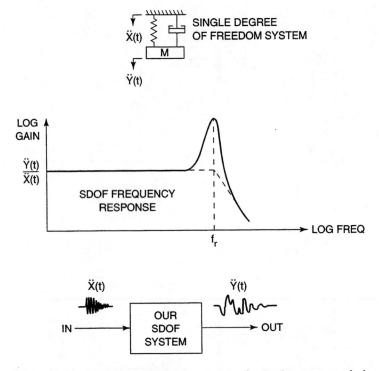

Figure 10.17   Single degree of freedom systems for shock response analysis

dynamic behavior for this system shows in the bottom of the figure with the system gain (shown as acceleration out/acceleration in—although the analysis can be done with any parameter) plotted versus frequency. The system's maximum gain is called its "Q" and occurs at the system's damped natural frequency, F_r. Shock response analysis in the aerospace industry is usually done with a Q of ten corresponding to a system with a damping ratio of .05, or 5% of critical damping. Figure 10.18 shows the frequency responses of several SDOF systems including one with 5% damping.

The cartoon at the bottom of Figure 10.17 shows the test article as a box with the single degree of freedom (SDOF) characteristics just discussed. A measured transient, usually a shock motion's acceleration, is analytically input to a SDOF model of your test article. The transient output is the response of that single degree of freedom system to the shock acceleration input. The peak acceleration of this system's output is noted as shown, and plotted at the resonant frequency of the SDOF system.

You can now state to the customer: If your device was resonant at F_r with a Q of 10, and was subject to the measured shock motion, the maximum acceleration it would experience would be $y(t)$ as shown. The question that follows is: Will the design survive this shock response at this SDOF frequency? In other words, what is the damage potential of this shock input to the system?

Other frequencies in the range of interest are covered by incrementing the SDOF system's resonant frequency and repeating the computational steps of the previous paragraph. We typically use 60 SDOF systems with a frequency spacing of 1/6 octave to synthesize the shock response spectrum to 10KHz. The overall shock response spectrum is the locus of the maximum peak responses of each of the 60 SDOF systems plotted versus their resonant frequencies. This shows for a real shock event with the shock input, in Figure 10.20. Thus, the shock response spectrum analysis is designed to answer this overall question: What is the damage potential of a measured shock input to a SDOF system with a certain Q whose resonant frequency varies over the test frequency range?

Test laboratories use several forms of shock response analyses. These vary depending on the section of the response transient used for the analysis. This shows in Figure 10.19. Here, we see an SDOF system subjected to an input transient acceleration in a certain axis that produces a measured response in the same axis. Both the input transient and the response show in the time domain in the middle of the figure. Typically, the response transient lasts much longer than the input transient. These high level shock events are caused by the sudden release of large amounts of energy. The suddenness causes the high frequency range of excitation amplitudes. Generally the shorter the transient is for a given energy release, the higher the frequency range of inputs. That portion of the response transient occurring while the input transient occurs is called the primary response. The response occurring after the input has gone to zero is the residual response. The entire response (primary + residual) is called, in the aerospace business at least, the maxi-max response.

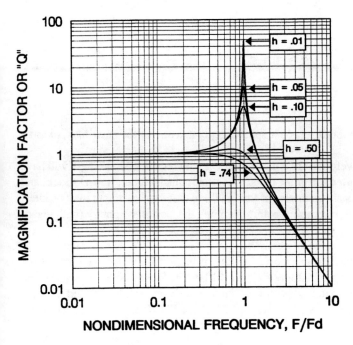

Figure 10.18 Single degree of freedom system frequency responses for various values of the damping ratio (h)

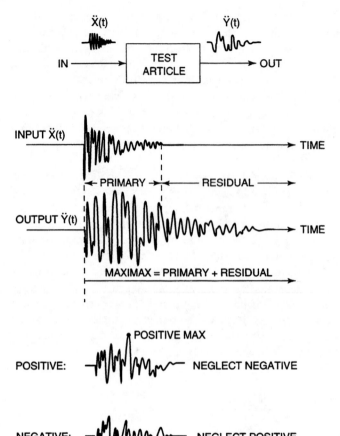

Figure 10.19 Definitions of terms for shock response analysis

Ninety-nine percent of the shock response spectra generated are for the maxi-max, or total, response.

Many excitation waveforms are not symmetrical with regard to sign. These will have higher amplitudes in one direction or the other, sometimes much higher. Impact shocks and those caused by explosives are examples of this. In these cases the customer may be interested in data from the shock response spectrum defining the lack of symmetry (with regard to sign and amplitude, not shape) in the excitation waveform.

In these cases, shock response spectra are generated for both plus and minus amplitudes. The response waveform from each SDOF filter is inspected for both the positive and negative peaks, as shown at the bottom of Figure 10.19. The loci of both the positive and the negative peaks are plotted in the spectrum. The final product is a positive and negative shock response spectrum as shown in Figure 10.20. If the SDOF responses are perfectly symmetrical in amplitude, these two curves will overlay.

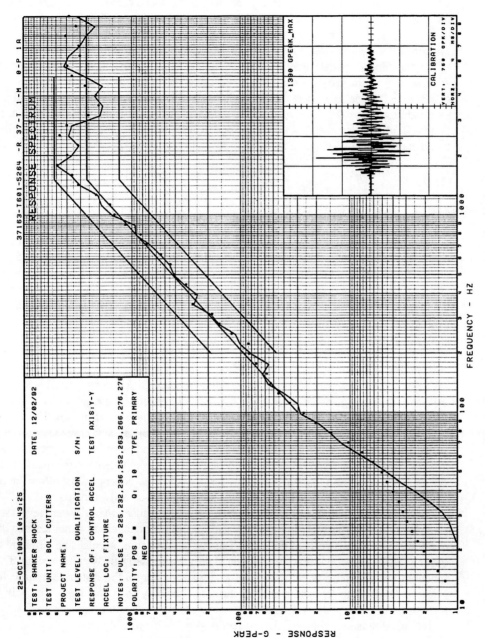

Figure 10.20 Shock response analysis from a shock test

Shock response analysis is not as intuitive as some other frequency analyses. Everyone has an opinion about whether a given answer is right or not. The designing measurement engineer who relies on these opinions for system validation commits a potentially fatal error. System validation must rely on bulletproof methodology. Creation of this bulletproof methodology and attention to the sufficient as well as the necessary conditions are another mark of the professional measurement engineer. I recommend that system validation be based on the analysis of classical pulse waveforms[9] (pulses, half sine, terminal peak sawtooth) for whom the shock response spectra are known, rather than shock test data. The system validation step for shock response spectra has no room for opinions and should be based on hard, preagreed upon, facts.

10.10 CONCLUDING REMARKS

Some readers will notice that I have not mentioned the subjects of sampling rates, windowing, and overlap processing in digitally sampled frequency analysis systems. Sampling rates are discussed in the chapter on sampled systems. The subjects of windowing overlap processing and their potentially corrupting affects on frequency analyses are discussed much more thoroughly in other texts and will not be covered here. The subjects are, however, important and must be considered in the design and operation of any digital frequency analysis system.

Last, I want to reinforce some remarks made earlier in this chapter on the subject of highly flexible, commercial digital frequency analysis systems. These systems are marvels of design and execution and are rapidly becoming fundamental, though expensive, tools for the measurement engineer. The analogy I use for these systems is that they are like a loaded gun in the hands of a child. Wait long enough and someone is going to be hurt. The amateur using these systems can get into trouble six ways from Sunday and never know it. Such systems should only be used with a fundamental understanding of both the process under investigation and the response of the frequency selective components in the measurement system to the data from that process. Only these two sets of understanding allow the data coming from the frequency analysis systems to be understood and validated.

As an example of this, we recently finished a 200 channel, major system level swept sine vibration test. This was a big, complicated, spectacular, dangerous, major league vibration test requiring two years of planning. The test required some special sine vibration frequency analysis functions that were unique to the class of hardware we were testing. These functions were included to our specifications by the vendor for the system (Syminex, Inc., Marseille, France) we bought to analyze the data—the only one in the world that will perform this particular analysis. Eight other vendors of major digital frequency analysis systems (United States and Europe), whose names you would probably recognize, flunked this analysis capability. They simply could not or would not deliver what we needed, or could not understand the analysis requirement.

After thoroughly checking out this system's ability to provide the valid analyses we needed, we still did not trust it. In order to prove beyond a shadow of a doubt that it was working as advertised, we recorded a fixed frequency square wave as a part of the 200 channel suite. This waveform has a very characteristic signature from the two types of frequency analyses required for the test. The analyses of this waveform verifies the measurement system's transfer function, linearity, signal-to-noise ratio, and the transfer function of the digital tracking filters—all in situ and in real time. This analysis was performed for every test run. It had to be right every time or we found out the reason why. Only when this analysis was correct was the rest of the data analyzed. Analysis of the data and preparation of the digital media required about 10 hours for each run of this test.

Why did we go to this trouble? We spent, after all, $150,000 on the analyzer and another $25,000 writing software for high speed electronic data transfer and network interfacing. Why didn't we trust it? We didn't trust it because this class of frequency analyzer is so flexible that its operation is risky, even to the professional. There are 5000 ways to set this class of tool up wrong, but only *one* way to set it up right for your application. How do you *know* you've got it set up right? Given this risk and the cost of testing, isn't it wise to check and validate the measurement system's performance on the fly?

This point is often missed by amateurs. They tend to see only the flexibility and miss risk and inherent complexity and the skill necessary to lessen them. They say things like "*Collecting data* is easy. What's the mystery? We've never had trouble with that before." Amateurs generally make statements like that just before irrevocably botching the filtering, sampling rate, and maybe the windowing and overlap processing control parameters and subsequently ruining the data. (By the way, that is a recent and direct quote from someone in my own firm.) The problem stems from ignorance of the fundamental contextual difference between *measurements* and *collecting data*. Collecting data is a button pushing exercise accomplished with little understanding.

In summary, the message is this. Amateurs think the subject of frequency analysis is simple, and can be trivialized to a button pushing exercise. Anybody can, after all, push the buttons. The proper and effective use of these systems requires that users thoroughly understand 10 to 15 nontrivial engineering issues and their interrelationships. These include the transfer function of the frequency selective component, analysis type, sampling ratios, averaging type, averaging times, windowing, single degree of freedom systems, sweep rate effects, RMS and peak detection, overlap processing, and much more. Each successful application requires a delicate balancing act among these technical facets of frequency analysis. The professional measurement engineer is characterized by understanding and balancing these facets all the time. Having any facet misapplied can ruin the data irrevocably. They are worth worrying about! How much you should worry depends on how much your data is worth to your enterprise. If your data is not valued, don't bother. Unless, of course, the reason the data is not valued is that no one bothers.

The result of the balancing act among the facets must be verified if the performance of the measurement system is to be *known* and not assumed. I guess that's a fundamental difference between professionals and amateurs in this business. Professionals *know* because they know the right questions to ask and validate the answers, while amateurs may or may not know the questions and *assume* the answers.

NOTES

1. Peter K. Stein, *Sensors as Information Processors,* Research/Development, June 1970, pp. 33–40.

2. D. L. Cronin, *Response of Linear Viscous Damped Systems to Excitation Having Time-Varying Frequency* (Doctoral dissertation), California Institute of Technology, Pasadena, CA, 1965. Work performed under NASA contract NAS-8–24–51.

3. Ibid.

4. Ibid.

5. Ibid.

6. Ronald V. Trull, Master's Thesis, Laboratory for Measurement Systems Engineering, Arizona State University, Tempe, AZ, 1969.

7. J. W. Miles and William T. Thompson, *Statistical Concepts in Vibration,* Shock and Vibration Handbook (Third Edition), McGraw-Hill, New York, 1988; ISBN 0–07–026801–0, pp. 11–1 to 11–14.

8. R. B. Randall, *Application of B&K Equipment to Frequency Analysis,* Bruel & Kjaer, Naerum, Denmark, 1977.

9. Sheldon Rubin, *Concepts in Shock Data Analysis,* Shock and Vibration Handbook, 3rd Edition, C. M. Harris editor; McGraw-Hill, 1988; ISBN 0–07–026801–0, pp. 23–1 to 23–30.

Chapter 11

Sampled Measurement Systems

Note how far into this book you've had to read before the subject of sampled measurements systems arose. There is a reason for this. The information that precedes this discussion is the necessary foundation for the discussion of sampled, or in the vernacular, digital, measurement systems. A measurements engineer has to *earn* the right to apply digital methods to a measurement problem. An engineer who does not understand the "analog" fundamentals noted in previous chapters lacks the technical context for solving the digital measurement problem. The subjects presented before this chapter need to be thoroughly understood before employing sampling techniques. The process of sampling is, in many important respects, a one-way street. Errors made in the sampling process are usually irrecoverable. Going the digital route is like "crossing the Rubicon."

The Rubicon is a river in Italy north of Rome and was the northern boundary of Italy proper and the southern boundary of the province of Gaul. In the Roman Empire the legions were outfitted by, and showed allegiance to, individuals. You kept *your* legions north of the Rubicon out of respect for the Roman senate and the empire. If a Roman nobleman crossed the Rubicon southbound toward Rome *with his legions*, it was a clear sign that he intended civil war. This is how Caesar initiated the Roman civil war of 49 B.C. against Pompey. "Crossing the Rubicon" implies actions that cannot be revoked. You cannot go *back* across the Rubicon.

Sampled measurement systems have that quality. You had best cross the digital Rubicon under complete and informed design control because you probably cannot recover from any problems you create in the process. The incredible plethora of table top, PC-based "systems" and components that have come on the market lately have exacerbated this problem significantly. They can make professionals very effective when used correctly. They allow amateurs to get into measurements trouble with great speed and ease, making them very dangerous when these systems are used cavalierly without understanding. These systems support the "black box" approach to measurements. This approach treats the measurement system as a black box, thus fostering use without knowledge. This approach replaces the validity checking process with, "It must be right. The computer said so. After all—it's digital."

Two other general situations, at least here in the United States, are very disturbing. The first is the general unfamiliarity with the real implications of Shannon's sampling theorem and its limitations. "Well, we just sample at twice the highest frequency then graph the results. Isn't that enough?" Second is an equally general unfamiliarity on the subject of aliasing and its implications and prevention.

So, how do you cross the digital Rubicon under complete and informed design control?

11.1 THE "GET IT RIGHT THE FIRST TIME" SAMPLING MODEL

Since the requirements for the reproduction of waveshape are more stringent than for frequency content, I'll take the tougher waveshape reproduction case as the baseline for this sampling discussion. In my experience, it is much more satisfying to measure an experimental parameter correctly and fully the first time around with no need for posttest "fiddling." Fiddling means anything you've got to do to the data to *make it valid* before handing it to your customer. It might be applying a digital filtering process, waveshape reconstruction, correction of transfer function errors, or a myriad of other "fiddling" processes. So, my second design assumption is that you want to sample the data correctly the first time with no need to correct after the fact. This is the "Get It Right The First Time" model. It takes the least time and allows the least risky approach. Posttest fiddling is both risky and inefficient. It takes personnel with high level skills and understanding of the phenomenon, the entire measurement process, and some digital signal processing to do this well. It is risky because it is easy, through limited knowledge or inattention, to do more damage than good to the data.

If you pursue the fiddling approach, surely the wizard you need to oversee the fiddling will be on vacation! I believe in doing things right the first time with no need for corrections. That is the initial sampling model you'll see here. Later, I'll discuss a design option you have, the "I'll Fix It Later" option, if you just can't get it right the first time and *must* fiddle.

A sampling model for any digital measurement application has two purposes of equal importance. The first is to prevent the creation of aliased frequency components in a specified frequency range. The second is to sample the input waveforms with enough fidelity to support the objectives of the measurement and, subsequently, the test. The sampling model is a *process* resulting in three design choices for any sampled measurement:

- a decision on the sampling frequency, F_s
- a decision on the anti-aliasing filter type
- a decision on the cutoff frequency setting for the filter, F_c

There are several inputs to the sampling model process:

- the valid data bandwidth, F_d
- are you required to recreate the waveshape?
- or the frequency content?

A second set of inputs includes errors inherent in the filter and sampler:

- aperture and jitter errors
- transfer function errors
- undersampling errors
- skew errors if nonsimultaneous sample and hold multiplexer A/Ds are used
- aliasing errors
- and a combined error that is a combination of the above five items

The sampling model is an engineering process with nine inputs and three outputs—a total of twelve engineering issues the measurement engineer must understand. All twelve issues must be under control or the data can be severely compromised in the sampling process. Such compromise can be irreparable—like crossing the digital Rubicon. If there is an issue in measurements engineering that is consistently misunderstood, or under-understood, this is it. This unfortunate state of affairs finds too many questions about how data are sampled answered by, "Well, it seems fast enough to us," or worse, "We've always sampled it this way."

The entire sampling model process shows in Figure 11.1. This process flow shows everything you'll probably ever need to know about sampled measurement systems on one page. There are three boxes across the top. The two on the left are labeled **"The Known."** These are the design issues that act as inputs to the process. The designing measurement engineer must *know* these parameters. Under **"The Known"** are two groups of inputs— **Goal Information** and **In Pass Band Error Budget.**

The **Goal Information** is the information that defines the goals of the measurement. The Goal Information tells you that the requirement is to guarantee something about the data from some lower frequency (usually zero for design purposes) up through the valid data bandwidth, F_d.

11.1.1 Valid Data Bandwidth, F_d

The valid data bandwidth, F_d, is the maximum frequency at which you must guarantee something about the data. This frequency is sometimes difficult for the designer to get from the customer. The customer may simply not know or have a poor estimate. Getting this frequency information is both necessary and crucial. You cannot proceed without it. If your customer won't give it to you, my advice is to invite them to return to your office when they understand the reasons for, and conditions of, their test better. The customer owes you this information.

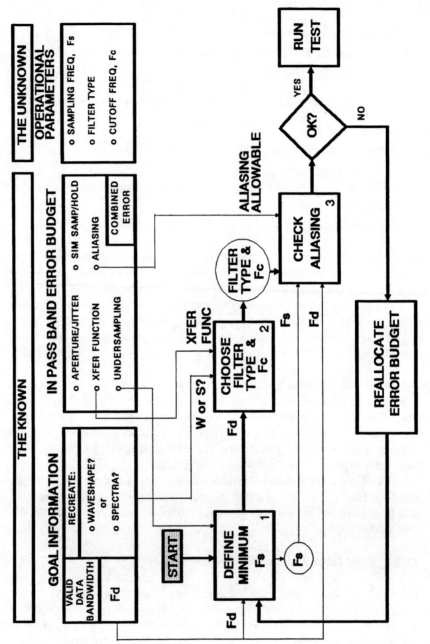

Figure 11.1 Sampling/aliasing model

There will be cases when you must be smarter than your customer. For an aerospace pyrotechnic shock test, for example, most customers would quote $F_d = 10KHz$ to you because that is the maximum frequency in most government shock specifications. Due, however, to the nature of the single degree of freedom filters used in the shock response spectrum computations, a little more valid data bandwidth, say 15KHz, is needed in the time domain to guarantee the shock analyses to 10KHz in the frequency domain. I would not expect even the most experienced customers to appreciate this level of subtlety without information before the fact. The designing measurement engineer has the right and duty to *unilaterally* increase the customer's estimate of F_d when justified.

When you get your best estimate for F_d, *double it* at least and use that for the valid data bandwidth. The downstream price for underestimating this number can be very high. Best be conservative here.

11.1.2 Waveshape or Frequency Content Reproduction?

The customer must tell you whether you are required to reproduce the waveshape or the frequency content of the measureands over a range of amplitudes. You already know the frequency range. As stated earlier in this book, the requirement for waveshape reproduction is generally tougher to meet since it requires a measurement system with the additional requirement of linear phase. The same holds for the sampling model. Waveshape reproduction requires higher sampling rates or methodology for waveshape reconstruction. It is simply more difficult to do.

The **In Pass Band Error Budget** includes several error sources occurring within the pass band defined from zero frequency to F_d. These errors come from the overall measurement system transfer function preceding the sampler, aliasing errors, errors caused in the sampling process itself, and a combined error representing their sum. This discussion considers only those important errors occurring in the anti-aliasing filter and sampler. This list, therefore, is by no means complete as a description of overall measurement system error. Pople[1] lists, for example, 64 error sources for an analog strain measurement using bonded foil resistance strain gages. If his example were expanded to a digital strain measurement system, the total would be up to at least 69 sources!

11.1.3 Aperture and Jitter Errors

Aperture errors are caused by the nonzero time over which the sampler measures the input voltage. If input voltage changes during this aperture time, there is an error in the measured value.

Jitter errors occur in multichannel sampling systems due to instability in the clock rates governing the sampling of each channel. The data channels are not sampled at the exact times you think they are, nor are they sampled at the exact times with which they have been time tagged by the sampling system.

There are several sources of information on aperture and jitter errors in sampled systems[2,3,4,5] that are much more thorough that I intend to be here. In my experience, these errors are usually very small compared to others in the error budget and can usually be neglected. However, as the designer you must justify this action with data and not opinion.

11.1.4 Errors Due to Nonsimultaneous Sample and Hold Analog to Digital Conversion

Almost any analog to digital converter you can buy today will have some form of voltage "hold" before the sampler. Vendors use various ways to implement this method whose intent is to hold the input voltage constant during the sample period. This prevents a reading error from what is sometimes referred to in A/D converter circles as "droop."

Another choice you must make in multichannel system design is whether to employ "simultaneous" sample and hold A/Ds if you are not deploying an A/D per channel. In a simultaneous sample and hold multiplexer A/D converter, all channels hold "simultaneously." When they are read sequentially by the single A/D converter, the values appear as if they were all read at the same instant in time—which is the point of the exercise. When the last channel has been read, all channels revert back into the "track" mode, following their input voltages, until the next simultaneous hold comes along for the next set of readings.

Not all multiplexer A/D converters have "simultaneous" sample and hold. The design choice is yours depending on your application. "Simultaneous" sample and hold multiplexer A/D converters cost more (20–50%) at procurement time. How do you decide if it is worth it in your application?

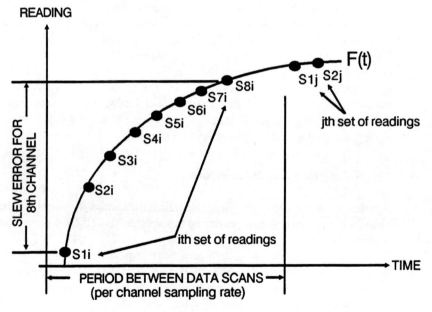

Figure 11.2 The skew problem with nonsimultaneous sample—and—hold MUX A/Ds

One of the qualities of a well-designed measurement system is that when channels are fed the same paralleled input waveform, the outputs match to design tolerances. You want all channels to give the same correct answer to the same input. This means that the outputs must be right for each channel and consistently right from channel-to-channel. Look at the example in Figure 11.2. Here, a hypothetical eight-channel multiplexer A/D converter has been fed the same varying input voltage, $F(t)$, on all channels. This is a non-simultaneous multiplexer A/D converter, so the output readings from the eight channels for the ith scan, S_{1i} through S_{8i} sampled sequentially, vary in amplitude for the same input because they were sampled at sequential periods in time. The first two readings of the second data scan, S_{1j} and S_{2j}, also show. This measurement error is "*skew*" in A/D terminology. In a multiplexer with simultaneous sample and hold, this skew problem does not exist since all channels are sampled effectively at the same time.

Simultaneity is absolutely necessary in certain measurement and analysis applications. Any time series that are going to be algebraically manipulated *with each other*— added, subtracted, multiplied, divided, cross correlated, have transfer functions determined—probably should be measured with simultaneous sample and hold A/Ds. A perfect example of a measurement in this class would be the calculation of dynamic major and minor principal stresses and directions from rosette strain gages on a structure during a shock test. Lack of simultaneous sample and hold under these conditions can lead to *spectacular* errors. Such errors are very hard to diagnose in some cases. A second analysis class demanding this feature is complex mathematical operations leading to the calculation of transfer functions and coherence, cross spectral properties, and cross correlations. This is why almost all commercial digital analyzers sold for use in dynamics testing, where these analyses are made on a daily basis, are designed with simultaneous sample and hold. The vendor simply legislates the problem out of the design in the most straightforward manner. Is this what you should do in your design?

A Straightforward Analysis Method.   Assume, as in the previous discussion, that you are interested in waveshape reproduction and that you know you're going to be doing some manipulations on the time series after acquisition. You want simultaneity in those time series. What kind of sampling model would you need if you chose *not* to deploy simultaneous sample and hold? What sampling ratio price would you have to pay?

As before, the highest frequency waveform that you have to guarantee is a full scale unit sine wave at F_d. The worst case for this problem occurs where the slope of $F_d(t)$ is maximum at zero and 180 degrees. This case shows in Figure 11.2. Assume that the zero crossing shown is at zero degrees. Remember, your design job here is to assure that all channels read the same to the same input. Assume again that you want the worst case skew errors for N channels to be not more than 1% of full scale. That is, the variation in reading between S_{1i} and S_{Ni} in the figure must be less than 1% of full scale. Then

$$\text{Sin}^{-1}(\text{allowed skew error}) = \Delta\Theta$$

This angle, $\Delta\Theta$, is the angle allowed between S_{1i} and S_{Ni} to assure the N amplitudes vary less than 1%. This is an interesting result since the allowed error here is .57 degrees. This means that the overall sampling ratio for the N samples is $N \times 360$ degrees/.57 degrees or $N \times 631$

samples per cycle at F_d. Sampling got fast quickly here! The overall sampling ratio for N channels then is

$$\text{overall sampling rate} = 360N/\text{Sin}^{-1}(\text{allowed skew error})$$

For the eight channel case in Figure 11.2, the overall sampling ratio is 5048 samples across all the channels per cycle at F_d! This is somewhat startling since we require sampling ratios of 10 or 20 for waveform reproduction for a single channel for the same class of errors.

This case for nonsimultaneous sample and hold Multiplexer A/D converters is generalized in Figure 11.3 for $N = 2$ through $N = 8$ channels. Here, the overall sampling ratio shows as a function of the maximum allowable skew error between the channels at F_d. At any lower frequency these errors are proportionally less.

The sampling price you must pay for nonsimultaneous sample and hold is spectacular if you have an application requiring time series simultaneity or anything close to it. My advice is to use simultaneous sample and hold regardless of your application. Be willing to pay the financial price up front at acquisition time as an investment in your later sanity. This is one less design option to worry about as you use your sampled measurement system. By paying the extra acquisition price you no longer have to worry about skew errors for most system design cases.

11.1.5 Transfer Function Errors

The measurement system will have its own transfer function error. This is the error for waveshape reproduction associated with the frequency response being not perfectly flat and

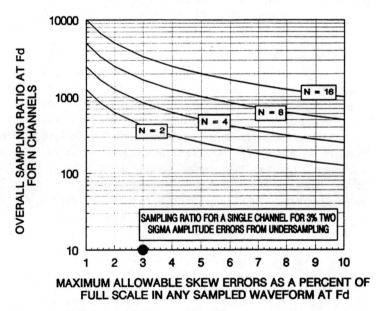

Figure 11.3 Amplitude errors caused by uncorrected nonsimultaneous sampling as a function of the number of sampled channels

phase response not being perfectly linear—an imperfect transfer function. In most measurement system designs the necessary anti-aliasing filter's transfer function predominates at the high end and can be considered the major contributor to this error. When considering errors in this section, remember that a .1 dB error in the frequency response amplitude is a 1% of full scale error at that frequency.

Although the issue of output-input linearity is not a legitimate part of the transfer function, we include it in this section in the interests of proper error bookkeeping.

11.1.6 Aliasing Errors

In sampled measurement systems, the sampling process itself forces images of the spectrum of the input signal to stack repeatedly along the frequency axis around the sampling frequency, F_s, and its harmonics ($2F_s$, $3F_s$, etc.).

High frequency components in the signal to be sampled can interact with the sampling frequency and be caused to "fold" back into a lower frequency range—the lower frequency range below F_d where your data lies. Such folded frequency components are pure error and are said to be *aliased*. They masquerade as valid frequency components—thus the term alias. If aliasing occurs, the data are corrupted irrevocably. You have crossed the digital Rubicon. After an alias has occurred, you have no knowledge of what frequency components have been aliased. All you do have in the frequency domain or the time domain is the sum of the aliased (invalid) and valid (nonaliased) frequency components. Since, in the general case, you have no information on what is an aliased component and what is not, there is no method to make a correction. You have crossed the Rubicon out of design control. Thus, *the first and major design goal for any digital measurement system is to, above all else, prevent the creation of aliased frequency components in the frequency range of interest and above an amplitude that you as a user can stand.*

I could spend the rest of this book just discussing the phenomenon of aliasing. A more thorough discussion of this phenomenon is found in the reference.[6] Suffice it to say that aliases are to be avoided like the plague in any measurement system.

If, however, you find yourself in the situation that you can *guarantee* that no high frequency components will ever exist in the sampled waveform above F_d, then anti-aliasing filters are not even needed and you can save the money and trouble. In this unique case, you may sample merrily without the possibility of an aliasing problem. Warning: If after making this design decision, you get surprised by high frequency components that appear mysteriously, you cannot recover. My advice is to make the *no anti-aliasing filter* decision with your eyes open to the risks involved. We never deploy sampled measurement systems for dynamic data without carefully designed anti-aliasing filters. *Never*.

11.1.7 Undersampling Errors

The worst case undersampling errors occur at the highest frequency of interest in the measurement, F_d. For a constant sampling frequency, these errors will be proportionally less at any lower data frequency. The maximum frequency in the valid data bandwidth, F_d, is the frequency you have to worry about and design for.

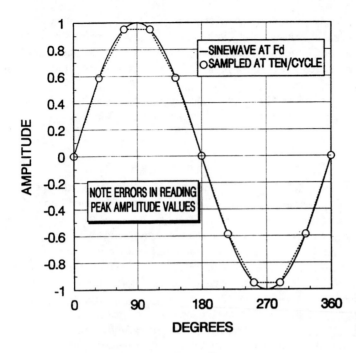

Figure 11.4 Example of worst case undersampling error for a sampling ratio of ten

Figure 11.4 shows a unit sine wave at F_d from zero to 360 degrees. The sampling ratio is the ratio of the sampling frequency to the sine wave frequency of interest. This sine wave has been analytically sampled at a sampling ratio of ten, or $F_s = 10F_d$. I've artificially assigned a zero phase shift between the sampling frequency, F_s, and F_d. Thus, you see the first sample at zero phase with an amplitude of zero. Eleven equally spaced samples show, one every 36 degrees. The eleventh is the first sample of the second period of F_d.

Look at the plus and minus peaks of the sine wave and its sampling. You see that there are two samples equally spaced around each peak. The readings at these samples do not equal the sine wave's peak amplitude—they are in error. They do not reproduce the wave shape. The case shown here is the worst case for undersampling errors for a sampling ratio of 10. Figure 11.5 shows a zoomed view of the positive peak. Here, you see the two equally spaced samples with amplitude of 0.951, or in error low by about 5%. In the absence or posttest data reconstruction, you've missed the sine wave's amplitude by 5%, which is not a trivial error.

You can further see that *any* other phase relationship between F_d and F_s would result in lower undersampling errors. Any other phase would force one or the other of these two samples to fall closer to the peak value than the case shown. This is why this is the worst case for this example. Do you control the phase between F_d and F_s in your measurement system? Can you tweak the phase of F_s to minimize this error at F_d? In the overwhelming majority of cases, you neither know nor can control this phase. It is simply not a design option you have.

Since this phase cannot be controlled, consider it a random type of error. If you tried

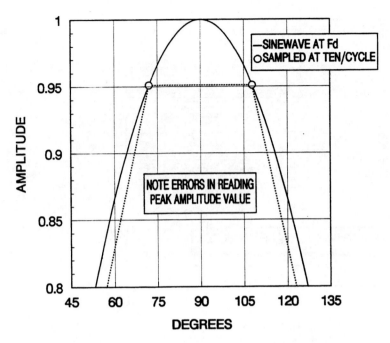

Figure 11.5   A closer look at worst case undersampling error for a sampling ratio of ten

this case 100 times with manual control of when you started your F_s sampler sampling your F_d waveshape, you would get a Gaussian statistical distribution of errors. You would get a distribution bounded with zero error on the low side and the 5% shown on the high side. I am now going to make a statement that will chill the heart of any statistician because it is statistically indefensible: For design purposes, consider the 5% maximum error value to be the three sigma (three standard deviations) value. In other words, 99.7% of all cases will show amplitude errors less than 5% for a unit sine wave at F_d for a sampling ratio of 10. From an engineering design standpoint, 99.7% is as good as 100%. I say that because, although statistically nonrigorous, it is *very useful* for design purposes. It gets us close enough for conservative designs.

Since these undersampling errors are Gaussian we can discuss them in terms of one, two, and three standard deviation errors sets. In this conservative design model, I assert that the three sigma error value is a useful design limit representing the largest error one could reasonably expect. It may be prohibitively expensive to design for the last .3% (99.7% to 100.0%).

The undersampling analysis shows in a generalized form in Figure 11.6. Here, the one, two, and three sigma error families are plotted versus sampling ratio. The case described above appears by following the sampling ratio of ten line vertically upward until it crosses the three sigma error set. The three sigma error is read horizontally from there as 5%.

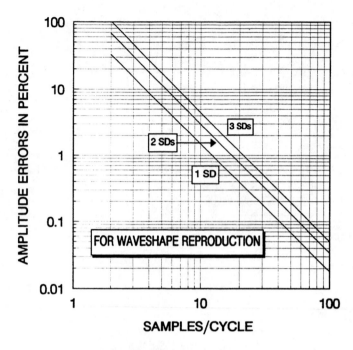

Figure 11.6 Undersampling errors
for various standard deviations (SDs)

This graph becomes useful as a design tool by using it in the inverse manner. In the design case, you know how much error you can allow from undersampling at F_d. You then choose the one, two, or three sigma error set you want to work with, and read down to the necessary sampling ratio at F_d. If your design error allocation for undersampling at F_d was 0.5% of full scale and you used the three sigma error family, you would define a sampling ratio of 22.

We use the two sigma error set for design purposes. This set assures that 95% of the undersampling errors will be less than that shown on the graph for any sampling ratio. I consider this conservative and sound design practice.

11.1.8 Combined Errors

This parameter is the "sum" of the previous six error sources. For a worst case error, the previous error terms can be added irrespective of sign. A root sum square approach may be used and is probably the more defensible case. How these error sources, and this is by no means an exhaustive list for an entire measurement system, should be put together in a cogent overall error statement is a technical area in some flux right now. There are no standards stating how this should be done. There is much talk and many papers given by amateurs and professionals alike, but no agreement or standard. Many treatments are not useful at all. The treatment in Jim Taylor's[7] books are as good as any in print, representing the first time error analysis and allocation has been presented as a measurement system design tool.

Editorial: You should have some method of error analysis in your hip pocket, if only for self defense. Certain contracts, at least in the aerospace business, have time bombs regarding error analysis for experimental measurements buried in fine print on page 302. They'll have a statement like, "All experimental measurements shall have tolerances of one tenth or less of the manufacturing tolerance of the parameter measured for design purposes." Usually these clauses are not passed down to the experimental group until much too late. Along with being largely arbitrary, and usually unwarranted and absurd, they are difficult to achieve. They drive the experimentalist to frustrating and expensive lengths to prove a tolerance at a level having no actual impact on the value of the product to the customer. Statements like these force you to prove that your experimental tolerance was 0.1°F on the temperature measurement that your were off on by 75°F! You measured a temperature invalid by 75°F—but you were accurate in doing so *and can prove it* as if this were important. If this occurs in your shop, you may have to analyze your way out of the predicament. Rationality lacks a place in this unfortunate process. The proper approach is never to let language like this get into a contract in the first place. If noted, your contracts people should immediately take exception to the language and discuss the matter with you.

11.2 THE SAMPLING MODEL PROCESS

As stated before, the assumption underlying the discussion of this process is that you want to measure the data waveshape with acceptable and defensible errors and low risk so that no posttest fiddling is needed. This is the *Get It Right The First Time* sampling model.

11.2.1 Define the Sampling Frequency, F_s

In Figure 11.1, the process begins by noting the requirements for the valid data bandwidth, F_d, and the allowable undersampling error. Knowing the allowable undersampling error, enter Figure 11.6, determine the standard deviation error set you want to work with, and read the necessary sampling ratio at the bottom of the figure. The sampling frequency, F_s, is simply

$$F_s = (\text{sampling ratio}) \times (F_d)$$

11.2.2 Define the Anti-aliasing Filter Type

The rules regarding waveshape reproduction are the same for the anti-aliasing filter component as they are at the measurement system level.

- all frequencies of interest within the flat range of the frequency response portion of the transfer function
- all frequencies of interest within the linear range of the phase portion of the transfer function
- all amplitudes within the linear range of the output/input characteristic (linearity)

Transfer functions.   A very useful way to look at the second criteria, phase linearity, is in terms of delay. The time delay of a measurement system component is linearly proportional to the first derivative of phase with respect to frequency, $d\theta/dF$. We want phase linearity with frequency for waveshape reproduction. Where the phase is linear, its first derivative, the time delay, will be a *constant* function of frequency. This shows in Figure 11.7 for an 8 pole Butterworth low-pass filter. This figure shows both the phase and delay for this filter as a function of the ratio of the frequency of interest to the filter cutoff frequency, F_c, where the response in nominally −3 dB down in amplitude. This filter has a very limited linear phase range from the lowest frequency to a frequency ratio of about 0.3. It shows nonlinear phase every where else. The delay starts out being constant at the extreme left of the chart and then immediately begins to vary in amplitude. It peaks at a frequency ratio of about 1.05, where the slope of the phase shift curve is maximum.

Showing phase in terms of its first frequency derivative, delay, is useful because delay can now be plotted logarithmically on the same plot as the magnitude as Figure 11.8 shows. Linearity judgments about the phase response can no longer be made with the logarithmic frequency axis. But, nonconstancy judgments about the delay are easily made. Waveshapes will be reproduced where the delay is constant.

Figures 11.9 through 11.12 show the transfer functions for four high performance anti-aliasing filters: (1) 8 pole Bessel; (2) 8 pole, six zero constant delay filter; (3) 8 pole

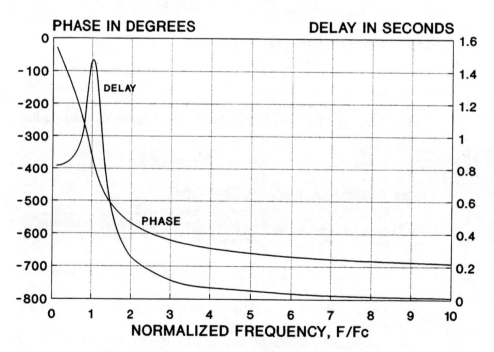

FREQUENCY DEVICES LP00

Figure 11.7   8-pole Butterworth (Frequency Devices LP00)

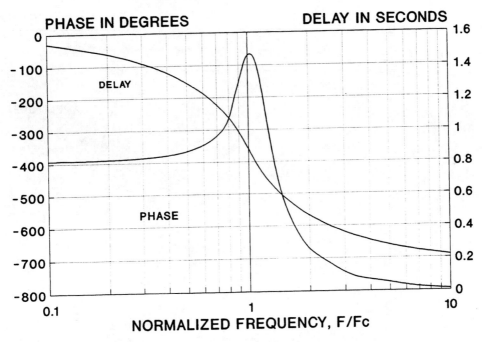

FREQUENCY DEVICES LP00

Figure 11.8   8-pole Butterworth (Frequency Devices LP00)

Butterworth; and, (4) an 8 pole, six zero elliptic filter. These data are courtesy of Frequency Devices, Inc.[8] These are the leading anti-aliasing filters of design choice for high performance systems. There are others available, such as the Chebyschev, but the generic design issues can be understood with these four. Filters of the same type from other vendors would show different internal circuitry, but very similar transfer functions. The imperfections in these transfer functions in terms of amplitude and phase within the passband are the transfer function errors referred to earlier.

The Bessel filter and the constant delay filter show constant delays for frequency ratios well past the cutoff frequency. This shows that they are designed to reproduce waveshapes within the passband. The price you pay for constant delay within the passband is roll-off characteristics that are not as efficient as other filters of the same order. The Butterworth and the elliptic filters show highly nonconstant delay within the passband of your data. These two filter types will corrupt the waveshape of your frequency rich data within the passband. They are lousy waveshape reproducers. They were designed to reproduce efficiently the *frequency spectra* of the input waveshapes. They are very flat within the passband, and drop like stones with radical roll-offs outside the passband. The price you pay for this radical roll-off slope *outside* the passband is highly nonlinear phase, and nonconstant delay *inside* the passband. If you want to reproduce only the frequency spectra, you do not require phase linearity. This is the general trade space you have with filter

FREQUENCY DEVICES LP02

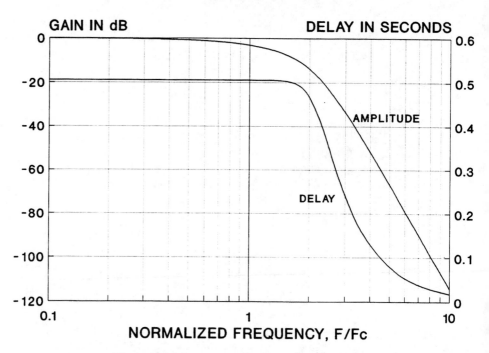

Figure 11.9   8-pole Bessel (Frequency Devices LP02)

choices. You are trading performance within the passband (phase linearity and amplitude flatness) for performance outside the passband (steep roll-offs). You must make an informed choice depending on your application.

Note the characteristic lobes in the amplitude portions of the constant delay and elliptic filters. These two filters were designed to be used with 12-bit A/D converters. They were designed so these lobes have amplitudes less 1/2 the least significant bit for a 12-bit A/D. The twelfth bit is worth −72 dB. Another 1/2 the least significant bit is worth another −6 dB, for a total of −78 dB. The designer has placed these lobe amplitudes at −80 dB or less so they cannot even be detected with a 12-bit A/D converter.

Figure 11.13 shows the comparison of the amplitude characteristics for these four filters to frequency ratios of five. The elliptic filter shows a much more efficient filtering characteristic for amplitudes outside the passband. The Bessel, constant delay, and Butterworth show almost the same terminal slope outside the passband. They just get there differently. I mention this because certain anti-aliasing models in the literature use only the terminal slope of a filter outside the passband to set the cutoff frequency. Note that these three filters would be treated in the same manner in such a flawed model. In reality, at any frequency outside the passband, their amplitude responses vary by as much as 40 dB—a factor of 100! Their anti-aliasing performance would, therefore, vary by a factor of *100* for

FREQUENCY DEVICES LP03

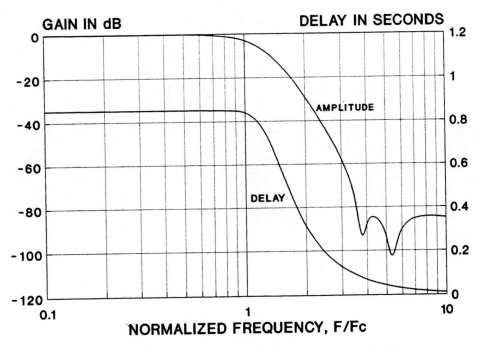

Figure 11.10   8-P, 6-Z constant delay (Frequency Devices LP03)

the same conditions. Clearly, how the filter gets to its terminal slope must be included in your sampling model. You must use the *entire* transfer function in your decision making, not just the terminal slope.

How these filters, in fact, get there shows in Figure 11.14. The filter amplitude responses are compared to a frequency ratio of one and from full scale, 0 dB, to −3 dB. Note that the Bessel, constant delay, and Butterworth filters all use the standard definition of upper frequency limit. Their amplitude responses pass through −3 dB at a frequency ratio of 1.0. The elliptic filter, however, is still going. It is not −3 dB down until a frequency ratio of about 1.15. All four of these filters have the same number of poles and zeros (the Bessel and Butterworth filters have their zeros at infinity), yet their transfer functions, therefore performances, are wildly different. A filter is not a filter is not a filter! You can see that this vendor is inconsistent within their own product line with three filters following the standard and one not following it. The message is that when discussing these subjects with your colleagues, the exact lexicon must be agreed upon to make sure you are comparing apples with apples.

Step (indicial) responses.   The step responses of the filters are additional, needed pieces of information that go in the design recipe. Figure 11.15 shows the step

FREQUENCY DEVICES LP00

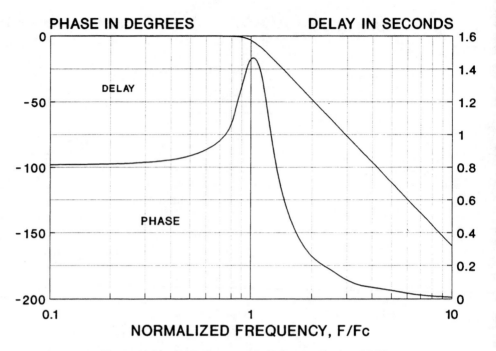

Figure 11.11   8-pole Butterworth (Frequency Devices LP00)

responses for the same four filters. These data are courtesy of Precision Filters, Inc.[9] These responses are also called the *indicial* response. It is the filters' responses in the time domain to a full scale, zero rise time step function at time equals zero. The responses have been nondimensionalized and normalized on the time axis.

The Bessel filter has the best behaved indicial response showing a 0.2% overshoot you can't even detect on this chart. This is a major advantage of this filter type. The constant delay filter shows an initial undershoot of about 4%, then the rise with an overshoot of about 4%. The Butterworth and elliptic filters both show overshoots of *15–20%* with subsequent "ringing!" The nonlinear phase of these filters causes this ringing. If your input signal is somewhat steplike, sampling within the overshoot time periods will result in large amplitude errors. In fact, the elliptic filter takes 13 times as long to settle to the same value of error as the Bessel filters! These indicial responses must also be considered in your anti-aliasing filter choice.

Upon understanding the transfer function and indicial response errors for this set of anti-aliasing filters, you can now make an educated choice depending on your application. For the case I'm presenting here, waveshape reproduction with no posttest fiddling required, the eight pole, six zero constant delay filter is the filter of choice from this set. It has a high terminal roll-off slope outside the passband, about 140–150 dB per decade for high

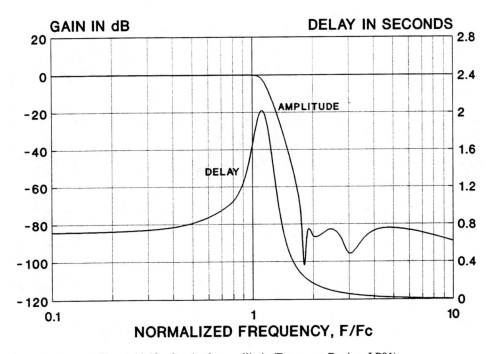

FREQUENCY DEVICES LP01

GAIN IN dB   **DELAY IN SECONDS**

Figure 11.12   8-pole, 6-zero elliptic (Frequency Devices LP01)

frequency rejection combined with superior performance in the passband and dead constant delay well past the cutoff frequency. Its indicial response behaves well for all but the most exacting pulse train waveshape reproduction tasks.

11.2.3 Choose the Filter Cutoff Frequency, F_c

The design choice for the cutoff frequency is based on: (1) the amount of anti-aliasing filter transfer function error your application can stand within the valid data bandwidth; and (2) the amount of attenuation you need outside the passband for otherwise alias-causing higher frequencies. The farther out you set the cutoff frequency, the smaller in passband transfer function errors, but the more high frequency energy will be passed. This is an eternal compromise faced by the measurement engineer daily. There is no free lunch.

11.2.4 Define the Allowable Aliasing Error

If there are high frequency components in your data that are suppressed insufficiently in amplitude, they will alias into your valid data bandwidth. The question is how much is too much? System level measurement system knowledge and knowledge of the phenomenon

FREQUENCY DEVICES 8th ORDER FILTERS

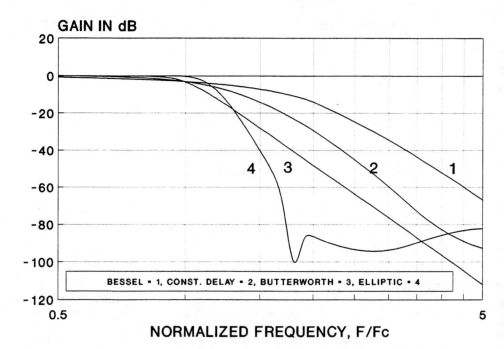

Figure 11.13   Filter gain comparisons (Frequency Devices 8th order filters)

are needed to answer this question. Somewhere on your design team this understanding must exist or you cannot proceed in a logical, defensible manner. Engineers who have newly entered this business via electrical engineering or computer science are at a sizable disadvantage here. These engineers in the absence of that knowledge, generally force anti-aliasing errors far below a reasonable level. This system suboptimization is usually expensive and unnecessary.

Author's note: I once wrote an outraged letter to the editor of a periodical on, among other things, PC-based data acquisition systems. The letter was about this subject—design suboptimization in the absence of proper overall system level knowledge. I ended up writing the expert column on data acquisition for that periodical. The moral is to be careful regarding what you complain about in print.

Where are the overriding error sources in your measurement system and what are the magnitudes of these errors? In the area of dynamic acceleration measurements for shock, vibration, and acoustics tests, the transducer calibration errors are at the 1–3% level (at best!) over the test frequency range. Even NIST can't do much better than that, and 99% of the time you're not even working at that calibration level. You're working with a transducer with, again at best, a secondary or tertiary calibration. Further, if the motion is even a little

FREQUENCY DEVICES 8th ORDER FILTERS

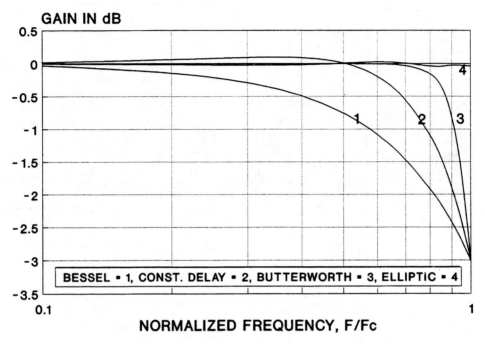

Figure 11.14   Filter gain comparisons (Frequency Devices 8th order filters)

complex, motion components in all three axes will exist even for uniaxial motion inputs. The best commercial accelerometers used for these tests exhibit from 1–5% cross axis sensitivity and a percent or two nonlinearity. The result is that even before your signals get out of the transducer, the errors are in the 2–5% of full scale range. Five percent is only −26 dB down from full scale. Experimental errors in the dynamics test business exist in the −26 to −40 dB down from full scale ballpark. Amplitudes below these levels are noise by definition. Why do some designers work very hard to suppress aliases down to the −70 to −80 dB level? I have never understood this absurd false economy. Here, there are uncontrolled, uncorrected experimental errors sources operable in the measurement system 30 dB *above* the allowable alias! This comes from a lack of knowledge of performance at the *system* level.

The allowable aliasing error is the maximum amplitude level within the valid data bandwidth above which you will not allow an aliased frequency component. It is a design parameter you set knowing the objective of the measurements you are trying to make. Aliased components may exist below this level, but you simply don't care. We typically design for aliases −50 to −60 dB down from full scale so we don't have to worry about them.

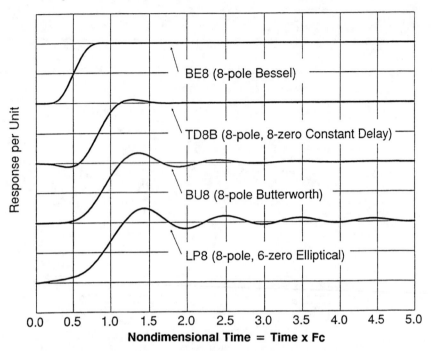

Figure 11.15 Comparison of step response for four low-pass filter types (Printed through the courtesy of Precision Filters, Inc.)

11.2.5 Check the Aliasing Performance

Now that all the performance parameters for your sampled system have been set, the anti-aliasing performance must be checked analytically. There are two cases to be considered in an anti-aliasing model. In the first case, your goal is the reproduction of the waveshape to the valid data bandwidth and no aliased frequency components above the aliasing allowable amplitude you defined in the previous step. This first method is the more demanding one and is discussed first. It also is the predominant case in our measurement system designs. If you only wanted to define one checking method for anti-aliasing performance, this is the one to choose because of its inherent conservatism.

The method shows graphically in Figure 11.16. All anti-aliasing models that I know of proceed from the following covert assumption, which I am now going to make overt. The assumption is that the spectrum of your data is flat, randomlike and extends to infinite frequency with no components *above* the channel's full scale. I'll discuss the risks in that assumption below. This is as good an assumption as you can make, within the valid data bandwidth. And, you have to assume *something* about the nature of your data before you proceed.

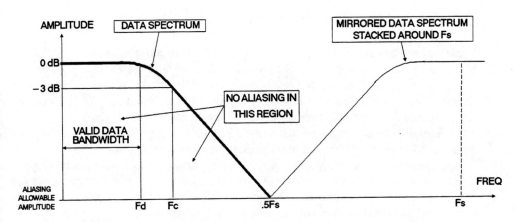

METHOD:
1. CONSTRUCT THE DATA SPECTRUM PASSED THROUGH THE ANTI – ALIASING FILTER IN THE LINEAR FREQUENCY DOMAIN.
2. NOTE THE ATTENUATION OF THE ANTI – ALIASING FILTER AT .5Fs.
3. THAT IS THE MAXIMUM AMPLITUDE OF AN ALIAS FROM A FULL SCALE OUT – OF – PASSBAND FREQUENCY COMPONENT THAT WILL ADD, AS ERROR, TO THE TIME HISTORY.
4. COMPARE WITH THE ALIASING ALLOWABLE AMPLITUDE.

Figure 11.16   Anti-aliasing check method #1

1. Draw (see Figure 11.16) the spectrum of your data after it has passed through the anti-aliasing filter you've chosen. Since the assumed data spectrum is flat, this is merely the amplitude transfer function of the filter as shown.

2. Note the amplitude attenuation of your filter at one half the sampling frequency, $.5F_s$. This is the maximum amplitude of an aliased frequency component from a full scale out-of-passband frequency component that will add—as pure and irrevocable error— to the time history below F_d.

3. Compare this amplitude to your aliasing allowable amplitude. If it is less, your sampling model is valid and you may proceed to test. If it is higher, then you must iterate the sampling model.

If your sampling model fails this test by producing aliased components above your aliasing allowable amplitude, you have only three design options open to you.

- Change the anti-aliasing filter type to one with a steeper terminal roll-off that will provide more efficient filtering outside the valid data bandwidth.
- Maintain the same filter type and cutoff frequency, but increase the sampling rate.
- If neither of these options works for your application, you must compromise, reallocate the combined error budget, and repeat the sampling model process with the new numbers until your model does pass.

11.3 THE "I'LL FIX IT LATER" SAMPLING MODEL

There may be design cases where you just cannot deliver the performance, on a per channel basis, demanded by the previous, bulletproof sampling model. In these cases, your back is against the design wall usually because of limited sampling rates. There is a second, more liberal sampling model that will get you out of this problem. I call it the *"I'll Fix It Later"* sampling model. The process steps in this model are the same as stated above.

Here is a plausible example that might drive you to using this sampling model. You are required to design a sampled measurement system with no, either series or parallel, analog mass storage capability. I'm not sure that this exclusion falls in the realm of good design, but let's include it for the sake of discussion. The system is to be used to monitor accelerations, strains, and pressures during acoustics testing of large aerospace structures over a valid data bandwidth of 10KHz.

You are required to measure the waveshapes of the measureands because there is meaningful risk of failures during the test and peak values at failures are of primary interest. Pertinent questions about the data include "What was the peak value of acceleration or strain at the moment of failure?" or "Was the dynamic pressure in the acoustics chamber peaking at the same instant?" These questions loudly call for waveshape reproduction over and above the assumed reproduction of the frequency content.

We assume this is a very large and complex structure, thus the data suite is 256 channels. What would the previous sampling model tell us about this measurement system? It would tell us that $F_d = 10$KHz and that the probable sampling ratio would be at least ten. The overall sampling rate would be, per channel

$$F_s = (F_d)(\text{sampling ratio}) = (10\text{KHz})(10) = 100\text{KHz}$$

The overall sampling rate for the 256 channels would be

$$\text{overall sampling rate} = (256)(100\text{KHz}) = 25{,}600{,}000 \text{ samples/second!}$$

Things get rapid in short order in this model! The real, full level acoustics exposure would take a nominal 60 seconds. It might be preceded and followed by two minutes each for calibrations and chamber runup and purging after the exposure, for a total of five minutes (3000 seconds). Therefore, the storage requirement is for 25.6 megasamples/second times 3000 seconds, or *7.68 gigasamples!* This is a huge amount of data with which to deal. In this kind of testing, the test conductor will be ready for a higher level run in an hour or two. What do you do, within an hour or two, with 7.68 gigasamples of data from the previous run to get ready to accept the 7.68 gigasamples from the next run? This is an interesting design problem in itself, but is outside the scope of this book.

The clear design problem in this example is the 25.6 megasamples/second sampling rate. What if you simply cannot provide, either for reasons of cost or performance, that level of sampling performance? You can find yourself in this same design problem with smaller channel suites, and lower bandwidths and sampling ratios. It depends on your hardware and your requirements. The problem is generic. The "I'll Fix It Later" sampling model gives the design maneuvering room you need.

In this model, more radical anti-aliasing filters are used so that the sampling ratio can be lowered. By radical, I mean a filter with spectacular out-of-passband attenuation performance paid for by nonconstant delay, therefore waveshape corruption, within the valid data bandwidth. The elliptic filter is the typical choice in this methodology.

By using the elliptic filter judiciously, both the cutoff frequency and the sampling ratio can be lowered. The sampling ratio can be lowered from the ten or twenty in the previous model to three to five in this model. A sampling ratio of five would get you down to 12.8 megasamples/second for 256 channels and this might be enough to produce a viable design. The price you pay is the sure knowledge that your multifrequency waveshapes are corrupted within the valid data bandwidth. But, at least, you're in the game rather than on the design bench!

Reconstruction of the valid waveshapes requires two steps of posttest fiddling, both of which are limited by noise levels. Such limitations must be thoroughly understood before attempting these methods. First, the data must be reconstructed in the time domain as if it had been sampled at sampling ratios of ten or twenty in the first place. We recommend Whittaker's method[10] using a SinX/X approach, first published in 1915. The method is valid because the lower frequency information being reconstructed is, in fact, already in the sampled time history via Shannon's theorem. This method has the advantage of producing an expanded time series that recreates exactly the sampled data values. Be aware that your time series file sizes will expand dramatically in this process.

The invalid waveshape of the reconstructed time series must now be corrected and the valid waveshapes recovered. This can be done in at least two ways. The information can be shifted to the frequency domain via a complex Fast Fourier Transform (FFT). Then, the anti-aliasing filter's inverse transfer function is applied to the FFT. The resulting corrected FFT is then inverse transformed back into the time domain recovering the valid waveshape. The key to this method is that each filter's transfer function must be known accurately to determine the inverse transfer function. In the time domain, the correction can be made by deconvolving the reconstructed, but still corrupt, time series with the channel's indicial response.[11] The key to this method is that each channels' indicial response must be known beforehand.

These correction methods require several qualities besides those the more conservative sampling model requires: (1) a staff skilled in digital signal processing; (2) the right software set to apply these methods in an efficient, risk free manner; and (3) an efficient and available computer platform in which to fiddle. For these reasons, I leave the "I'll Fix It Later" sampling model to the experts with the correct toolset and understanding.

11.4 TWO SAMPLING CAVEATS

11.4.1 Frequency Content Reproduction

The first caveat is that if you are lucky enough that your requirements are for frequency content reproduction alone, you have even more design leeway for sampling. This would be the case if Fast Fourier Transform, autospectral density, or one-third octave analyses were all that were required. For frequency content reproduction applications, sampling ratios of *just more than two* are necessary. The anti-aliasing checking method for this case shows in Figure 11.17. Note that aliases are allowed in the textured triangular area on the chart in this sampling methodology. These aliases occurring outside the valid data bandwidth can be removed with digital filtering if required. The same criteria for waveshape reproduction apply to digital as to analog filters: flat frequency response, linear phase, and output/input linearity.

Most commercial frequency analyzers operate with default sampling ratios of 2.56 for firmware reasons. This is why these analyzers must be used with great care when you're trying to use them for waveshape reproduction. We do so in some of our pyrotechnic shock work at sampling ratios of 10.24 for 10KHz data, allowing > 100KHz sampling rate. To do this, the analyzer must be operated on the 40 KHz analysis range (40KHz × 2.56 = 102,400 samples per second) to produce a waveshape valid to 10KHz. In this analyzer, four times as much valid data bandwidth is needed to produce valid waveforms than is needed for valid frequency content reproduction!

11.4.2 Signals Higher Than Full Scale

The second caveat is a usually covert one regarding sampling on dynamics tests. Earlier I made an assumption that the frequency content of the sampled signal could be described with a flat spectrum to infinite frequency, with no components above full scale. This as-

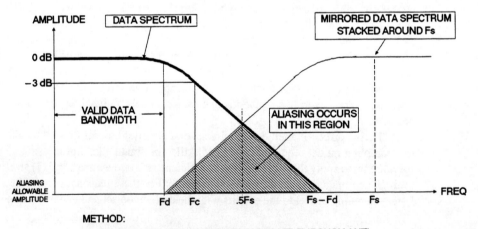

METHOD:

1. CONSTRUCT THE DATA SPECTRUM PASSED THROUGH ANTI – ALIASING FILTER IN LINEAR FREQUENCY DOMAIN.
2. CALCULATE Fs – Fd. THIS FREQUENCY WILL FOLD TO Fd.
3. NOTE ATTENUATION OF ANTI – ALIASING FILTER AT Fs – Fd.
4. THAT IS THE MAXIMUM AMPLITUDE OF OF AN ALIAS AT Fd FROM A FULL SCALE OUT – OF – PASSBAND FREQUENCY COMPONENT.
5. COMPARE WITH ANTI – ALIASING ALLOWABLE.

Figure 11.17   Anti-aliasing check method #2

sumption is made, covertly or overtly, everywhere. Whether it is valid depends on your application.

In pyrotechnic shock work, for instance, the usual transducers of choice are piezo-electric or piezoresistive accelerometers that are internally highly resonant as shown in an earlier chapter. If high frequencies in the shock motion excite these resonances, *apparent* acceleration motions 10–100 (called Q in the earlier chapter) times full scale may be produced in the transducer. These high output levels will boom through the anti-aliasing filtering capability and produce aliased components at levels 5–50 times your design anti-aliasing allowable amplitude. You can almost guarantee that your data within the valid data bandwidth will be ruined when this happens. If these levels exceed the channel's linear input limit, problems of frequency creation by nonlinear operation will also occur as mentioned in an earlier chapter. This problem must be anticipated and provided for in your system design, usually by increasing the data channel's range and valid data bandwidth before the test to include the resonant response within the linear range, thus handling the linearity problem.

When you get surprised by these resonances causing unacceptably high aliasing levels, your only recourse is to know conclusively that this occurred, and quietly consign the data to the trash can.

This overranging problem, caused by excited transducer resonances and some other effects, can happen at any frequency. They are not peculiar to shock testing. They can just as easily crop up in a 30 Hz rate gyro transducer. The problem is a generic one in measurement system design.

11.5 CONCLUDING REMARKS

The subject of sampled systems is becoming more and more prevalent in measurement system design. The first reason for this is the alluring quality the data can have when it gets into a digital format. A new vista of possibilities opens for the measurements engineer when this occurs. The second reason is the amount of new digital systems hardware that continues to explode into the marketplace. Problems that were not solvable at reasonable cost 5 years ago are solvable now.

Not too many years ago, commercially available data acquisition system applications software was of a very limited nature. That's why everyone wrote their own software. The third reason is that decent data acquisition packages, designed for engineering users and not programmers, are coming into the marketplace and becoming viable design alternatives. That is the upside.

The downside is that many engineering students coming out of universities see the engineering universe as something on the other side of a keyboard. If they can't solve the problem from the keyboard, they are not interested in solving the problem! This, in my opinion, is a serious and limiting flaw in an engineer's skillset. It is a fatal flaw in a measurements engineer.

As such, these new engineers tend to think that absolutely legitimate, necessary, and effective analog methodologies are somehow passé or dated. This limited scope of thinking is dangerous. The measurement engineer has no inborn right to a digital system. That right must be *earned* by the development and demonstration of measurement systems level thinking and understanding. This statement is worth repeating: A digital system in the hands of a user without that system level understanding is like a loaded gun in the hands of child. Someone or something (usually the data) is going to get hurt.

NOTES

1. J. Pople, *Errors and Uncertainties in Strain Measurement;* in Strain Gauge Technology, A. L. Window and G. S. Hollister, editors; Applied Science Publishers, Englewood, NJ, 1982, ISBN 0–85334–118–4; pp. 209–66.
2. James L. Taylor, *Computer-Based Data Acquisition Systems;* Instrument Society of America, Research Triangle Park, NC, 1986, ISBN 0–87664–921–5; pp. 95–112.
3. James L. Taylor, *Fundamentals of Measurement Error,* Neff Instrument Corporation, Monrovia, CA, 1988.
4. *Handbook and Selection Guide for Computer-Based Data Acquisition and Control Systems;* Neff Instrument Corporation, Monrovia, CA, 1988; pp. I-9 to I-19.
5. A. Sanchez, *Understanding Sample-Hold Modules;* The Best of Analog Dialog, 25th Anniversary Edition; Analog Devices, Inc., 1991; pp. 51–54.
6. Ibid.
7. See notes 2 and 3.
8. Frequency Devices, Inc.; 25 Locust St., Haverhill, MA 01832, 617/374–0761.
9. Precision Filters, Inc.; 240 Cherry St., Ithaca, NY 14850, 607/277–3550.
10. Samuel D. Stearns and Don R. Hush, *Digital Signal Analysis (Second Edition)*; Prentice Hall, Englewood Cliffs, NJ, 1990; ISBN 0–13–213117-X; pp. 75–79.
11. Patrick L. Walter, *Deconvolution as a Technique to Improve Measurement-system Data Integrity;* Experimental Mechanics, Society for Experimental Mechanics, Bethel, CT, August 1981; pp. 309–14.

Chapter 12

Measurement System Operational Methods

Over the years we have identified several operational methods that absolutely support the effective acquisition and presentation of valid data. We have institutionalized them into our working culture. They become the operational guidelines for all new measurement system designs. The list is an evolving one, so the information below is a snapshot of our experience base in this area. Some of these operational features will seem "See Spot run!" simple to you. So be it. Thorough practice and craft sometimes look simple. Perhaps it should.

A college professor of mine defined an expert as someone who had made every mistake in the book—*once*. By that definition, if you make a mistake twice you are a fool. By that definition, in turn, I am a fool as are all my colleagues! We've all made certain critical mistakes more than once. These operational methods evolved in the crucible of those operational mistakes. They are the result of positive corrective action driven by some of the burnt tail feathers I've mentioned before. These are simply *what works* in the operation of effective measurement systems.

I can discuss these issues in two ways. I can discuss them in terms of design features in a dry litany of, "Whatever you do, design in these features." Or, I can discuss the operational methods that support the effective presentation of valid data and tell you why they are important. I choose the latter course since I can include some horror stories that drove the improvement of the operational method.

Some of these operational methods represent a decidedly minority view in this business. Most organizations do not, unfortunately, take this much care of their data. These are organizations that expect to routinely lose 10–30% of the data on any test by their own admission. I think they pay a heavy and hidden price for not taking that care. These operational methods and design features are a discernible difference between an amateur and a professional. You should look for that difference in the workplace.

12.1 GENERAL MEASUREMENT SYSTEM OPERATIONAL METHODS

12.1.1 Independent Span Verification

This single subject is the most important in this chapter. I cannot overemphasize its importance. That's why I'm going to spend a little ink on it. Assume you are providing measurements for a test where a very light, very expensive spacecraft deployable structure is being put in a "Zero G" condition. Long vertical cables support the structure's weight such that the structure reacts to its deployment kinematics as if it were in an orbital zero G condition. Assume that there are 14 such lines, each with a full bridge load cell in series with the structural load. Each line is then carefully tensioned with dead weights over pulleys to take up the correct portion of the structure's weight.

A senior measurements technician assumes that the bridge completion cards in the load cell signal conditioners are all in the proper full bridge configuration because he was the last person to use them. He sets them up, applies the correct excitation, balances correctly, applies the correct metrology shunt calibration for each channel's span, and reads the loads as the loading process begins. Five of the load channels are clearly off by a factor of *two* in the wrong direction when the dead weights are applied. Worse—quality assurance personnel catch the error. (Note: no measurements professional *ever* allows quality assurance personnel to catch an error. This is an insult of the highest order and an admission of failure to a professional.) The errors are, fortunately, caught before irreparable damage can occur. *If these errors had remained undetected, this expensive flight spacecraft structure would surely have been destroyed.* This horrifying situation happened to experienced professionals in our organization. It really got our attention and told us that there was clearly something missing from our operational method. What actually happened? What was missing?

In fact, the senior technician was not the last person to use the data channels. The person who had used them put five of them in an external half bridge configuration and didn't return them to the standard full bridge configuration after his test (Mistake #1). The senior technician did not check them before use (Mistake #2). Upon investigation, we found that the signal conditioners had a design flaw. The signal conditioner, with a full bridge transducer connected and a half bridge completion card installed, would deliver the excitation voltage, balance and even shunt calibrate! The problem is that the half bridge completion network shorts out half the external full bridge transducer. Since only the external half of the transducer bridge is working, it takes twice the load to cause the expected output signal! You shunt and span the channel for a 25 pound full scale load—but it takes 50 actual pounds to make the channel read 25 pounds. The resulting span is clearly off by a factor of two and in the wrong direction to boot.

What really happened here *generically?* Measurement channels, in improper configurations, were spanned with proper techniques using legitimate metrology calibrations resulting in improper spans. But, the spans *were not verified* after the fact using independent means (Mistake #3). Independent means are those not associated with the primary calibration method.

The generic question you have to ask yourself is how do you *know* your data channels are set on the right span before you run? The following responses are not the correct answer to the question:

- I can see the gain switch setting on the bridge signal conditioner.
- I can see the full scale engineering units setting on the charge amplifier.
- I can see the volts per inch switch setting on the X-Y plotter, oscilloscope, fiber optics oscillograph or ink recorder.
- Or worse, the computer said it was on the right range.

How do you *know?* You know by performing an independent span verification on every data channel that verifies the span setting is correct (or not) before you run.

An engineering unit calibration is one where real engineering units are input to the measurement system *in situ.* Examples include dead weight loads for load transducers, +/−1 G tip calibrations for DC coupled acceleration transducers, dynamic pressure from piston phones for microphones, actual deflection inputs using micrometers or gage blocks for any DC coupled deflection transducer, angular inputs for rotary deflection transducers and in-clinometers, angular rates from a rate table for rate gyros, or actual pressure into a differential pressure transducer in a wind tunnel model.

In each of these cases, real calibrated engineering units are input and the data channel is forced to read the right value. This is the bulletproof calibration method and allows highest level of certainty. "I put in 25.00 pounds of dead weight and the measurement system read 25.03 pounds." No one will argue with that method—at least at zero frequency and with established linearity. In-situ engineering unit calibrations are the best method for the setting of system spans.

Calibrations where transfer standards are used to set the channel's span are one level lower in certainty. In our horror story, the transfer standard was the metrology shunt calibration for the load cell. Other examples of transfer standards are dialing in of the box charge sensitivity and range on piezoelectric measurement channels, or the dialing in of the manufacturer's gage factor for strain indicators. Often, the situation forces you to use the transfer technique. You are simply not going to have 100,000 pounds of calibrated dead weights in your pocket to input to your 100,000 pound load cell. Where do you get a cali-brated 50 watts/cm^2/hr to calibrate your absorbed flux radiometer, or 1000°F for your ther-mocouple, or 500Gs for your accelerometer? In these cases, some sort of transfer calibra-tions are mandatory.

Since the use of transfer calibrations is almost mandatory in any test laboratory, you do not *really know* the system's span (set using the transfer standard) unless you verify it independently. That is why the independent span verification method is so valuable. We use it on all tests on every independent gain path in the measurement system. Our consistent experience is that, even with highly motivated and trained professionals working on mea-surement systems they control, independent span verification finds span problems in 5% of the data channels on average. Independent span verification allows us to find these prob-lems and fix them before the test runs. It costs a modest amount of labor to perform these checks, some of which are serial to the test effort.

It is, in fact, cheaper to blindly set the ranges, not verify them, and run. Most people in this business do exactly this. If we find and fix span problems 5% of the time using independent span verification, what do you think the rate of undiscovered span problems, not checked with this method, is in the industry as a whole? You can bet it is significantly larger than 5%! How much are these undetected, uncorrectable errors worth to you in your test laboratory?

Returning to the horror story noted above, what should have happened? The channel should have been excited, balanced, and spanned using the metrology calibration data—exactly as it was. At this point, the transducer cable should have been removed from the transducer and plugged into a millivolt/volt transducer simulator set at zero unbalance (we use either the BLH 625 or the Vishay 1550 simulators for this purpose). The simulator is the *independent* check. The load cell has a nominal full scale sensitivity of ±3.00 mV/V. That unbalance value, 3.00 mV/V, is then input from the simulator. The channel *must* read the full scale value within a tolerance or the channel is incorrectly spanned despite, and independent from, its switch or gain settings. When the span is conclusively verified, reconnect the transducer, rebalance if necessary, recheck span via the shunt calibration, and run the test.

Table 12.1 is a page from our published document on independent span verification. It states by transducer type: (1) the method by which the fine span is input to the setup process; and (2) the method of independent span verification. If the primary method of setting span is an engineering unit calibration, then independent span verification is not needed—you know the channel span is correct—and the table entry is "not applicable." If the primary method of setting span is a transfer standard, then the table entry is the necessary independent span verification method.

The purpose of the independent span verification is to catch human and machine errors. It absolutely catches the X2, X3, X5, or X10 gain error caused by operator oversight, poor visibility, inattention, or mispatching. It will also catch the recalcitrant automated signal conditioning system that tells you it's set on the 3000G range when it is really set on the 100G range because it did not accept your computer's range command. I'll say more about this failure later in this chapter. Independent span verification is not designed to catch 0.2% span errors. You must find those with other methods if they are important to you.

12.1.2 Noise Level Documentation

Your data simply cannot be validated without noise level documentation. Your customers have the right to ask you for it. You should institutionalize noise level documentation runs in your operational procedures. As an example, we have a practice for vibration testing of recording data from before the dynamic exciter's power amplifier turns on until after it turns off after the run. A potentially destructive excitation glitch can occur whenever these very high power amplifiers are on. It would be very embarrassing if the exciter system glitched and we were not recording the structural responses and control accelerations. The recording immediately before the exciter starts also acts as a pretest noise level documentation. The recording immediately after the excitation ends is a posttest noise level documentation. These two recordings tell you nothing about the noise levels *during* the run. These are documented with check channels.

TABLE 12.1 INDEPENDENT SPAN VERIFICATION METHODS

Transducer Type	Measured parameter	Set-up Method	Independent verification
Current Probe	Current	Calibration Coefficients	Series Ammeter
LVDT	Displacement	Displacement Calibration	N/A
Microphone	Pressure	Piston Phone	N/A
Piezoelectric	Acceleration	T-Insertion	Box Sensitivity
	Force	T-Insertion	Box Sensitivity
	Pressure	T-Insertion	Box Sensitivity
	Moment	T-Insertion	Box Sensitivity
Potentiometer	Angular Displacement	Angular Calibration	N/A
Rate Gyro	Angular Velocity	Rate Table	N/A
Resistance Temperature Transducer	Temperature Absorbed Flux	Calibration Coefficients	Transducer Simulator
Servo Accelerometer	Acceleration	Tip Calibration	N/A
Inclinometer	Angle	Sine Block	N/A
Wheatstone Bridge (1–4 active gages)	Straine, Force, Pressure, Torque, etc.	Mechanical Input	N/A
	Same as Above	Gage Factor	Rcal or Transducer Simulator
	Same as Above	Rcal	Transducer Simulator
Wheatstone Bridge (piezoresistive)	Acceleration	Tip Calibration	N/A
Wheatstone Bridge (Sim. from Potentiometer)	Angle, Displacement, Pressure, etc.	Mechanical Input	N/A
Wheatstone Bridge (Sim. from Potentiometer)	Angle, Displacement, Pressure, etc.	Rcal	Transducer Simulator

The key here is to record enough data to support comparison with whatever data reduction you're going to do during the run. For instance, if the data reduction was power spectral density analysis with 64 averages, this would define an averaging time depending on the analysis bandwidth. We would record at least this much time after the test with no structural excitation so the equivalent analysis of the noise levels could be performed.

On major structural tests, we allow the measurement system to record data periodically over the 24 hours immediately prior to commencing the test. This defines the diurnal noise levels in the measurement channels. In these cases, you usually see periodic, low level drift in the measured noise associated with building temperature variations. Even in air conditioned test laboratories, the building's wall temperatures will change, radiating more or less infrared energy to your test article, changing its temperature enough to see in the strain and deflection data. In certain *very* sensitive structural tests, these data allowed us to pinpoint when the slopes of the diurnal noise levels were minimized. That's when you test. Unfortunately, it's usually from 2 to 5 a.m.!

The other key to documenting noise levels is the judicious creation and monitoring of check channels. They can save your measurement's life. When things go well, they document that they went well. When things go wrong, the noise level channels can help lead you to the problem. On transportation tests with FM/FM tape recorders, we always record a shorted channel to define transportation-induced tape speed variations that are inseparable from the otherwise valid data. On thermal vacuum, structural and dynamics tests we include check channels identical with the actual transducers but responding to everything other than what we want to measure.

On major dynamics tests with complicated data reduction schemes, we record a constant diagnostic waveform, whose analysis we know, in parallel with the test data and through the entire system. If the analysis of this referee waveform is right, then it is highly likely the analysis of the test data is also right. A typical referee waveform for low frequency sine vibration testing would be a 10 Hz repetition rate square wave that gives a clear signature under dynamic analysis.

12.1.3 Analog Signal Visibility

A major drawback of modern digitally controlled signal conditioning systems is that you cannot see the waveform at any point in the measurement chain until it pops up on the screen. This is a fatal design error on a vendor's part.

We design our measurement systems with lots of analog visibility via test points in the measurement chain—particularly in the low level portions of the chain where the signals are most vulnerable to noise level corruption. If you're designing by integrating components from several vendors, this is easy to do. If you're buying a standard turn-key system from some vendor it may be impossible.

In the absence of analog data visibility, you simply do not know what is happening to your signal in the complex mixmaster of a measurement chain. You do not know the ingredients in the recipe. You only know the result. No chef in any restaurant in the world would operate like this. You shouldn't either.

12.1.4 Use of Diagnostic Instruments

The *analog* oscilloscope is the most important instrument in any test laboratory. If you could only have one instrument to work with, the analog oscilloscope would be the one. I say analog oscilloscope because most modern oscilloscopes sample and suffer from the frailties of any sampled system. Analog oscilloscopes are getting harder and harder to find. You should stock up!

We include a suite of low and high level diagnostic instruments in every measurement system we deploy regardless of the valid data bandwidth. You'll see at least a mechanical engineer's oscilloscope and a digital multimeter in any system we deploy. (Look at the photo of the Thermal Data Acquisition System in the chapter on software.) A mechanical engineer's oscilloscope is one with about 1 MHz of bandwidth, but with single-ended and differential amplifier sensitivities of at least 100 microvolts per inch. These instruments are absolutely invaluable when you are trouble shooting. When this occurs, and it will, you don't want to be looking around for your diagnostic toolset. You want it right there in the rack ready to go.

12.1.5 Real-Time Verification of Data Storage

In systems using FM/FM or digital tape recording, you must be able to see data coming off tape while the device is recording to be able to *know* the device *is* recording. Laboratory tape record/reproducers have the extra set of heads allowing this. You monitor these heads while the test is in progress to make sure that what is coming *off* tape is what is going *on* tape.

Some "modern" tape record/reproducers do not have this capability (certain TEAC digital and analog cassette recorders and Metrum's RSR 512 digital cassette recorder are examples of this). The only indication you have when you're recording is that the record light is on. Would you trust your career and your customer's data to the record light? You cannot see the data on tape unless you stop recording, rewind, and playback. But then it's too late! Any tape record/reproducer without the capability to verify data storage during recording is a toy.

12.1.6 Transfer Function Verification

We verify the entire transfer function of dynamic data channels on a regular basis. the transfer function measurement concurrently verifies the three criteria for waveshape reproduction: flat frequency response, linear phase response, and output/input linearity via the coherence function.

We recently participated in a major offsite dynamics test with four contractors and three government agencies and we were responsible for the strain and acceleration data from a certain part of the test article. We requested, as a standard practice, complete transfer function verifications on our data channels. These transfer functions showed that all thirteen (!) 28 track tape record/reproducers (responsibility of another contractor) were delivered with the wrong electronics cutting the valid data bandwidth in half! No other contractor or government agency planned to perform transfer function verifications for two reasons: (1) "That's a waste of time and we're on a tight schedule"; or (2) "What's a transfer function?"

If these transfer functions had not been checked, 364 data channels on a *$20 million* test would have been run with insufficient bandwidth. I would not want to take the call from the paying customer on that one!

12.1.7 Dry Runs

Plan dry runs on every nonstandard test. A dry run is one in which all test environments occur in their proper sequence except the planned input to the test article. Data should be taken during the dry run on all channels. The perfect example of this is when a transient dynamics test requires high speed photography. The photographers come in and set up their cameras and lights. Test time comes and, ten seconds before the firing, the cameras and lights come on popping every circuit breaker in the county, shutting off the entire test system. This has happened in almost every test laboratory in the world. You solve this problem with dry runs. Include them in your test planning.

Warning: No photographer should ever be allowed to plug any piece of his equipment into power circuits supporting the measurement systems on a test. This equipment invariably takes large amounts of power and can introduce inductive noise levels into the power lines.

12.1.8 Establishment of Flexible Single Point Ground Reference

Measurement systems should have a single point ground reference to keep ground loops and other noise levels under control and minimized. This is easy to say and, sometimes, difficult to do. I also recommend that the grounding and shielding configuration for your measurement system have some flexibility built in to handle changing conditions. It is highly likely that the power grounding configuration you design for will change at some time in the future. It is further likely that this change will not be under your control. You may not even be aware of it. Suddenly, your measurement system has killer noise levels that were not there the day before. Flexibility in grounding and shielding can help prevent you from chasing your tail when you're trying to isolate and fix the very subtle causes of a 200 microvolt noise level.

Be aware that if you bring in your facility electricians to help you work the problem, they may not even perceive you have a problem. *Perfect* power to an electrician may be *abominable* power to a measurements engineer.

12.1.9 Where to Put the Gain and Filtering in Your Measurement System

A basic unwritten ground rule in measurement system design is put necessary gain and filtering as close to the transducer as you can. This practice tends to minimize the system's sensitivity to later noise sources and maintain a proper signal-to-noise ratio across the valid data bandwidth. The ultimate expressions of this are transducers with built-in excitation if required, gain, and filtering. Unregulated DC power goes in and high level signal voltages come out. This is a very alluring capability.

There are at least two problems with this approach. For non-self-generating transducers, it removes control of the excitation waveform from the designer. The designer can no longer tailor the excitation waveform to control self-heating, verify power-off noise levels, or support information conversion.

Self-generating piezoelectric transducers with internal signal conditioning remove from the designer the ability to perform the crucial T-insert functions, unless specifically designed for that function. There is no longer any opportunity to perform necessary independent span verification or in-situ transfer function measurements. We prefer to put the gain and filtering just outside the transducer in external signal conditioning so that we have maximum visibility and accessibility at the right points in the measurement chain to perform the important functions noted above.

12.2 METHODS FOR QUASI-STATIC SYSTEMS

12.2.1 Component-by-Component Zero Balance

For static systems, you need an ability to balance your system as defined in the chapter on the Wheatstone bridge. The condition you want to create easily is zero output for zero input at every stage in the measurement chain. This also creates the desired overall condition of zero output for zero input.

The creation of this condition allows zero shifts at any stage of the measurement system to be verified and monitored. It allows you to answer the question, "Is the zero shift I see at the output of this measurement system due to what I want to measure (good) or something else (not so good)." It also allows you to pinpoint where the problem is occurring in the system.

If you do not use this method, you lose all zero information when you change ranges. This is not good in a static measurement system designed to keep absolute track of zero.

12.2.2 External Zero Reference—The Star Bridge

Assume you've got some internally strain gaged bolts loading a spacecraft separation system. You want to load the separation system in your facility and read the static load via the strain gaged bolts. You then ship the spacecraft to the launch site where your people will read the bolt's load on a totally separate measurement system, usually a strain indicator, to verify the system preload.

How do you zero the second strain indicator before you connect the preloaded bolts? In generic terms, how do you establish a measurement system's zero so that it matches a previous system's zero, when you have no opportunity to put the transducer in a zero mechanical input state? How do you transfer an absolute zero from one measurement system to another? You cannot unload the bolts because you lose the information you are seeking.

A similar generic question is how do you verify the long-term zero stability of a measurement system channel in a case where the measureand may not be stable with time? What is long-term change in the measureand and what is measurement system zero drift?

The answer to these questions is the use of an external zero reference. For full bridge transducers, the device is a Star bridge and shows in Figure 12.1. The Star bridge is a configuration that absolutely guarantees a stable zero output when it replaces a full bridge transducer. Its balance point cannot drift even if its resistors do drift. You can see that there is no possibility for a voltage to develop between the output B and C corners of the Star bridge. Each resistor has a value of half the transducer impedance. In this manner, both the excitation supply and the readout device are loaded properly.

The Star bridge is used as follows. A transducer is excited, balanced, and spanned on measurement system #1. With the transducer in a zero input condition, the measurement system reads zero output (any residual unbalance at zero input is suppressed). The transducer is disconnected and the Star bridge substituted. The zero offset is recorded (this offset was previously balanced out). The transducer is reconnected and the test proceeds.

At measurement system #2, the loaded transducer is connected and the system spanned using the same shunt calibration information as used for system #1. Now the loaded transducer is disconnected and the Star bridge connected. The system zero is now adjusted to give the same offset as when the star bridge was connected to measurement system #1. Now the span and zero of the original measurement system are recreated in the second system. The loaded transducer is now reconnected and the actual load read directly.

If you are transferring zeros externally among measurement systems with the same balance networks or voltage injection circuits, the Star bridge resistors do not have to be matched, nor do they have to be precision resistors. You can build one of these for $20 in parts depending on your standard transducer connector. We wire them in the backs of con-

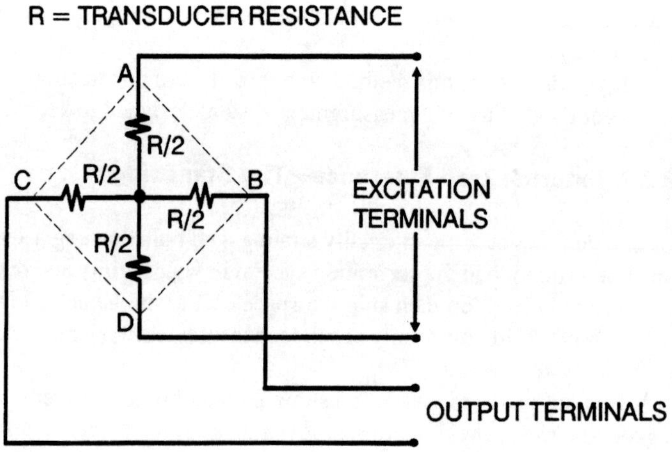

Figure 12.1   Star zero reference bridge

nectors simulating the transducer connectors and pot the entire assembly into a brass cylinder. Every technician gets one. If you're in the more complicated situation of transferring external zeros among systems with markedly different balance networks, then the resistors should be matched precision resistors (+/−.05% resistance match, and 50 ppm/yr resistance stability). The cost just went up to $50 apiece. That is dirt cheap when you consider what troubles the use of the external zero reference can prevent for you.

A second use of the Star bridge is in testing the common mode rejection capabilities of your signal conditioning. Note that the potential at the central node of the Star bridge is half the excitation voltage and that zero current flows through the resistors leading to points B and C. Thus, the two output terminals are at the transducer's common mode voltage with zero differential input. This is the perfect condition for testing a differential common mode rejecting device. Simply vary the common mode voltage by varying the excitation voltage and noting your system's output. Whatever you see is noise.

12.2.3 Bipolar Multilevel Shunt Calibrations

Whenever possible you should shunt calibrate using multiple shunt levels of both polarities. This documents system linearity as well as system span. The shunt calibration information exists on the *change* in level between one shunt and another—not on the level itself. Thus, even when a shunt calibration is imposed on a channel, data is still coming into the measurement system on top of the shunt calibration level. You lose no data in this process. You need have, therefore, no reservations about using shunt calibrations during a test run as long as the data remains within the system's linear input limit.

12.3 METHODS FOR DYNAMIC SYSTEMS

12.3.1 Excess Bandwidth

Whenever possible, provide at least double the bandwidth called for in the data request. In certain cases in transient measurements, we provide eight times the requested bandwidth! This will allow your measurement system not to be surprised by the inevitable surprises in the test. You do not need anyone's agreement to do this.

12.3.2 Recorded Calibration Signals for Dynamic Measurement Systems

Best practice is to record calibration signals allowing transfer function verification. Sine waves of various frequencies across the valid data bandwidth can be used for this purpose. Several amplitudes should also be recorded to document system linearity. A fixed repetition rate square wave with harmonics in the right places can also be used. The square wave has the advantage of being self-checking for linearity since each harmonic has its own proper amplitude that can be checked easily. Further, frequency creation by nonlinear operation can be easily spotted spectrally.

12.4 METHODS FOR AUTOMATED MEASUREMENT SYSTEMS

12.4.1 Never Trust an Automated Measurement System

An automated measurement system is one where the signal conditioning and filtering is under computer control. These systems are usually deployed without a knob or test point anywhere. You're lucky to get an on-off switch. The vibration control and data reduction market is flooded with these systems from half a dozen otherwise reputable vendors.

Users who trust what a computer says about the configuration of an automated signal conditioning system do so at their own risk. A competitor of ours has a major automated measurement system they use to support vibration and shock testing. All its charge amplifiers and anti-alias filters are computer controlled. Not only does this competitor not perform independent span verifications, they believe what the computer says as gospel without checks of any kind. We consider this absolutely foolhardy. All they really know is the configuration they told the system to be in on a channel-by-channel basis. They know *nothing* about its actual configuration at test time because they do not verify. They type the piezoelectric transducer's box charge sensitivity and the channel's full scale range in from the keyboard—and run. Very dangerous. Very scary.

Here is a better example. We use top of the line digitally controlled charge amplifiers in our own dynamics testing. They have absolutely the best transfer functions around for laboratory dynamics testing. We recently used them on a major pyrotechnic shock test with about 100 data channels. After setting the ranges (we thought) and acquiring the shock data, we found several channels severely clipped. This made no sense since the measurement locations were several acoustic interfaces away from the shock initiators and the levels should have been fairly predictable. The measured levels did not pass the sanity test. What happened?

Nine of the charge amplifiers did not accept the controlling computer's ranging command. Our directions were to run these channel's on the 3000 G range and the controlling software told them to go to the 3000 G range. The nine channels remained on the 100 G range where the previous T-insertions occurred. The vendor's control software does not report this anomalous condition before the run. What they report to you is a useless summary of your *directions* to the system. No information whatsoever is provided about the system's *response* to your directions. As a result, we ran 1500 G shock data through data channels ranged for 100 Gs with predictable results. The errors were discovered after irretrievably losing the data, when a range verification program was run.

The obvious solution to this problem is a system performing automated T-inserts (independent span verifications) *after* the ranges are set by computer command. This absolutely verifies ranging prior to the run.

Never trust computer controlled signal conditioning in measurement systems. They will turn on you and devour you when you least expect it. *Always* check independently and verify.

12.4.2 How Much Data Storage Room Have I Got Left?

It amazes me how many vendors field digital data acquisition systems that do not tell you how much data storage room you have left *in the test configuration you are going to run.* Some systems tell you the entire acreage in bytes. Who cares about bytes? What you want to know is how much recording time you have in this test configuration (channels X sampling ratio X valid data bandwidth). And, computer, please tell me in useful units like seconds or minutes. We will not buy a system that reports this crucial information in bytes, blocks, or any other programmerese unit not understood by everyone on the test crew.

12.5 CONCLUDING REMARKS

Operational methods are to the measurement system as the written music is to the piano. Proper and conservative operational methods are an integral part of effective measurement system design. The best measurement system design in the world can be easily negated by using inappropriate or shoddy operational methods. Unless the music is played well, the best piano in the world makes only noise.

Professional operational methods may even be more important on a day-to-day basis than the details of the design. These methods (and the list in this chapter only scratches the surface) can sometimes allow a marginal measurement system design to perform admirably in the hands of a professional. The reverse is generally not true. A great measurement system, misoperated for whatever reason, will produce only expensive and questionable junk.

The test I mentioned in Section 12.1.6 is a case in point. The test cost $20 million dollars and lasted 40 milliseconds. We were the *only* contractor that got data of unquestionable validity on that test. The fact that we got valid data is attributable as much to proper and proven operational methods—from which we would not back away—as it was to a solid measurement system design.

Chapter 13

Data Validation Techniques

Suppose for a moment that you have just finished a test with your newly designed, bright and shiny measurement system. Finally, you have the first piece of data for the customer to peruse. It looks like a squiggly line on paper, as much experimental mechanics data does. Further, suppose that before the test you did everything you knew how to do to assure yourself and your customer that the data you were about to take would be valid.

Up walks the customer. He peers intently at your data for a while and says, "That is very interesting. However, my hardware simply cannot be acting that way. Your data is obviously wrong. Please call me when you've figured out what is wrong since I am in a hurry on this test program."

Gulp. All measurements engineers have had this interaction happen to them. If the reader is in the measurements business and this has not happened to you, it will.

At this point there are just two things you can do. You can throw the customer out of your office and invite him back when senses and the power of logic have been regained. This, for reasons of public relations and follow-on business, is frowned upon. The second choice is to prove, after the fact and beyond a shadow of an engineering doubt, that your data *is* valid. This is as far as the measurements engineer need go most of the time. A customer who maintains that your data is invalid, in the face of proof to the contrary, is simply admitting that he does not understand what the data is telling him about the test process. He does not understand that, in fact, his hardware *can* act that way. This is no longer a measurements problem. This is a data understanding problem for the customer. At this point, the measurements engineer's responsibility shifts from providing demonstrably valid data, to helping the customer to understand the message the data is giving about the hardware.

This chapter is about the *provision of that proof to the contrary, that proof of validity.*

This proof comes through a process of posttest data validation. This is a most exciting area in measurements engineering because it is where the power of perceptive computer software can make a real difference on the test floor. When data validation methodologies are captured in software processes, these conclusive tests can be automated in digital data acquisition systems. Data is only declared valid in these cases after it has passed these codified tests for validity. This usually takes most of the subjectiveness out of the data validation process. It also affords the test team a way to capture the corporate measurements experience base in this area. I'll discuss these aspects more in the later chapter on knowledge-based systems.

This chapter will provide you some simple, yet ironclad, methods for the determination of data validity. These methods are as close as we are ever going to get to "back of the envelope" engineering, which is a very good way to do business. If you can explain a problem and/or solve it on the back of an envelope, you surely understand it.

We'll approach the subject of data validation techniques with the following analogy. It is fun as well as apt. After the listing of the data validation questions noted below, some proven methodologies will be discussed in more detail.

13.1 MEASUREMENTS SYSTEMS SHOULD BE LIKE SPIES

You are the spymaster.[1,2] One of your spies is undercover observing some people doing something you want to know about. These people are secretive and do not want you to know about it. They do not know the spy is there. You know the spy is loyal and capable of observing and supplying objective information about the process. Do you believe the information supplied? Given those caveats, the information should be trustworthy. You still do everything you can to verify the supplied information. This is the process of data validation.

Here's another case. The same loyal, capable spy is observing the same people doing the same something. This time, they *know* the spy is there. Do you believe the information supplied?

It is highly likely that the information acquired was compromised because the folks *knew* the spy was there. Maybe they even lied to the spy on purpose—called disinformation in the intelligence business. In any case, the mere knowledge of the spy's presence affected their interactions and, at least, colored the information. The information is no longer objective news, but has become a subjective editorial. Is the information trustworthy? Probably not. You will still have to verify it by another means or somehow correct it, presuming you know what is wrong with it.

Spymasters' lives are difficult. Their professional lives are about the question, "Do I believe the information, or not?" But the spymaster has a major advantage over us. It is, at least, *possible* to acquire objective knowledge about the process of interest to the spymaster under the right circumstances: undetected, capable, loyal spy.

It is, by definition, *impossible* for your measurement system to acquire perfectly objective information about the process you're interested in. As stated in an earlier chapter, the laws of physics say it is impossible to acquire information about a process without transferring energy with the process. That transfer of energy changes the process and, thus, the acquired information. Physics does this *by definition and without exception.* You do not get a vote in the matter. The process now lies to your spy, your measurement system, giving it false, subjective information. So why measure at all? Aren't you doomed? You measure because by proper system design and management of the energy flowing between the process and your system, you can get close enough. It may not be easy or cheap, but you can. The problem is that your boss and the project manager think it *is* easy and, therefore, *should* be cheap.

So, here comes some information from your measurement system during your test process. You know the process has been changed by definition. Assume that you've done a good design job and the amount of the process change you've caused is insignificant in an engineering sense.

You must still check up on your spy. Good spymasters check up on everything. What are the questions you ask about the information supplied so that you can assess it to determine its worth? These kinds of questions must be asked and answered if you have any chance of convincing yourself your data is valid—and not disinformation.

Now we need some experimental data from our spy. Figure 13.1 shows a time history from a test. It does not matter what the measured parameter is, nor does it matter how the amplitude and time axes scale. It could be grams or tons, microseconds or years. These questions are universal and generic. A time history is appropriate because all your data starts out in the time domain anyway. It may later get transferred to the frequency domain in your application. But it starts here. If it's invalid here, it's likely invalid everywhere.

13.1.1 Data Validity Checking Questions to Ask Your Spy

The balloons in the figure refer to 12 questions that come to mind about *this* data set. I'm sure there are more, but 12 are good enough to start with. If you get the right answer to these

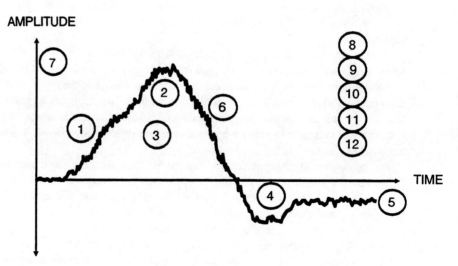

Figure 13.1   Your data's final exam

12, it is highly likely that your data are valid. But, you don't get the right answer unless you ask. Also please note that for another data set with different characteristics, these questions might be different. The questions fit the application.

As I go through the questions, I'll also refer to the measurement system design issues that control the answers to the question. Each of these design issues is discussed in this book. If you did your design correctly, you already have the right answers. Please note that most of these questions refer to things occurring *before* the first digital anything in your measurement system—the A/D converter. Consider the A/D converter home plate. You have to get to analog first, second, and third bases before you can score a digital run!

These are the questions I would ask regarding the validity of this waveform.

1.   Is this really the rise time of the measureand, or is it the rise time of the measurement system?   You want to measure the measureand's rise time. The measurement system will have its own characteristic rise time. This is the time required for the system's output response to a step input to go from 10% of its final amplitude to 90% of its final amplitude. This 10–90% time is a function of your design and has nothing to do with the measured input. Your data acquisition system simply cannot respond any faster. If your data acquisition system's rise time is one-tenth of the measured rise time, then you've made a 1% error in the measured rise time. You've slowed it down. If the system's rise time is one-third the measured rise time, the errors are about 6%. Remember, in any properly designed sampled system there are anti-aliasing filters to limit the bandwidth *on purpose*—thus increasing the system's rise time (making the system slower). These are opposing design criteria—fast rise times versus valid anti-aliasing performance.

- governed by measurement system's high frequency transfer function with caveats for resonant transducers

2.   Is this really the pulse peak, or was it affected by the measurement system's frequency response, phase response, or linearity?   To what degree is this waveshape distorted by the system? The system's frequency and phase response taken together define the transfer function. When input/output linearity adds to the recipe you've largely defined the system's ability to reproduce waveshapes. Was the transfer function adequate to measure this parameter? Was there enough high frequency response to handle the high frequencies in the waveform? Was there enough low frequency response to handle the low frequencies? Did the system have the right phase response to prevent phase distortion? Did the system exhibit frequency creation (shudder) because of nonlinear behavior?

- governed by the measurement system's entire transfer function and input/output linearity characteristics

3.   Is this really the pulse shape, or was it affected by the data acquisition system?   Same discussion as above.

- governed by measurement system's entire transfer function, input/output linearity characteristics, and transducer/phenomena isolation ratio

4.   Is this really a negative phenomena (cavitation, bounce-back, rare-faction, stress-reversal, rebound, etc.), or is it undershoot in the data acquisition system?   Data acquisition systems with a transfer function not extending down to zero frequency will undershoot to a pulse input—output goes *negative* when the input returns to zero! Depending on the low frequency transfer function, some systems will undershoot and then "ring," putting nonexistent amplitudes into the information.

I read a paper published in a respected measurements-related symposium proceeding in which a plot appeared of absolute pressure that actually went negative (−160 psi?) violating all known laws of physics. Given the waveform, it's likely that what I was looking at was data acquisition system undershoot. This is very embarrassing for the author who was trying to demonstrate a totally different aspect of the experiment. The author lost all credibility about his real issue by publishing patently and obviously invalid data that should have been caught in the first place.

- governed by data measurement system's low frequency transfer function

5.   Is this permanent set in the measureand, or is it zero shift in the measurement system?   Did the test article yield? Or was the transducer overloaded and yield? Was there some permanent offset in the measurement system? Did something get too hot? Did some cables get fried? Did some system electrical component go into "saturation" due to either too much high or low frequency energy and cause a system zero shift? Did transducer insulation resistance change?

- governed by data acquisition system transfer function, linearity characteristics and transducer material properties

6.   Is this high frequency really in the measureand, or is it ringing in some resonant component, carrier noise, or "ripple" created during data handling?   Note that the high frequency content increases when the event begins. Can anything be concluded from this? The record ends before this high frequency content ceases. What can you conclude from this? The basic thing you conclude is that you didn't take enough data! How do you know the underdamped transducer is not resonating due to high frequency input from the measureand? Is this frequency the result of an alias created by (shudder again) your sampling methodology? If you're using a carrier system, is the carrier leaking through into the output data? What about triboelectric effects in your cables?

- governed by transducer transfer function, carrier system design, anti-aliasing and sampling method, noise levels from phenomena

7.   How do the uncertainties in all the measuring instruments, and handbook data used to interpret the result, combine to affect the result?   Maybe this is transient strain data and the structural design group wants to calculate stress. How does the design group's use of the handbook values for material properties and assumed geometry affect the stress data? Should you use an error propagation model? Does the model even work for dynamic events? What is the overall experimental uncertainty in the data? Does it even matter whether you have an 8, 10, 12, 14, or 16 bit A/D on this job? What about noise levels and how they affect the result?

- governed by error propagation model, inherent uncertainties in using reduction constituent equations, noise level documentation methods

8.   Is this really the result of the pulsed phenomena you want to measure, or is it the result of another part of the test environment?   Maybe what you want to measure is dynamic pressure in a hydraulic system. Is this waveform actually pressure, or is it the result of temperature, or magnetic fields on the data acquisition system? How do you know? Are you prepared to prove it?

- governed by the design's selectivity in environmental responses and noise level reduction and documentation methods and procedures

9.   Is this really a result of the fundamental transduction process you've designed into the system, or is the waveform the result of the desired measureand's affects on the cabling system?   Were the data cables subject to the same transient pressure environment as the transducer was? How do you know you aren't looking at transient capacitive affects in the cables? This one is nasty because it's a noise level caused by what you're trying to measure, and occurs in the same time frame and portion of the frequency spectrum as the data you want.

- governed by the design's selectivity in environmental responses, cabling system design, and noise level reduction and documentation methods

10.   How much of the waveform is self-generated voltage?   You're measuring the dynamic pressures with strain-gage-based pressure transducers. The response you want is that from the resistance change in the strain gages due to pressure. How much of what you see is nasty self-generated thermocouple voltages due to transient temperatures? These correlate in time and frequency with the measureand and cannot be filtered out. This could be a problem.

- governed by physical transduction processes within the transducer, its mounting, and its sensitivity to environments other than the measureand

11.   How much energy did the measurement system transfer with the process during the measurement, and was it significant?   In order for the supposed pressure transducer to function, its volume in front of the diaphragm must change. When the diaphragm deflects due to pressure this volume changes. The deflecting diaphragm, in turn, loads the internal flexures on which the strain gages are mounted. This change in strain is, in fact, what you really measure. The original and necessary change in volume *requires energy* from the process, thus changing it. How much is too much?

- governed by the complex physics at the transducer/phenomenon boundary

12.   Was the measurement system designed to minimize the 11 previous effects in the first place, and to document them unambiguously during the test in the second place?   In the answer to this question lies the realm of the professional measurement system designer and operator.

- governed by all the above and a lot more!

A basic set of validation techniques appears next. These are the methods used to answer the 11 questions. They can act as a checklist for you in the design of your particular data validation method set. These are generic and can be applied to almost any experimental data. In some cases, they reiterate topics already covered. In others, they are new information.

13.2 ISOLATION RATIO CONSIDERATIONS

The first posttest question the measurements engineer should ask to validate the data is, How much have I changed the process itself merely by measuring it? These questions were addressed in some detail in the chapter on non-self-generating transducers. Any time you make a measurement, energy flows across the process/system boundary, changing the process.

The isolation ratio is a measure of how much energy will flow across the boundary. The isolation ratio is the ratio of the acceptance ratio of the measurement system to the sum of that acceptance ratio and the process's emission ratio. If the isolation ratio is 1, there is no energy flow and the data is error free.

You want a system isolation ratio that is as close to one as possible so the offending energy flow is minimized—given reasonable financial and schedule restraints! The approach is to choose a measurement system with an acceptance ratio so large compared to the process emission ratio that the errors are negligible.

13.3 VALIDATION OF THE WAVESHAPE

The next step involves the inspection of the output waveshape itself. The rules for waveshape reproduction are:

- all frequencies of interest must exist within the flat region of the amplitude portion of the system's transfer function
- all frequencies of interest must exist within the linear region of the phase portion of the system's transfer function
- all amplitudes of interest must exist within the linear region of the systems' output/input characteristic

Does your data follow the three rules? Knowing (1) the amplitude limits imposed by the linearity rule, and (2) the frequency limits imposed by the flat amplitude and linear phase rules, you can now inspect the recorded data and determine validity. Do amplitudes exist outside the linear range? Do frequency components exist in the waveshape outside the limits imposed above? If the amplitudes and frequencies in your data fall within the limits imposed by your design, then the data are unaffected by the measurement system transfer function and the data is valid—so far.

Two further checks need to be made. Each is associated with a different end of the system transfer function.

13.4 RISE TIME CONSIDERATIONS

The rise time of a measurement system is a performance characteristic associated with the high frequency end of a monotonic system transfer function. Rise time is not defined for nonmonotonic systems—systems with resonances. Beware of this when reading component specifications. Vendors of resonant components, such as transducers, are notoriously loose with their definitions in this specification area. The rise time shows in Figure 13.2 and is the time required for the system's output to go from 10% to 90% of its final amplitude

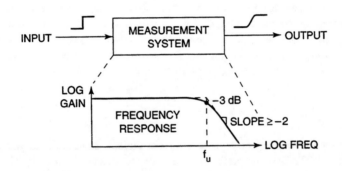

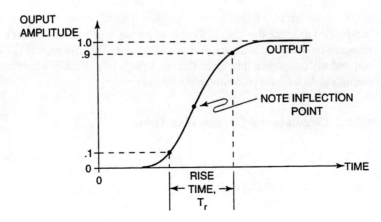

Figure 13.2 System rise time definition

when subjected to an input step function. This waveform must show a point of inflection, where the second derivative with time changes sign, during the rise. This definition only applies to systems having high frequency roll-off slopes of minus two or more (such as -3, -4 or -8). A system with a high frequency roll-off slope of minus one will show an exponential rise to a step input and such a waveform has no defined rise time. Almost all modern measurement system components show transfer functions with more complex roll-offs than minus one, so the discussion is germane.

Why would the rise time of a measured phenomenon be important to your customer, and thus to you as the systems designer? The rise time of a phenomenon has to do with the rate at which energy was transported around, within, or progressed through, the phenomenon itself. This phenomenology is of prime importance to some investigative areas: dynamics, shock, impact or blast phenomena, combustion processes, and high strain rate testing. It may, therefore, be important for your measurement system to reproduce validly the rise time of the phenomenon of interest. How do you prove you are doing this?

Two pieces of system information are needed to support this performance check. First, you need to know the measured rise time of the phenomenon—this you get from your data. Call this time t_{rm} (t for time, r for rise, m for measured). The second parameter you need to know is the system's rise time, t_{rs} (s for system).

You can do the following simple analysis before the test if you know the expected rise time of the phenomenon. This is usually very difficult information to get from your customer. Good luck! A knowledgeable customer could easily be off an order of magnitude on a prediction of his phenomenon's rise time. More often, this check occurs after the test as a part of the data validation process.

13.4.1 Measure the System Rise Time

The measured rise time, t_{rm}, comes from the test data. How do you get the system's rise time, t_{rs}? The first and preferable method is to measure it directly. Here, and as early in the measurement system as possible, input a pure step and note the system's rise time in response to the input. Be sure to use a function generator with a very short rise time (lots of bandwidth) to make this check. This is an example of more is better! Check the system's response on the readout device you use in the test. This rise time is t_{rm}. The system cannot respond any faster than this, since this rise time is a function of your system design and has nothing to do with the measured phenomenon.

13.4.2 Calculate the System Rise Time

In the second method, the system rise time is calculated using an approximation.[3] For a monotonic system component (signal conditioner, amplifier, demodulator, FM/FM tape record/reproducer, FM/FM multiplexer, etc.), the relationship between the rise time and the upper -3 dB, f_u, frequency is:

$$t_r \times f_u = 0.35 +/- 10\%$$

Further, as components cascade into systems, the resulting rise time is approximately the root sum square of the individual component rise times

$$t_{rs} = (t_{r1}^{2} + t_{r2}^{2} + t_{r3}^{2} + \ldots\ldots + t_{m}^{2})^{.5}$$

for components 1 through n. The upper frequency limits can be determined from each component's specifications for -3 dB points in the use condition. This is the method to be used if this analysis is desired as a part of the system design function before the test.

13.4.3 Compare Measured to System Rise Time and Assess Error

In either case, you now have enough information to assess the error in the measured rise time caused by the nonzero measurement system rise time. Obviously, a system with a zero rise time would have no effect at all on the measured rise time. Calculate the ratio of the measured rise time to the system rise time. The larger this ratio is, the smaller the error. The effect shows in Figure 13.3. The message from this graph is clear. As the ratio of the rise times approaches two or three, the errors begin to increase almost exponentially. At a ratio of one, the error is infinite—the system displays its own rise time at its output and you have no idea what the phenomenon's rise time was. At ratios of five or greater, the error is 2% or less. Errors of 2% in rise time are, in my experience, usually acceptable.

These error numbers can be used to correct the measured rise times. The system's rise time can only lengthen the measured rise time, so all corrections are downward in magnitude (tending to lessen the measured rise time).

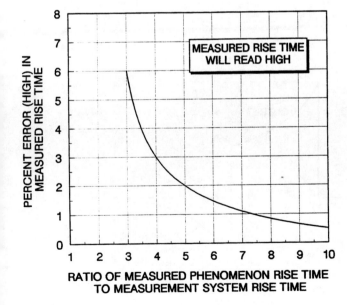

Figure 13.3 Effect of measurement system rise time on measured phenomenon rise time

The bulletproof design criteria for rise time measurements, when this analysis occurs for design purposes, is to design for a system rise time one-tenth as long as the expected phenomenon's rise time.

13.5 UNDERSHOOT CONSIDERATIONS

Undershoot is a measurement system characteristic associated with a low frequency transfer function that does not extend to zero frequency. This system type includes all piezoelectric systems, condensor microphones for sound pressure level, and velocity-based seismometers for example. Undershoot is not a consideration in any measurement system that is DC coupled. If you are designing or using a DC coupled system, you may go on to the next section.

The systems in question all exhibit low frequency roll-offs. Systems that exhibit this behavior will undershoot to a pulse input as shown in Figure 13.4. This undershoot is the

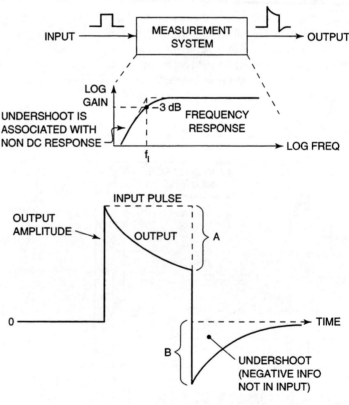

Figure 13.4 Undershoot phenomenon

problem presented in the data validation process. The degree of the problem is a function of how "long" the pulse is compared to the system low frequency behavior that is fixed by design. You can see, in Figure 13.4, such a system with a pulse at its input. The output, as shown earlier in the book for first order or higher systems, will decrease as shown. At some later time, the input pulse returns to zero. When the pulse is at a constant value, the output has decreased by an amount A. When the input returns to zero, the output undershoots by an amount B. What you lose at A, you undershoot at B (B = A), giving a clear error from the measurement system. There is no negative information in the input waveform, but there is negative output. You can see here that the longer the pulse, the worse the undershoot problem, as illustrated in Figure 13.5.

The nature of the undershoot problem, therefore, is the determination of, *"How long is too long?"* This determination can either be done before the test by design, or after the test as part of data validation. It is very easy to do with piezoelectric measurement systems because they can be T-inserted in situ. Simply insert, with the transducer connected, a step voltage in the form of a zero based square wave. The repetition rate should be slow enough to give your pulse train enough "on time" to show you the decay and eventual undershoot. We find that with most piezoelectric systems one to two pulses per second works best. The amplitude of the inserted voltage should give your system its full scale output. The designer or operator of the system must decide how much is too much in this validity check. In this example, I'll use a 5% undershoot error as the allowable criteria. You might choose less or more in your design application.

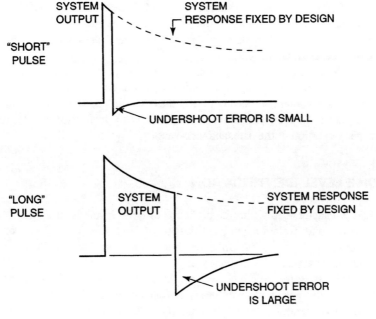

Figure 13.5   Effect of pulse length on undershoot error

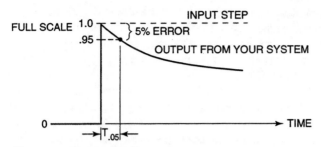

① INPUT FULL SCALE STEP

② NOTE DECAY IN OUTPUT. MEASURE THE TIME, $T_{.05}$, FOR OUTPUT TO LOSE 5% OF FULL SCALE

③ COMPARE WITH YOUR DATA   **Figure 13.6**   Undershoot analysis

You input the square pulse train and observe the system output as shown in Figure 13.6. This observation must be made on a DC coupled instrument such as an oscilloscope or you will become terminally confused! Measure the time it takes for the output to decay to 95% of the input level—a 5% loss in amplitude. Call this time T seconds. You can now state: For your measurement system, for a worst case input (a square pulse) of up to T seconds, undershoot-caused errors will be less than 5% of the input pulse amplitude. Any pulse duration less than T seconds will show undershoot-caused errors of less than 5%.

In reality, the errors you will have made here should be less than the 5% stated above (or whatever your real criteria is) because the square pulse is a worst case input. Your data, by its nature, will likely be more "gentle" than the square pulse—so the undershoot errors should be less than this case.

The last thing needing mention in this section is what part of the waveform you choose to validate any negative excursion in your data (assuming an initial positive excursion—the argument works the same in the other direction). In the simplest case, you do this analysis on the longest positive excursion in the waveform if you are trying to validate a negative portion of the data, and vice-versa.

13.6 NOISE LEVEL IDENTIFICATION

Earlier chapters dealt with the problems of noise levels in measurement systems and the procedures for isolation and identification of those noise levels based on the general transducer model of Chapter 7. Understanding this powerful model allows you to identify noise levels and the path by which they enter your measurement system. If you are using a non-self-generating measurement system, the ability to control the minor input or excitation, gives you design control over the separation of non-self-generated data from self-generated noise. This identification and separation handles several questions noted in the first part of this chapter.

You always have to have acceptable and documented noise levels before your test is run. A procedure for this was also provided earlier in the same chapter. As a general statement in the area of noise level control and documentation, the following is a good guide. Always identify and document noise levels. They are as important as the data. They should be treated statistically only as a last resort, as the use of statistics implies lack of understanding of the deterministic noise coupling mechanisms. Data, without noise level documentation, cannot be validated.

13.7 USE OF CHECK CHANNELS

There are times in experimental mechanics measurements that the use of check channels is mandatory. The primary example of this is when the measurement system uses self-generating transducers of any kind. Here, the designer does not have control over the minor input to the system, the excitation, since self-generating systems do not require this input. This is the single major detriment to the self-generating measurement. At the transducer level at least—whatever comes out, comes out! The major tool for noise level identification at the qualitative level in these cases is the use of check channels.

In the chapter on information conversion when using non-self-generating transducers, I noted that this method could not separate path 2 and 4 responses—the nonself-generated responses to the undesired and desired environments. The use of check channels is one method to separate these responses in a qualitative sense at least.

A check channel is a measurement channel that is as identical with a complete, active channel as possible, with a single exception. That exception is that it is not sensitive to the major input, the measureand. It is subjected to everything else in the environment!—but not what you are trying to measure.

Check channels are carefully designed and installed on the test article to be noise level finders and checkers. They have no other function. Noise levels from check channels are qualitative, not quantitative, in nature. The noise levels from check channels cannot be used to correct the data plus noise on any other channels. The reason for this is that the check channels data and associated active channel data do, in fact, come from different channels. These channels will respond in a quantitatively different manner to the noise causing environment. You, therefore, do not have the right to make a quantitative subtraction of their responses—thus qualitative, but not quantitative.

The information from check channels is used to bound the severity of the problem on the active channels. The information allows you to make a statement like the following that usually bounds the noise level problem: It is highly likely that the noise levels we see on the check channels we have installed on your test article are similar in nature to that occurring on the associated active channels. Your job, as a measurement system designer/operator, is to make that statement true by proper design of your check channel installations.

One way to identify professional measurements people when you are working with them is to look for tell-tale check data channels. Amateurs will tell you they never use them—"Why would you want to do that? It takes up a channel! And besides, we never have any problems." Professionals, who know that the measurements universe is perverse by definition, use them regularly to bound their measurements noise level risk when those risks cannot be measured directly. We have deployed digital systems working in very difficult long-term measurements conditions that have 27 active channels and 20 check channels—a check to active channel ratio of almost one for one!

13.7.1 Resistance to the Use of Check Channels

We participated on a recent major transient dynamics test run in a vacuum environment. There were four other contractors and two government agencies involved in this very complex test with over 500 data channels to be measured. We had a number of check channels simulating quarter bridge strain, full bridge axial and bending strains, full bridge load, and piezoelectric acceleration on board the test article. These check channels allowed us to bound the noise levels that were impressed on the other active channels for which we were responsible. Ours was the *only* organization that demanded and planned check channels on the test. We were also the *only* organization that could, in fact, stand up to the posttest data validation process.

In my experience, there is a dismal and pervasive lack of understanding of the simple power of check channels in some organizations. Many instrumentation organizations, that seem otherwise professional, show a profound lack of this understanding. They run complex, risky tests consistently without check channels. When asked how, in the absence of the crucial data check channels can provide, they validated their data, what you hear is foot shuffling, waffling, and attempts to change the subject.

Foot shuffling and waffling in this context sound like this:

"Does this data look right to you?"

"We're not really sure about these curves."

"We've never had problems before with this stuff."

"Why would that be a potential problem?"

"No. We don't use check channels. We generally expect to lose 30–40% of the data channels on a test anyway." (I actually heard an engineer from another company, a company whose name you would recognize, say this in a meeting attended by our joint customer. As a manager in this business I personally get upset when we lose *1%* of the data on a test! When we lose 1% of the data on a test, we're doing a postmortem to find out why.)

"We were too busy."

"Noise? What noise?"

"Oh, somebody else is responsible for that."

Perhaps they think that check channels, sometimes called dummy channels, are for dummies. In fact, check channels are for smart engineers who know enough about what they are doing to admit: (1) there are uncontrolled noise level risks involved in any test; and that (2) they care enough to identify and bound those risks for the customer.

13.7.2 Check Channel Capacity

Check channel capacity on tests must be guarded zealously by the measurement organization. If, for instance, you were going to run a test on a 100 channel measurement system, you would never share the 100 channel capacity with your customer! You would tell them they were allowed 90 channels of data. The other ten remain in your back pocket for use as diagnostic check channels. This is neither perfidy nor lack of cooperation by the measurements organization. This is the necessary provision of valuable insurance to protect and validate the precious data—the measurements organization's *only* product.

13.7.3 Examples of the Use of Check Channels

Figures 13.7 and 13.8 show examples of check channel information for measurements we routinely make on spacecraft level thermal vacuum tests. These check channels operate in 2-wire resistance measurement systems using both metal foil and thermistor resistance temperature transducers. Any change in the lead wire resistance in a 2-wire resistance measuring system will result in an apparent error in temperature.

Here, you can see the resistance data taken on two precision resistors in the thermal/vacuum environment being read as though they were active temperature channels. The only environment they do not react to is temperature because their thermal coefficients of resistance have been chosen so small as to be negligible. These data show 7300 data scans over a 6.5 day test schedule. Figure 13.7 shows data from "OHMSCHK1," a 350 ohm precision resistor simulating the foil temperature sensor (Measurements Group EB-TG-250BK-500 Balco foil sensors). The reading has been corrected for the room temperature lead wire resistance. You can see the resistance drop off at about 0.5 days as the lead wires cool down in response to the chamber walls cooling to $-300°$ F. The reading remains stable until the chamber walls warm back to room temperature at 6.0 days. The resistance change is about .25 ohm, equivalent to an uncorrected error of .38°F.

Figure 13.8 shows similar data for "OHMSCHK2," a 1014 ohm precision resistor simulating negative coefficient thermistors (Yellow Springs, Inc. 440003A) also used in this type of testing. The lead wire resistance change shows the same characteristic as the previous figure for the same reason. However, due to the higher sensitivity of the thermistors, the errors are about 0.01°F. Since all lead wires were insulated, and had similar routing and radiation exposure, the errors noted by these dummy channels tend to bound this non-self-generated response (resistance change) to an undesired environment (radiative heat transfer to the cool chamber walls).

Figure 13.9 shows a self-generated noise levels from the system noted above with 27 active channels and 20 check channels. This system measures a number of bending moments and torques on some delicate hardware during integration. The problem is that the moments at which we want overload warnings to go off are of a rather small magnitude. The zinger is that after the system's initial electrical and mechanical zero is established, it is *18 months* before it can be reestablished to check the system zero shift! We have taken exquisite measures in the system design to be able to track a number of these self-generated drift noise levels.

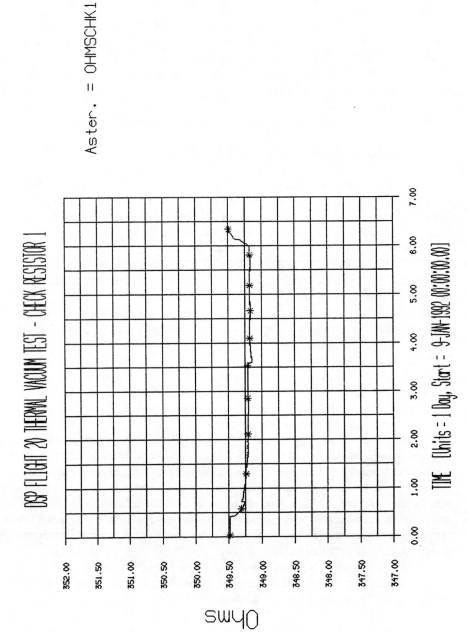

Figure 13.7 Check channel from a thermal vacuum test

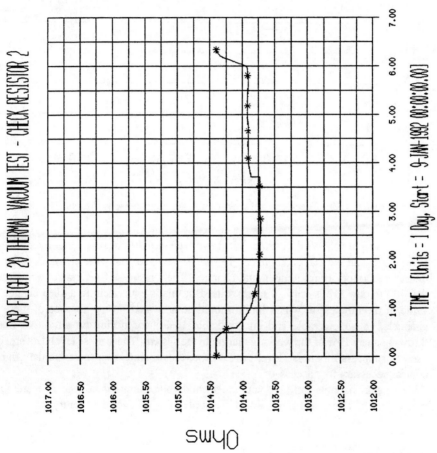

Figure 13.8 Check channel from a thermal vacuum test

325

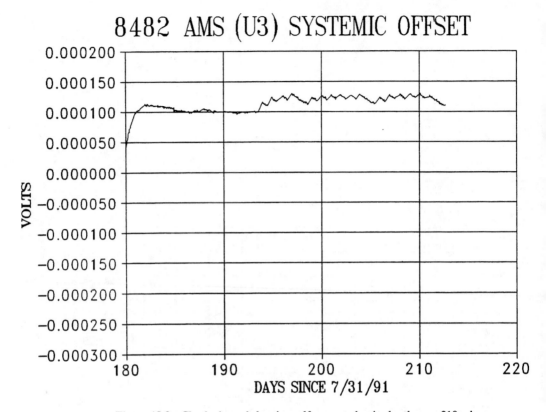

Figure 13.9   Check channel showing self-generated noise levels over 210+ days

This figure shows the system drift in voltage into the A/D converter over a 210+ day period. You can see that on about day 195 we began to see daily variations associated with diurnal laboratory room temperature changes. The system self-generated zero shift from day 180 to day 212 is only 70 microvolts! But this is equivalent to a 10% change in the value of the moment at which limits should be activated. In the absence of these check channels, an undetectable and uncorrectable 10% error occurs on all active channels, throwing away 10% of the working margin in the system. These noise level data are, in fact, used to suppress this noise level in real-time, thus maintaining the extremely tight zero control necessary for success.

Check channels can save your career and your client's data. It amazes me that this practice has not been institutionalized in test planning all over the country.

13.8 THE DATA MUST MAKE SENSE

Valid experimental data makes sense in the physical world. In the reverse, this relationship is valuable to the measurements engineer in determining the validity of the data. The relationship is reversed by asking the question, "Does this data make sense in the real world? If so, then the data are likely valid." For the measurements engineer, the data validation test thus becomes one of sensibility—does the data pass the sanity test, does it make sense?

Usually, these tests are based on the laws of physics that govern the process under investigation. Understanding and use of these laws allows the measurements engineer a sure foundation on which to test the data. No one is going to argue with the laws of physics. I've included examples of "Make Sense?" checks from the static and dynamic testing worlds.

13.8.1 Example from the Static Test World

A simple example would occur in a static test of a spacecraft structure mounted upon a trusswork that adapts it to the launch vehicle interface. The physical laws invoked are those of statics. In a developmental static test the trusswork would generally be strain gaged to measure the tension and compression loads in the truss members. During the test, loads would be input to the spacecraft structure, simulating launch conditions, and reacted out through the supporting trusswork. These input loads would typically be measured accurately using load cells. The "Make Sense?" validity check would involve calculating the six degree of freedom loads and moments reacted out through the trusswork with the vector loads and moments being input. This simple check serves two functions: (1) it serves as the overall sanity check, and (2) it serves to assess the total measurement errors being made in the reaction loads.

13.8.2 Examples from the Dynamic Test World

Suppose you had just run a test where you had measured the time history of deflection of a test article as shown at the top of Figure 13.10. If you were to measure the slope of the time history of deflection at every data point and plot that, you would have the middle plot. This is the velocity time history for the test article. If you were to measure the slope of the velocity time history at every data point you would get the bottom plot, the acceleration time history. You have, of course, just graphically differentiated the deflection time history twice.

You also can play this game in the other direction. The net integrated area under the acceleration time history is the velocity time history. The net integrated area under the velocity time history is the deflection time history. In practice, there are several limiting caveats with performing these integration and differentiation steps, usually concerning the contaminating presence of noise in the time history. These are simple reflections of the laws of physics called rigid and flexible body dynamics. Real test articles must follow these rules.

These particular rules are perhaps easiest to grasp when applied to pyrotechnic shock testing. In this testing a structure is subjected to shock loads and acceleration associated with the extremely fast energy release from a pyrotechnic device such as a bolt cutter, a pin puller, or a panel cutter—something that goes bang loudly. In a pyrotechnic shock test on the ground the test article begins and ends at zero velocity. Before the shock, the test article is sitting there in the lab. After the shock, the test article is still there in the same position. Its net velocity and deflection changes over the test period is zero—by definition.

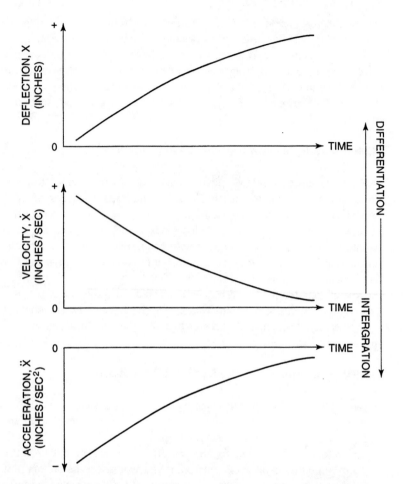

Figure 13.10 Deflection, velocity, and acceleration as functions of time

Since there is no net velocity change, the integrated area of the acceleration time history must be zero. Observe the time history at the top of Figure 13.11. Here, all the plus areas tend to equal the minus areas and your eye tells you they should add to zero. This acceleration response is possibly valid for a pyrotechnic shock input. The time history at the bottom of the figure is, however, invalid because the area under the curve simply cannot integrate to zero—there is a lot more positive area than negative area. This response is physically impossible for a zero velocity change pyrotechnic shock input.

Consider a drop shock test. The test article drops onto a hard surface with a controlled kinetic energy and the response is measured. Here, there *is* a net velocity change. The test article begins the test immediately before the shock at some finite velocity and ends the test at zero velocity. The physical implication is that in-line shock accelerations must not integrate to zero. They must integrate to the net velocity change. The conclusions for the wave-

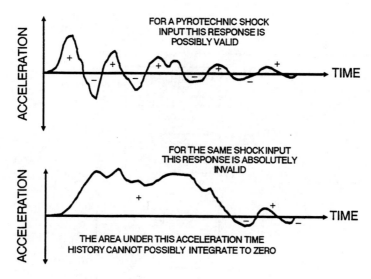

Figure 13.11   Valid and invalid pyrotechnic shock acceleration time histories

forms shown in Figure 13.11 are reversed in this case. For a drop test the upper acceleration time history is not possible since the area under the curve clearly can integrate to zero. The lower curve is possibly valid in this case since its integrated area is clearly nonzero, as it should be.

Pyrotechnic shock data has the unique and appealing quality of being internally self-consistent. There is generally enough information in the shock transient itself coupled with the applicable laws of physics to validate the data without resort to other sources.

Author's note: In measurement engineering the data speaks, in terms of its validity, for itself. Validity checking uses the data itself, the known performance characteristics of the measuring system and the laws of physics. Nothing else is needed. If the measurement engineer needs to go to the customer and ask, "How does this look to you? Could this be right? How does it compare to your prediction?," he/she has flunked the course. When the customer becomes the arbiter of data validity, the measurement engineer has failed.

Figure 13.12 shows the process flow charts we use for automated pyrotechnic shock data validity checking. These validity tests are applied to the sampled time histories of response acceleration measured on the test article. The process is as follows:

1. Measure the pre-event AC noise levels and the DC level.
2. Suppress the pre-event DC level. You need to do this so that subsequent zero shift in the data can be found if it is there. This is clear evidence of a problem.
3. Compute the plus and minus peaks of the transient as a percentage of full scale.
4. Perform a ranging check. Is the data between 5% and 100% of full scale? Yes? Go on. No, document a ranging problem.

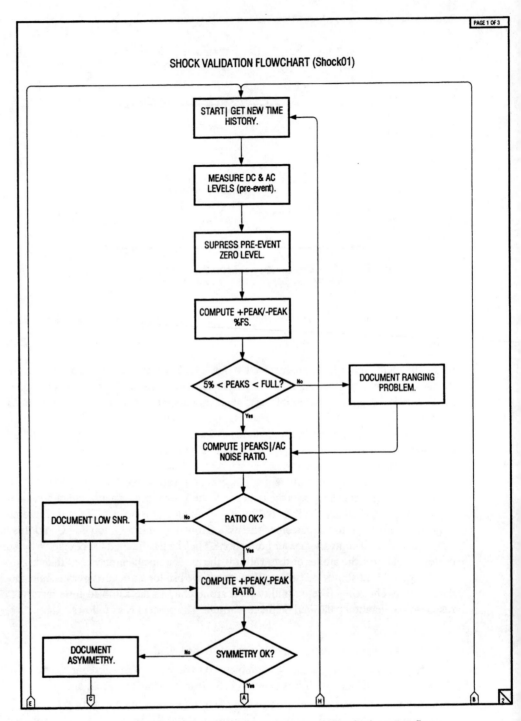

Figure 13.12 a,b,c Automated shock validity check process flow

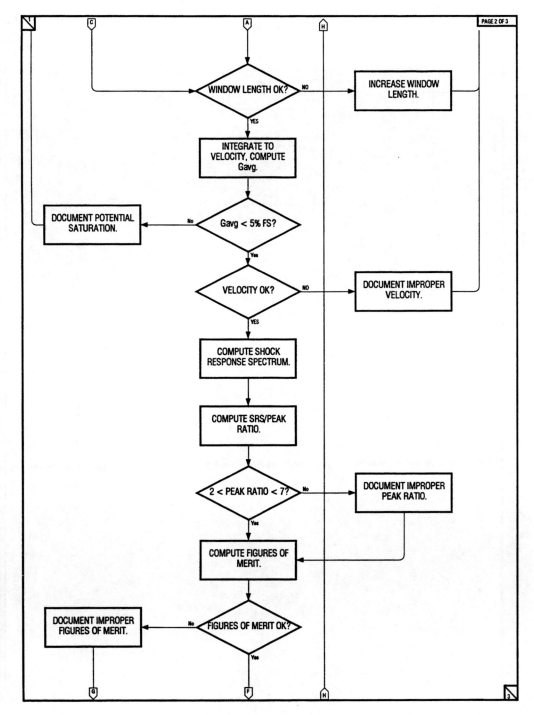

Figure 13.12   Continued

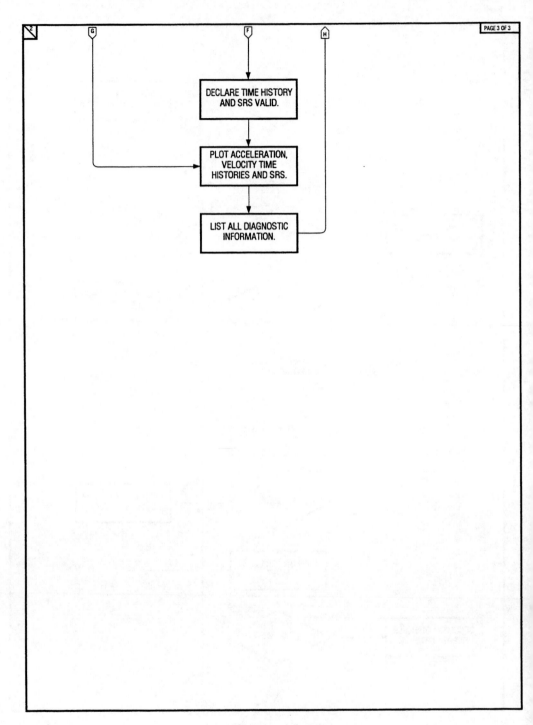

Figure 13.12 Continued

332

5. Check the data levels versus the AC pre-event noise levels and calculate the signal to noise level ratio. Document.

6. Compute the plus maximum peak to minus maximum peak to document symmetry.

7. Check window length. Has the transient decayed into the pre-event AC noise levels by the end of the window? If not, reacquire the transient with a longer window.

8. Only now do you integrate to velocity and compute the algebraic average acceleration, G_{avg}. Is G_{avg} more than 5% of full scale? If yes, then document a channel saturation problem and bail out. There is no point in taking saturated data any further in the process. It goes in the trash can.

9. Inspect the velocity waveform. Is it OK? If not, document improper velocity and bail out.

10. Only after passing the previous nine tests do you commit to computing the shock response spectrum. Compute the spectrum.

11. Compute the ratio of the SRS maximum to the maximum peak in the time history. Test for valid limits for this ratio. Document any problems.

12. Compute other figures of merit. Test them versus limits for known validity. Document any problems.

13. Only after passing the previous tests do you declare the time history and the SRS valid. Plot and archive all data.

14. List all diagnostic information.

15. Do it all over for the next response.

Tabulated results of automated shock validity checking show for six acceleration channels in Figure 13.13. Note the pass/fail summaries listed in the far right columns showing how many validity tests a time history failed. These data were chosen, by the way, to provide some invalid data demonstrating the feature.

These kinds of automated validity checking procedures can really make a difference on the test floor. They take most of the subjectivity out of the validity checking process. This is particularly true if you have involved your customer in the development of the methodology. Then it is their process as well as yours.

13.9 WHAT DOES IT ALL MEAN TO YOU?

The need for data validation techniques is not limited to either analog or digital measurement systems. The methods and procedures are similar in either case. Data is either valid or it is not. The rules for validity checking do not depend on this artificial analog versus digital distinction. However, in digital measurement systems the power of automated data validity checking can reach its full potential when these checks are implemented in software. The tricky part here is casting the knowledge of the measurement engineer into software. In this sense, automated validity checking requires an expert system. At the first level at least, it does not require a learning system. The validity checking rules can be cast effectively into

RESPONSE OF	PLOT #	+PEAK	PEAK XFS	-PEAK	PEAK XFS		SNR (dB)	GAVE	GAVE %PEAK	GAVE %FS	DECAY %PEAK	+SLOPE	-SLOPE	PEAK RATIO	SYM	FOM1	FOM2	FOM3	ERR CNT	P/F
ACCEL 1(X)	9554	1788.	17.9	-2517.	25.2	OK	45.5	-12.0	0.48	0.12	8.4	0.82	0.84	3.5	1.41	12.3	6.6	13.1	0	PASS
ACCEL 2(Y)	9555	1916.	38.3	-1016.	20.3	OK	49.2	-45.9	2.39 F	0.92	8.4	0.94	0.93	3.4	1.89 F	44.1	28.3 F	51.6 F	5	FAIL
ACCEL 3(Z)	9556	3498.	35.0	-3113.	31.1	OK	51.8	4.1	0.12	0.04	7.9	0.90	0.89	4.4	1.12	2.4	1.2	4.4	0	PASS
ACCEL 1(X)	9557	1816.	18.2	-2513.	25.1	OK	44.3	-6.0	0.24	0.06	7.6	0.82	0.85	3.3	1.38	6.3	3.2	6.6	0	PASS
ACCEL 2(Y)	9558	1656.	33.1	-1151.	23.0	OK	48.3	12.9	0.78	0.26	15.8 F	0.93	0.92	4.1	1.44	14.2	6.6	14.7	1	FAIL
ACCEL 3(Z)	9559	3350.	33.5	-3535.	35.3	OK	48.9	22.7	0.64	0.23	8.9	0.93	0.97	4.2	1.06	10.7	5.1	24.2 F	1	FAIL

Figure 13.13 Tabulated results from automated shock data validity checking software

software rules. The potential application of neural network software in validity checking is still in its infancy—but may turn out to be significant.

If you are responsible for the digital hardware/software portions of the data acquisition system, there is a major opportunity you share with the systems designer. That opportunity is to create a set of validity checking tools in the software that is aware of two things: (1) the system's design and performance capabilities; and (2) the measured data output. The data output can then be automatically compared with the system's inherent design capabilities, and validity quickly checked in software in near real-time.

For example, the system could be aware of its own rise time capability (this answers question #1 in the section on spies). It could automatically check what it measured against what its capabilities are, and an assessment made of the error in the measurement. The rise time could even be corrected automatically.

The system could be made aware of the transfer function's magnitude and phase versus frequency (questions #2, 3, 4). It could then determine the frequency content of the data via FFT techniques, and compare that content to the transfer function's capability. The assessment of the transfer function's effect on the measured data could then be made automatically. A correction is even possible.

The system could be designed to track its own internal electrical zero to determine if it had shifted (question #5). This data could be used to assess overall system zero shift and isolate its cause. Did it come from the phenomena or transducer, or from the electrical parts of the system via saturation? We have already implemented systems operating on this principle.

The system could be designed to isolate and document certain noise levels on the fly (questions #8, 9, 10). Such noise level data is necessary to validate the data output. Unless you can answer the question, "How much of this information is noise?" you can't validate. Your customer has the right to ask that question and you must be prepared to answer it.

If you are responsible for the overall design of a measurement system, these are a subset of the questions and areas of concern that must be addressed in your design. Remember, they are application specific—there might be a different set of questions for your application. But there are a fundamental set of questions that demand answers in your design. Notice that a significant portion of the overall concern must be handled before you ever get to a digital component. The data problems created by lack of concern for these issues may be impossible to handle once the transition to the digital realm occurs. You may have already crossed the Rubicon and cannot compute your way out of the problem. Even worse, you may not even know your system has the problem.

The second major opportunity is to instruct your digital hardware/software person to include the right set of effective validity checking tools in your system design. Such tools asking and answering the right questions go miles in convincing your customer that you and your design/test team know what you're doing and that your data are valid.

In closing this section, I need to tell you what we do when, despite all our care, we still cannot prove the data is valid to our own satisfaction. What do you do as a last resort? What do you owe your customer in this case?

You owe Figure 13.14. This figure shows a PSD plot that we could not validate because of the suspiciously high low frequency response. This was caused by a loose acceler-

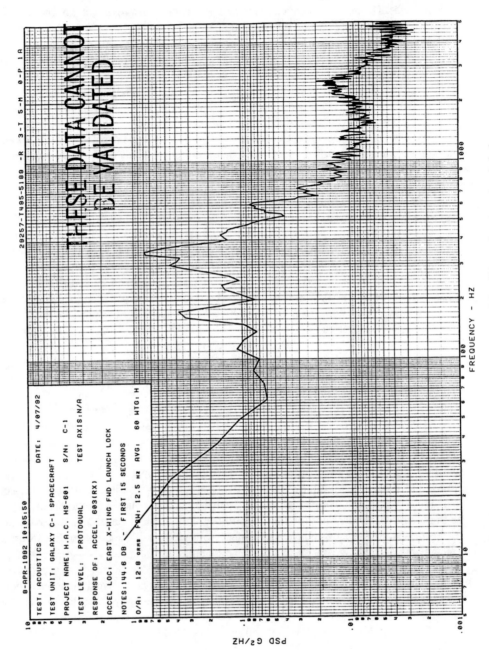

Figure 13.14 Autospectral density analysis of invalid acceleration signal during an acoustics test (loose connector)

336

ometer connector buried in another company's spacecraft on a fixed price acoustics test we ran for them. Note that this plot has been stamped "THESE DATA CANNOT BE VALI-DATED." We have rubber stamps in each laboratory to handle this case. And we use them now and then too. We cannot prove these data are valid, since they are highly suspect given the known loose connector. We have informed the customer of that with the stamped message. The stamp says, "Customer—use these data at your risk. We cannnot prove it is valid." My policy is always to tell the customer the absolute truth about the worth of the data product—even if we don't like the message. You owe your customer that truth.

NOTES

1. This general data validation issue was originally presented by Peter Stein (Stein Engineering Services, Inc., 5602 E. Monte Rose St., Phoenix, AZ, 85018–4646, 602/945–4603). I've expanded it here with his blessing.
2. These ideas first published by the author in *Data-Acq Systems Aren't Perfect, So Ask the Right Questions during Data Analysis*, Personal Engineering and Instrumentation News, PEC Inc., Rye, NH, December 1991, pp. 73–77.
3. Peter K. Stein, *Measurement Engineering Volume I*, Stein Engineering Services, Phoenix, AZ, 1964, p. 134.

Chapter 14

Knowledge-Based Systems for Test Data Acquisition and Reduction

Author's Note: In this publication,[1] the term data acquisition system is used interchangeably with measurement system. The audience for this publication was not aware at the time of the hierarchical difference noted earlier in this book. Although this paper was written from an aerospace perspective to an aerospace audience, the measurement system design philosophy presented here is absolutely generic. Please note the style of this chapter differs somewhat from other chapters of this book. It was written as a technical paper for presentation at the premier aerospace testing symposium in the country and is printed here verbatim.

Modern computer-based measurement systems for static, dynamic, and thermal testing featuring high sampling rates and large channel capacity are capable of literally swamping users with information. Further improvements in speed (here defined as analyses or plots per hour) may only increase this problem rather than lessen it. Such "speed" improvements appear to be a major thrust of some commercial and in-house data acquisition system designers. What is really needed by *users* are improvements in the *utility* (usefulness) of the information provided leading to improved understanding and real-time decision making during testing.

The Measurements Engineering Department of TRW's Space and Technology Division is designing and building test data acquisition systems which are based on the requirements from the users of these systems and the customers for the information. These data acquisition systems are designed to provide significantly increased data utility at rates which support optimum assimilation and understanding when it is important—during the test. They are, in a sense, based on the knowledge bases of the designing measurements engineer *and* the customer. The key design features of two major "Knowledge-Based" data acquisition systems are described with the attending system design philosophy.

A method for effectively obtaining the important design requirement specifications from the customer for the data is also described.

14.1 THE MYTH OF SYSTEM "SPEED"

For the purposes of this paper, we define the "speed" of a digital data acquisition system as the rate at which it can produce report-ready hardcopy (PSDs, FFTs, shock response analyses, time histories of temperature and heater powers, page listings, plots of stress/strain/deflection versus load, etc.), call it plots per hour. Let us not kid ourselves—we still produce data on paper! During a recent 34 day systems level (author's note: full spacecraft) thermal/vacuum test the thermal data acquisition folk generated 21,000 pages of plots and listings which included only spacecraft temperatures, heater powers, and absorbed radiometer flux. This represents a stack of paper six feet tall and was generated by a rather slow automated thermal data acquisition system. Is this the right way to do business in the future?

When considering the design of a digital data acquisition system, there is a general and alluring tendency to think that if a little system speed is good, a lot more must be better! Vendors of data acquisition hardware and front-end data processors are making amazing strides in these areas. Systems based on these types of front ends can generate literally hundreds of plots per hour. If 200 plots per hour is good, then 500 plots per hour must be *better!* If we can generate a six-foot stack of paper with a relatively slow data acquisition system, think of what we could do with a *really fast* one!

The attached cartoon, Figure 14.1, illustrates the point. There is a tendency for the systems designer/owners of the Mk.27 Super Fast Automated Data Acquisition System (lovingly known as the Mk27SFADAS) to increase its performance on the basis that if 200 PSDs per hour is good, then 500 per hour would be terrific!

This approach forgets the poor guy on the right—the customer responsible for assessing the data and the test. This is the guy who has to look at, interpret, and understand the 21,000 pages of thermal data or the 500 PSDs per hour. This poor fellow is in the hot seat under intense pressure not to make a mistake or miss any important issue in the test—and he is being literally *buried* in data which may or may not help him in real-time high leverage decision making. The project type on the right is familiar to all of us.

The "FASTER MUST BE BETTER" approach comes from the unspoken assumption that the utility of the data produced increases forever with increases in system speed. In this (design) model, 500 PSDs per hour is more useful than 200 PSDs an hour, simply because they are there. This is shown in Figure 14.2 as the linear line heading for the ceiling.

A design approach which we feel has a significantly improved probability of really making a difference is called the "FASTER MIGHT BE BETTER APPROACH," or "FAST IS FAST ENOUGH." This approach takes into account the human being(s) who is the assessor, the interpreter, the assimilator, and the judge of the data and the test. This design model focuses on providing data as fast as is really needed and in forms which assist real-time decision making and problem solving during the test. This is illustrated by the

Figure 14.1 Is faster necessarily better?

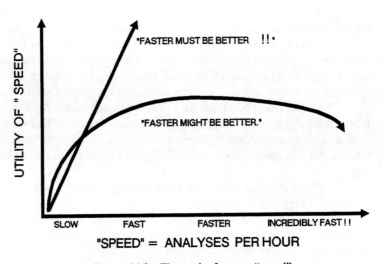

Figure 14.2 The myth of system "speed"

curved line in Figure 14.2. In this model, the rate of increase of utility goes down after a certain critical speed is reached. In this model it makes no sense to spend money increasing speed from 200 to 500 plots per hour if 200 is all the customer can assimilate. Spend the company's hard earned money giving yourself tools to conclusively identify invalid data so it never enters the design assessment process and your customer tools to assist him in that process. That is what Knowledge-Based data acquisition systems are about.

14.2 KNOWLEDGE-BASED SYSTEMS: WHOSE AND WHAT KNOWLEDGE?

We wish here to use a less than rigorous definition of Knowledge-Based systems. It is, however, a definition which is very useful in designing data acquisition systems which really work effectively in the laboratory. (Author's note: We fully admit that this definition is a mechanical engineer's definition—a measurement system user's definition, and not the more rigorous software engineer's definition.)

By a "Knowledge-Based" system, we mean a data acquisition system whose design is securely founded on the knowledge and experience of the design and using engineers and the customer community for the data. That, on the face of it, sounds like common sense at best, and motherhood at worst. It is surely the former. Experience has shown that vendors of data acquisition system hardware, software and integrated systems understand the hardware and software of their systems. They do not understand what makes a *real difference* in your test laboratory. They are not test people—they do not understand what testing is about. They understand what their products are about. It is not their fault because they are playing by a different set of rules than we are. A commercial vendor has a totally different goal in mind when designing a data acquisition system or component (unless it is a one of a kind system for you). That goal is driven by the marketplace and is to sell lots of stuff to a wide variety of customers. By this set of rules, you design systems which are characterized by extreme flexibility and minimal efficiency. Your goals are diametrically opposed in a major aerospace test laboratory. You want efficiency because 90% of the time data acquisition system requirements are predictable and repetitive and you are under continuing and severe pressure to cut the cost of doing business while maintaining the quality of your data product.

Knowledge-Based data acquisition systems here are characterized by: (1) extremely close attention to the expressed needs of the data customers; (2) an ability on the part of the designer to translate the data customers' needs into appropriate functional requirements (to see into their problems and understand the underlying physical principles); and (3) two unique and additional sets of (generally) software tools. The first set of tools is designed to perform real-time validity checking on the data; the second, to enhance the data's utility to the user.

14.2.1 Validity Checking Tools

These software tools are designed to discover and identify data which are invalid and prevent their use in test data assessment and the subsequent design verification processes.

These tools are based on sound measurements engineering principles and the laws of physics. Examples of a complete set of validity checking tools are discussed for a new, major automated data acquisition system for shock, vibration, and acoustics testing with a focus on shock data. These tools should be designed into a system and not be optional to the operator. The data must pass through the validity checkers before being declared valid and released.

14.2.2 Utility Enhancement Tools

These software tools are designed to allow the data user to *focus* on the data, or problems with the data, which are important *now* to test problem solving—while the mass of "non-critical" data are being tended to appropriately. A complete set of these tools are discussed for the dynamic data acquisition system noted above and for an Advanced Thermal Measurement System supporting systems level thermal/vacuum and thermal cycling tests.

14.3 THE INFORMATION UTILITY HIERARCHY AND SPACE

In order to generalize our thinking about Knowledge-Based systems (regardless of what they measure) and how they should fit into an overall test laboratory strategy, we have devised the Information Utility Hierarchy. This hierarchy defines the increasing levels of test data utility: How useful is this data *now* in the conduct and assessment of this test and this design? This has been a very useful tool for us in keeping ourselves on the design track. The hierarchy is shown in Figure 14.3. The levels are as follows in increasing order of utility:

LEVEL 4: INTEGRATE THEORY AND EXPERIMENT
  ▨ real – time comparison and validation of test data
    with theory; possible theory modification on – line

LEVEL 3: INSIGHT
  ▨ process is intuitively and quickly understood;
    the "AHA!" experience

LEVEL 2: OBSERVATION
  ▨ process objectively observed

LEVEL 1: VALIDITY
  ▨ data faithfully describes process

LEVEL 0: IGNORANCE
  ▨ no data available

Figure 14.3   The information utility hierarchy

Level 0: Ignorance

No data is available.

Level 1: Validity

Data acquisition systems operating at this level provide data which is demonstrably valid. It clearly and faithfully describes the process under investigation. This is the most basic job of the designer/user of any system, digital or otherwise. Do not pass go and do not collect $500 if your data is not valid.

Level 2: Observation

At the Level of Observation, systems provide data in a form which allows the process to be objectively observed in an overall sense. The data acquisition system is not allowed to give its "opinion" about the data. This is a fairly high level at which to operate.

Level 3: Insight

At the Level of Insight the data acquisition system provides information in a form allowing the process to be intuitively and quickly understood.

Insight, n.: the act or result of apprehending the inner nature of things or of seeing intuitively

Intuition, n.; immediate apprehension or cognition

These definitions should be very pleasing to any engineer. We have all had experiences at the Level of Insight. This is the *"AHA!"* experience. Knowledge-Based systems are designed to support the *AHA!* experience and operate, as a minimum, at the Level of Insight.

Level 4: Integration of Theory and Experiment

At this highest level, data acquisition systems are designed to allow the integration and/or comparison of experimental data with theoretical predictions. A data acquisition system operating at this level allows the customer to leave the test laboratory with valid test data, a verified and valid test which, without doubt, did or did not meet the test objectives, and a verified and updated analytical model.

In summary, most modern digitally based data acquisition systems operate at level 2, the Level of Observation. Knowledge-Based systems are designed to operate at levels 3 and 4 providing information as well as data.

An example is in order to appreciate the quality of the difference between Knowledge-Based and non-Knowledge-Based data acquisition systems. We are developing an Advanced Thermal Measurement System (ATMS) to support future systems level thermal tests. A convenient way to appreciate the shift in thinking required to develop these systems is to look at the Information Utility Space.

The Information Utility Space for ATMS is shown in Figure 14.4. Here, the Information Utility Hierarchy is plotted versus a time epoch and a parametric input which is temperature in this case. The time includes: the previous past or historical data from a previous similar test or spacecraft flight; the present past or historical data from the test presently being run; "now" or the data which has just been taken; and the future. System cognizance of the previous past and the future represent new capabilities. We will add a cryogenic

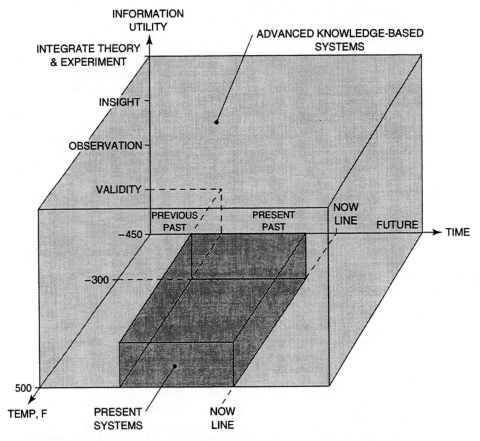

Figure 14.4 Information utility hierarchy for an advanced thermal measurement system

temperature measurement capability so the temperature axis goes from 2°K to about 500°F, the upper limit for typical spacecraft temperatures.

We surveyed our seven leading competitors about their capability in thermal data acquisition as part of our homework on this effort. The systems of all eight companies fell within the volume noted as "Present Systems." All of these systems operate from −320°F to 500°F on the temperature axis and in the "present, past," and the "now" on the time axis. As previously stated, they also operate at the Level of Observation.

Knowledge-Based advanced systems, at least in the thermal area, must significantly expand this operating volume as noted in the figure. Our ATMS will operate from cryogenic temperatures to 500°F and will include cognizance of the present, past, and the future. Knowledge-Based systems will operate at level 4, the level of integrating theory and experiment. A comparison of the relative volume is analogous to the size of the utility-of-information leap we expect to make in this area.

14.4 THE DATA ROOM AUTOMATION SYSTEM (DRAS)

Let us begin by saying that we all hate the acronym DRAS but could not come up with a better one. So DRAS it is!

DRAS is an automated, data base driven, digital data acquisition system supporting vibration, shock, and acoustics testing in all our major facilities. All necessary analyses, with the exception of swept sine data acquisition/reduction, are in-place and running. Swept sine will come later. (Author's note: by the time of printing swept sine analysis will have been implemented, completing the DRAS suite.) It is a totally automated system driven from a DEC Micro-VAX II processor. DRAS supports up to four (Author's note: now five) dynamics tests on a noninterference basis and features preprocessors for each test. The system runs on a "no pencil" basis with all recording/playback of test data, and recording/display of test logs, data logs, run logs and report-ready hardcopy under operator control via CRTs at the four (five) test stations. All analyses are archived in report ready format for retrieval on a gigabyte optical disc which will store 100,000 analyses (Author's note: the system presently has over 45,000 analyses archived. We generate 10,000–15,000 dynamics analyses per year). Extensive sorting capabilities are provided so that even a single unique analysis can be easily located and output to a laser plotter. DRAS allows this laboratory to operate at a maximum rate of 24,000 analyses per shift per year.

There are two basic ground rules for hardcopy outputs for any Knowledge-Based data acquisition system. Rule #1: Output hardcopy plots in a manner in which the data stands alone. This means that if a single unclassified PSD were found in the parking lot after three years, there would be enough information on that analysis to totally identify the test during which that analysis was generated. Data must be useful to your customers years after we think it is obsolete. Rule #2: Format your plots so they speak to your customer and address information important to him—not you.

Figure 14.5 is an example of an analysis which breaks both rules. This happens to be a 2 KHz PSD from the digital output of a reputable multichannel FFT analyzer. Look at the information it gives in addition to the PSD itself. It says "SETUP GRP SPEC LWR VW" as well as "DG +30 DB WTG HA 1V" and "AVG 11.1 DELTA P:9.10EU." This, in fact, is information of critical importance to the measurements engineer providing the data. It is of no use whatsoever to the customer nor is it generally understood by him. A Knowledge-Based system such as DRAS outputs the same information as shown in Figure 14.6. The title block includes that information which is of interest to the measurements engineer *and* the customer. It is, in fact, in customerese. It has been designed to exactly overlay analyses which were done on analog and hybrid systems before DRAS was even conceived and which will likely never exist in digital form. This is a clear and simple example of increased data utility.

14.4.1 DRAS Validity Checking Tools

DRAS has certain of its validity checking tools running now and will have a fairly complete set when the Phase II implementation is complete. The DRAS validity checking matrix will look like this when complete:

DRAS VALIDITY CHECKING MATRIX

Validity Check	Swept Sine	Random Vib/Acous.	Shock
A/D ranging	X*	X*	X*
Cable noise	X		
DC offsets	X*	X*	X*
Span verifications	X*	X*	X*
In-line responses	X*	X#	X*
Cross axis responses	X*	X*#	X*
Q magnitude	X*	X	X*
Sweep rate effects	X*		
SRS initial slope			X*
SRS peak ratio			X*
Time history integration			X*

* = operational
# = used for random vibration only—not used for acoustics

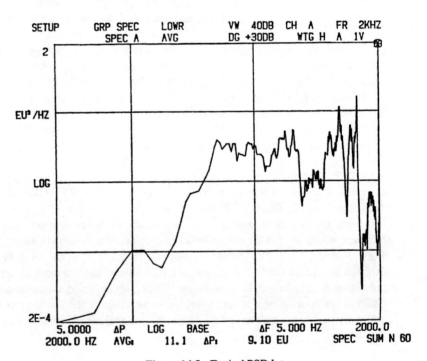

Figure 14.5   Typical PSD lot

A/D Ranging Check.   This check is run during the digital sampling process and notes on the plots whether A/D full scale limitations were reached at any time.

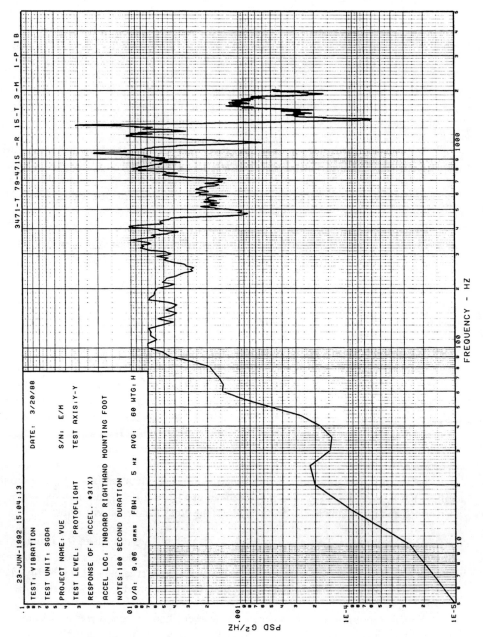

Figure 14.6 Same PSD plot in Knowledge-Based system

347

Cable Noise Check.   This check is associated with swept sine testing and is designed to locate loose transducer connectors prior to commencing sweep. After the test article has been brought to level, but before the sweep commences, all accelerations will have FFTs performed looking for the high low end spectrum characteristic of loose connectors. Summary results are displayed for the operator before the sweep commences.

DC Offset Check.   Reports any DC offset in the sampled time histories prior to the imposition of the dynamic stimulus on the test article. Out of tolerance results are displayed for subsequent corrective action.

Span Verification.   Dynamic calibration signals are measured at calibration time prior to and after the test run. Out of tolerance results are displayed for corrective action.

In-Line Responses.   This check is used for uniaxial swept sine and random vibration tests and occurs after all analyses for a test run are complete.

The archived analyses are automatically checked to verify that in-line responses match the control input below the first structural resonance. Any responses not meeting this criteria are flagged. This (check) finds misinstalled or mispatched transducers.

Cross Axis Responses and Sensitivity Checks.   Both checks occur after all analyses for a test run are complete. The first check verifies that cross axis acceleration responses are less than control inputs below the first in-line resonance. This also finds misinstalled or mispatched transducers.

The second check is based on the acceleration transducers' nominal cross axis sensitivity. A typical (high impedance) commercial piezoelectric acceleration transducer has a maximum specified cross axis sensitivity, usually 3%. In the presence of an arbitrary cross axis acceleration field of 100 units with zero units in the sensitive axis, no more than 3 units of acceleration will be reported by the transducer. It has just created 3 units of acceleration in the presence of zero, clearly an error. In a triaxial acceleration field measured by a triaxial set of accelerometers, the upper bound of this (noise) acceleration can be calculated as a function of frequency and noted on the analysis as shown in Figure 14.7. The check says that "below this acceleration for this transducer in this acceleration field, the data are invalid." For a uniaxial acceleration measurement in a triaxial field, this analysis cannot be performed and no statement can be made regarding cross axis sensitivity errors.

Q Magnitudes.   This check is made after analyses are complete. It calculates the structure's measured "Qs" with reference to the control input and checks them versus realistic upper and lower limits for this variable. If a Q of 11,000 is calculated, it is highly likely that there is a setup error on the test. Out of limit Qs are reported to the operator.

Sweep Rate Effects.   This check is made after filtered swept sine analyses are complete for a test run. The check is made using Trull's method[2] by calculating the amount of frequency shift and resonance broadening caused by the interaction of the nonzero sweep rate and the measured resonance bandwidth (see Figure 14.8). Resonances which are significantly affected by sweep rate are reported to the operator.

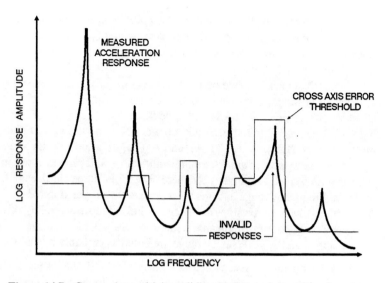

Figure 14.7   Cross axis sensitivity validity check (requires multi-axis measurements)

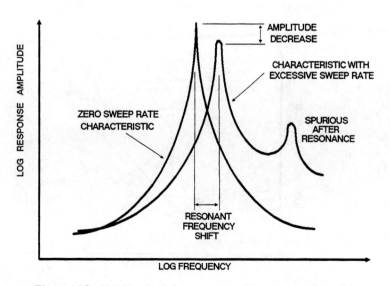

Figure 14.8   Validity check for sweep rate effects (sine testing only)

Shock Response Spectrum Initial Slope.   Reproducing the waveshape of a (pyro-technic) shock event is one of the most challenging jobs of the measurement engineer. The rules one must follow to perform this feat are some of the most stringent in experimental mechanics measurements.

The first check is made on the leading slope of the shock response spectrum in the midfrequency range, from 100–500 Hz nominally. Smallwood[3] shows that a pyrotechnic shock response spectrum (SRS) will have three distinct regions of slope leading up to the high level responses found generally above 1 KHz as shown in Figure 14.9. These frequency ranges have slopes of +1, +2, and +3 (+6, +12, and +18 dB/octave) respectively. In practice, these responses are discernible only with a well-ranged transient with correspondingly low noise levels—but they are there. The initial slope check calculates the SRS slope in the midfrequency range and checks it against the proper slope of +2. It is impractical to perform this check in the slope = +1 region due to the susceptibility of the analysis in this frequency range to measurement system noise. SRSs which deviate significantly from the proper slope in the midfrequency range are noted as probably invalid.

Shock Response Spectrum Peak Ratio Check.   This check calculates the ratio of the maximum peak of the shock response spectrum to the maximum algebraic peak in the measured shock time history as shown in Figure 14.10. For pyrotechnic shocks the valid range for this ratio is defined as from 2 to 6. For impact shocks (simulating pyrotechnic shocks and specific to a particular shock testing machine) the valid range is from 2 to 4. SRSs whose peak ratios deviate from these ranges are noted as probably invalid.

Time History Integration Checks.   Time history integration checks are run on each acquired shock time history (typically 4096 points). The time history is integrated and the net area (area = G-milliseconds) is calculated. For a zero delta velocity shock this area must theoretically integrate to zero. If it does not, there is net velocity remaining after the shock.

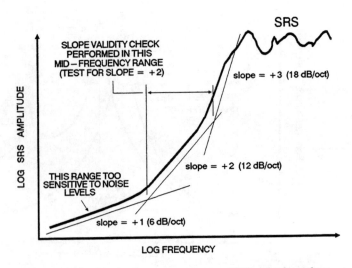

Figure 14.9   Shock response spectrum—validity check on slope

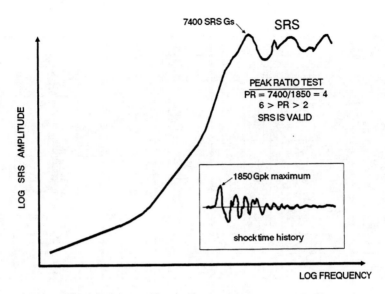

Figure 14.10   Shock response spectrum peak ratio test

To the degree that the time history does not integrate to zero, to that degree you have error in your measurement. Such error will be directly passed on in the SRS. A simple example is shown in Figure 14.11. Note that this SRS and waveshape fail all three validity tests. The leading edge, preshock, DC level is measured and suppressed in the integration. The integrated value is shown in the top margin as INT = 64.14. Close inspection of the time history shows a zero shift in the transient after the arrival of the shock. The time history is invalid because of this artifact. The integrated value is much too high and failed the integration test.

We have investigated a number of integration based algorithms for shock validity checking. One of the simpler of these is shown in Figure 14.12. A figure of merit (FOM) was calculated using a comparison of the positive and negative areas of the time history. For a perfectly measured shock, these areas must be equal, their ratio being one. This figure of merit was calculated for a large set of shock data from a pyrotechnic shock test with a known subset of invalid data due to a setup error. This is a perfect data set for checking the effectivity of these figures of merit. Here, the number of time histories are plotted by their figures of merit as defined in the equation at the top. For a perfectly measured shock, with no zero offset in the time history prior to the event, this FOM is zero. Note that if this large data set were tested versus an FOM of 60, 90% of the invalid data would fall above this limit and 95% of the valid data would fall below this limit. Similar and more discriminatory integrative techniques will be employed in the Phase II implementation of DRAS.

14.4.2 DRAS Utility Enhancement Tools

A useful set of utility enhancement tools is presently being developed for this data acquisition system. They are defined by the matrix on page 353.

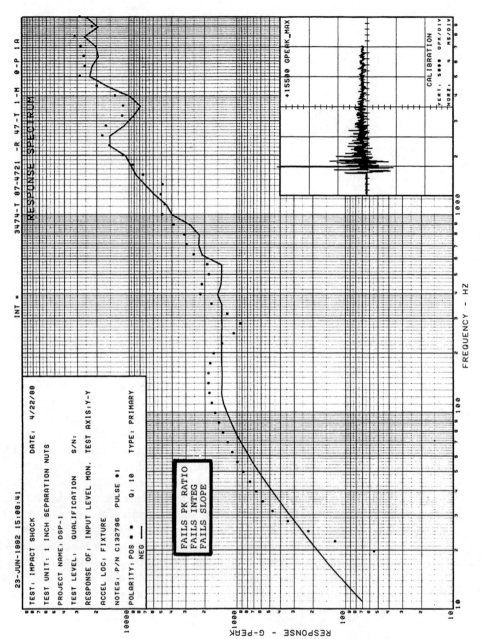

Figure 14.11 Shock response spectrum that fails validity tests

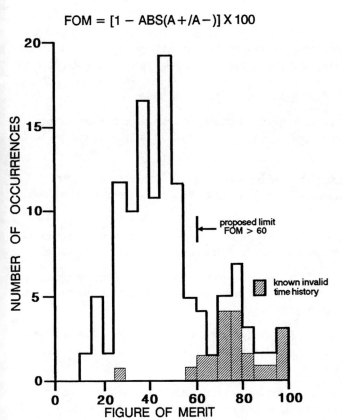

Figure 14.12   Validity checking results for a zero delta-V pyrotechnic shock

DRAS UTILITY ENHANCEMENT MATRIX

Utility Enhancement Tool	Swept Sine	Random Vib/Acous.	Shock
Analysis sorting	X*	X*	X*
Peak finder	X*	X*	X*
Overall listings	X*	X*	X*
Transfer func. corrections	X	X	X
Subtraction of spectra	X*	X*	X
Overlay tolerance bands	X	X*	X*
Waterfall displays	X	X*	
Zoom capability	X*	X*	
Integration/differentiation		X*	X*
Overlay analyses	X*	X*	X
Compute next input level	X	X	

* = operational

Analysis Sorting. This feature is based on a commercial VAX-based data base manager and allows powerful and fast sorting methods to be used on archived data plots. Even a single plot can be easily retrieved and replotted for hardcopy. Typical Boolean sorting tools are provided allowing sorting on project, test article type, job account, operator, test data, etc. For example, a request might be, "Give me all analyses run on Project 4000 black boxes during October of last year" and out they would come.

Peak Finder. This routine inspects all similar analyses from a given test run and lists the peak frequencies and amplitudes for each analysis by transducer number in decreasing order of amplitude.

Overall Level Listings. This routine inspects all similar random analyses (either PSD FFT, or 1/3 octave) and lists the overall RMS levels over the test analysis bandwidth by transducer.

Transfer Function Correction. This utility allows correction of dynamic time histories and analyses using the measured transfer function of the measurement channel.

Subtraction of Spectra. This utility is used to subtract (or add, divide, or multiply for that matter) spectra. The primary use of this utility is to subtract the spectra of noise levels from spectra of noise levels plus data.

Overlay Tolerance Bands. This utility allows the creation and drawing of tolerance bands on analyses and greatly enhances the comparison of analyses from location to location and from run to run.

Waterfall Display. Waterfall displays are very useful when tracking problems during a test run which vary with time. Many commercial analyzers have this feature.

Zoom Capability. This capability is used to increase the resolution of the analysis in certain frequency ranges in order to look at spectra very closely and (to) separate responses very close in frequency.

Integration/Differentiation. These routines are used to move spectra among the acceleration, velocity, and displacement regimes.

Overlay Analyses. This routine is used to overlay similar analyses from any run, or runs, of a test. Further, DRAS will be capable of overlaying similar analyses from previous tests. It will be possible, for instance, to directly overlay what happened on the Flight 3 acoustics test with analyses from the Flight 1 acoustics test. Any similar archived analysis will be available for this feature.

Compute the Next Input Spectrum. After a low level sine/random vibration or acoustics run, DRAS will be able to compare the response levels from each transducer as a function of frequency with limit levels allowed in that location in that direction and compute (based on system linearity and constant damping) the input spectrum for the next higher test level. Since the system's damping will actually increase with level, the DRAS prediction of the next level will be conservative.

DRAS is an evolving system which was designed to be a Knowledge-Based system from the beginning (Author's note: that is why software is soft—to support evolution). Our experience is that it is very difficult to take a non-Knowledge-Based system and evolve it into a Knowledge-Based system. The reason for this is the nature of the controlling software. Our experience is that non-Knowledge-Based systems do not have the appropriate structure nor software hooks to allow efficient transition to a Knowledge-Based system. We have had to start over with the appropriate design in both of the cases discussed in this paper.

14.5 THE ADVANCED THERMAL MEASUREMENT SYSTEM (ATMS)

The Advanced Thermal Measurement System is presently being designed to support future systems level thermal vacuum and thermal cycling tests on large Shuttle-class payloads. The functional specifications for ATMS are based on our perception of what the future is going to look like over the next 20 years—well into the twenty-first century. That perception is based on on the view that systems level thermal testing is going to be characterized by the following factors.

(Note: The Advanced Thermal Measurements System was implemented as the Generic Thermal Measurements System—GTDAS. It is referred to as GTDAS elsewhere in this book. We renamed it because certain senior managers do not like the term "advanced." To them it spells overrun on cost. "Generic" is much less threatening and closer to the point. A simple renaming did the trick.)

14.5.1 Increased Technical Difficulty

There are several factors causing the difficulty of systems level testing to increase. Spacecraft heat dissipation is increasing from the 1–2 Kw class to the 10–20 Kw class bringing with it more complexity in the thermal control subsystem design. Future spacecraft will be flying with active cryogenic cooling systems for optical payloads. Heat pipes are going to proliferate in future thermal systems testing causing testing and verification problems due to our 1G test environment. It may not be possible to test all heat pipes on the ground in 1G. How do you verify performance? Optical instruments and benches will require much closer thermal control and verification of gradients to insure precision performance.

14.5.2 Increased Technical Risk Driven by Program Plans

There is a growing tendency on the part of some programs to drop major optical or science payloads into the spacecraft integration and test flow later and later so that overall program time can be shortened. At the same time these major subsystems may be delivered with less and less predelivery thermal testing, again in order to save cost. This puts the integrating contractor in a unique position. Such contractor is responsible for the overall system thermal performance but (may not be) responsible for the payload thermal design. In these cases, the systems level thermal vacuum test may be the *only* opportunity to exercise the system and discover thermal design, manufacturing, or analytical model errors. These problems *must* be identified and isolated during this test. There is no other opportunity. This increases the risk associated with not finding a problem in this test.

14.5.3 Emphasis on Reducing Cost and Schedule

There is continuing high pressure on programs and functional areas to reduce the cost of doing business and the (cycle) time required for the program. A capability that can take several days out of a systems level thermal test is now very important.

In summary, the future looks paradoxical for testing organizations. Test complexity and associated risks are going up when, concurrently, tests costs and span time are under pressure to come down. These forces tend to multiply the difficulty of the problem the professional tester is asked to solve. The Advanced Thermal Measurement System is a direct result of thinking about new ways to approach the future solutions to these problems.

14.5.4 Needed ATMS Capabilities

Our customers communicated to us with some clarity their ideas for new capabilities needed for the future. They are listed here at a generic level.

1. Increased Information Utility
   Need ways of observing experimental data which are more perceptive, allowing easier and more certain problem solving.
   Need an ability to *focus* on problems rather than the mass of nominal data—find the thermal needle in the haystack.
   Need expanded time horizons:
   - FUTURE: predictive capabilities
   - NOW: what just happened
   - PRESENT PAST: this test's history
   - PREVIOUS PAST: previous test's history
   Need comparison of experimental data with analytical predictions for static and transient models

2. Increased Data Suite
   500 test temperatures
   300 flight heaters
   150 test heaters
   50 radiometers
   50 analog voltages
   500 telemetry channels
   450 derived channels
   2000 channels (1050 H/W, 950 S/W)

3. Need Cryogenic Temperature measurement capability

4. Need Increased Software "Robustness"

robust, adj.: strongly formed or constructed; requiring little maintenance.

Put tools for software maintenance in the hands of the users to the maximum extent feasible—maximum robustness

Increased Data Utility and an Expanded Time Horizon (Level 3).   The tools necessary to provide needed improvements in these areas are, in reality, so similar that they can be discussed together. Further, these tools replace no features which are presently in use. They are additional features which allow more perceptive viewing of data and creation of information—information being at a higher level than basic test data. The basic services which allow system operation at Level 2, the Level of Observation, include: measurement of appropriate parameters (at an appropriate rate!), capture of spacecraft telemetry data, color limit checked CRT displays in page and graphic formats, ability to create hardcopy plots and listings in report-ready format, all at the test site. These features are assumed and will not be discussed further.

Ability to Operate in the Future (Level 3).   ATMS will have several capabilities to operate in the near-term future. These are predictive capabilities allowing a view of the thermal future given no further intervention in the system's thermal state.

Certain types of commonly used CRT page displays are amazingly imperceptive. These are the displays characterized by packing as much limit checked data on the screen as possible, so you don't have to worry about a lot of pages. These displays can really hide a problem in the near-term future. An example is shown in Figure 14.13 (with limit status shown by the carats). There is no rate of change information here. What parameter, if any, is in trouble? Which one will go out of limit next? Is there a pony in there somewhere? Note that this display was presented (to the system operator) at 5 a.m. on a Sunday morning. None of us would be at our sharpest at that time. A problem here is easy to miss.

ATMS will operate in a mode in which the amount of time remaining before a limit exceedance is automatically predicted "on the fly" for all limit checked channels. A (mock) CRT display of this function is shown in Figure 14.14. A precision prediction is not needed here. What is needed is to prevent a problem which will occur in minutes. A four point traveling linear extrapolation of limit exceedance time is shown. The delta time in the future is the important parameter. A linear extrapolation is shown but other methods are under consideration which inspect the second (time) derivative of temperature and its sign and based on that sign apply a more conservative fitting algorithm—an "intelligent" algorithm if you like. ATMS will also generate and update a global limit exceedance display as shown in Figure 14.15. Here, those channels which will go out of limit soonest are displayed in ranked order by time. A CRT display would, of course, be in color. For the channel nearest its limit exceedance time, TM037, it can be seen that it will go out of limit in about four minutes and is presently seven degrees from its upper limit.

ATMS will also have an ability to predict the condition of future thermal equilibrium. This is especially important when testing structures which are massive with large thermal time constant in thermal balance tests. If one knows the future time and temperature of

PROGRAM = DSP   **FLIGHT = 14 T/V**   **DATE = 12/21/90**   **TIME = 0500**   **P = 03**   **SCAN 255**

TEST TEMPERATURES (DEGREES F)

T304	-45.3	T305	-54.3	T306	-50.4	T307	-41.8	T308	-41.6
T309	-47.9	T310	-48.7	T311	-41.0	T312	-44.1	T313	-48.4
T314	-54.2	T315	-46.7	T316	-44.6	T317	-49.9	T318	-52.8
T319	-43.3	T320	-41.4	T321	-47.2	T322	-56.3	T323	-47.3
T324	-45.3	T325	-50.2	T326	-49.9	T327	-45.1	T328	-46.8
T329	-51.7	T330	-15.0	T331	-44.1	T332	-48.5	T333	-53.1
T334	-20.3	T335	-43.8	T336	-48.5	T337	-55.2	T338	-44.0
T339	-45.3	T340	-60.9	T341	-56.3	T342	-63.5	T343	-42.5
T344	-42.5	T345	-44.7 <-	T346	-43.2 <-	T347	0.6	T348	36.6
T349	36.6	T350	33.9	T351	26.7	T352	39.4	T353	43.1
T354	5.6	T355	5.7	T356	21.7	T357	21.4	T358	-19.0
T359	-19.2	T360	-25.3	T361	-25.3	T362	24.8	T363	24.5
T364	26.7	T365	26.5	T366	13.8	T367	14.0	T368	15.6
T369	15.4	T370	8.9	T371	13.3	T372	53.0	T373	52.4
T374	37.3	T375	46.2	T376	30.6	T377	15.6	T378	32.1
T379	17.5	T380	24.4	T381	-10.4	T382	32.1	T383	1.6
T384	27.3	T385	24.7	T400	47.6 <	T401	44.5 <	T402	46.8 <
T403	52.7	T405	47.4 <	T406	23.6 <-	T407	48.8 <	T408	42.8 < <
T409	34.6 <	T410	70.4	T411	45.4 <	T413	56.1	T414	45.1 <
T415	51.5	T416	50.9	T417	48.9 <				

Figure 14.13   Example of a very imperceptive display

thermal equilibrium, an informed decision can be made of whether to wait for that condition to occur. When time constants of days are involved, significant test time may be saved.

A planned set of applications and displays are shown in Figures 14.16 14.17, and 14.18. As shown in Figure 14.15, if a structure exhibits exponential behavior, and if the time and temperature of the initial conditions are known, the equilibrium time and temperature can be easily calculated and estimates refined as the condition progresses. Figure

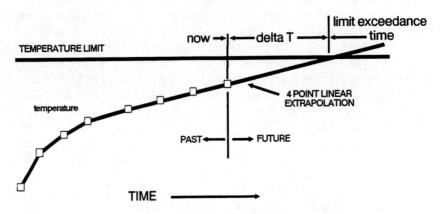

Figure 14.14   Future prediction of limit exceedance

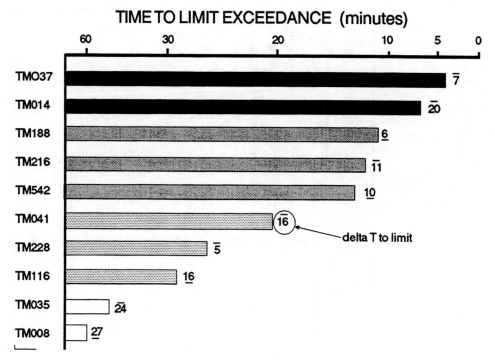

Figure 14.15   Ability to predict future limit exceedance

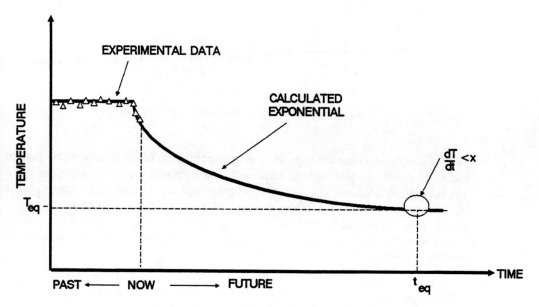

Figure 14.16   Ability to predict future thermal equilibrium

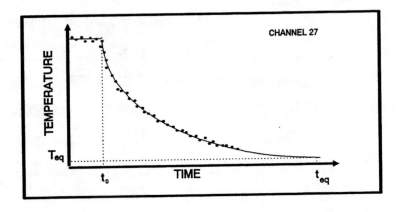

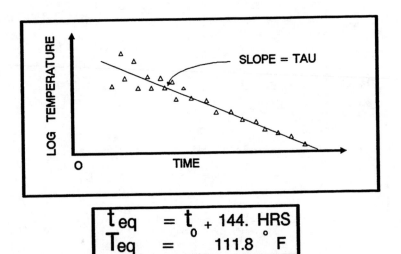

$$t_{eq} = t_0 + 144. \text{ HRS}$$
$$T_{eq} = 111.8 \ ^\circ F$$

Figure 14.17   Ability to predict future thermal equilibrium

14.17 shows an exponential temperature response at the top of the display. In the bottom half of the display we plot the same response but with temperature plotted logarithmically. The slope of this characteristic is the thermal time constant. Once this slope converges the time constant is known and equilibrium time and temperature can be plotted as shown. A global display of thermal equilibrium conditions is shown in Figure 14.18. Here, equilibrium time and temperature are plotted for a global set of parameters showing conditions all over the test article.

Ability to Operate in the Previous Past (Level 3).   ATMS will be able to store not only all data from an on-going or present test, but also the data from a previous test what we call a previous past. Comparisons by parameter will be easily done from the keyboard.

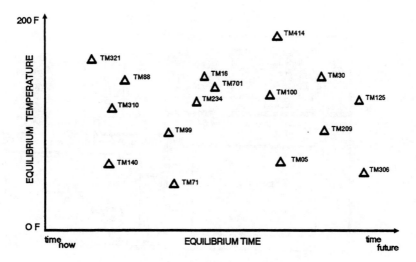

Figure 14.18   Global thermal equilibrium display

Figure 14.19 shows a simple overlayed display of a test phase for a temperature, TM027, with the same phase for the same parameter from a previous test. This prevents the customer from having to show up at the test site with the 21,000 pages from a previous test for comparison in real-time!

Ability to View Thermal Data Spatially (Level 3).   A capability to view test article temperatures spatially and graphically will be included in ATMS. An example of this dis-

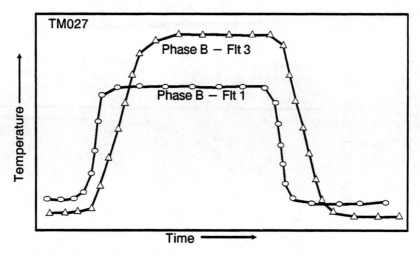

Figure 14.19   Ability to compare present past and previous (historical) past in real-time

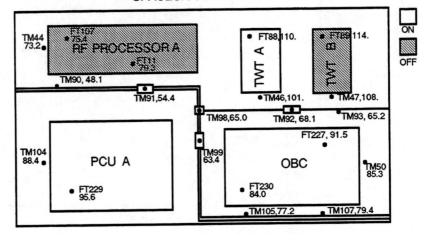

Figure 14.20   Spatial display of panel status

play capability is shown in Figure 14.20 for a hypothetical spacecraft panel. Multiple views would be available. This is not a new idea by any means; several aerospace contractors already have this capability on-line. It is just too good an idea not to implement.

Ability to Integrate Theory and Experiment (Level 4).   ATMS is being designed to facilitate the easy comparison of experimental data with theoretical predictions. A potential display for this feature is shown in Figure 14.21. Experimental data is shown overlayed with both a static and transient prediction for that parameter.

In summary, ATMS will be a Knowledge-Based system having as complete as possible set of tools providing significantly increased information utility to the users and the

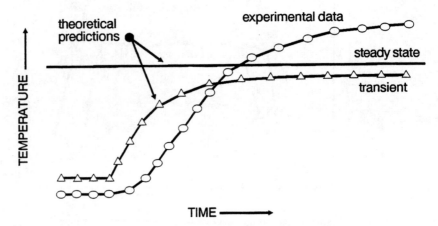

Figure 14.21   Ability to compare theory and experimental data in real-time

thermal customer. Further, ATMS is being designed to operate at Level 4 where theory and experiment can be integrated during the test itself.

14.6 METHODOLOGY FOR DISCOVERING YOUR CUSTOMER'S NEEDS

We have found that the process of eliciting the customer's needs can be a difficult and frustrating one. Most involved players are busy working today's and next quarter's problems. They may not have time to focus on solutions to problems we may not yet have even noticed. Yet, that is what *vision* is all about.

Certain techniques of group dynamics, however, can be very useful in this elicitation. Nominal Group Technique (NGT) has been found to be a powerful tool for arriving at group consensus regarding what is needed in the future. Nominal Group Technique is a structured method of "brainstorming." We use this technique with our customers to elicit their consensus of what capabilities are really needed for the future. You go into an NGT session with a question like, "What new capabilities are needed in future data acquisition systems to reduce test risk, cost and schedule time?" You come out of the session with your customer's answers to that question ranked in their priority of importance. This is priceless data to a systems designer. Such a session takes 2–3 hours and is effective with groups of 10–20 people.

We follow up with individual conferences to flesh out the top level details of the customer's ideas. This is done for the top ten raked issues developed in the NGT session. If you can handle you customer's top ten ideas for the future, that is generally enough! These sessions are followed by the development of a written functional specification for review by all concerned parties. The detailed design specifications and design follow from the agreed-upon functional specifications. This approach has been followed for both DRAS and ATMS.

14.7 SUMMARY

Both the Data Room Automation System (DRAS) and the Advanced Thermal Measurement System (ATMS) have been designed from the outset as Knowledge-Based data acquisition systems. DRAS is operating as one presently (Author's note: both systems are now up and running). As such, we assert that they are on the leading edge of data acquisition system design for aerospace testing.

Knowledge-Based systems such as these are designed with the experience and knowledge bases of the designers, test conductors, and customers for the test and the information clearly in mind. Such a design base cannot be acquired from commercial vendors—they do not understand your particular problems. They understand their problems. Systems, such as these are designed to support complex testing by providing significantly increased perceptiveness and acuity into the test process on the floor during the test.

These capabilities really can and do make a difference. They do not fall into the myth of system speed—that faster is necessarily better. Fast is fast enough. What really makes a difference is the provision of capabilities which allow better real-time decision making during the test process itself.

Author's Note: Our philosophy about Knowledge-Based systems is being expressed in more measurement systems as time goes by. In certain cases, it makes more sense to apply Knowledge-Based principles to systems that are not otherwise Knowledge-Based. This might happen because the system was deployed before Knowledge-Based principles were defined, or because it is just too expensive to upgrade. In these cases, Knowledge-Based improvements can make a real difference. It is not, after all, an all-or-nothing game.

We have a digital data acquisition system operating on a project whose purpose is to monitor several hundred quasi-static loads, moments, and angular deflections on some large hardware items during integration. The customer is interested in the time histories of these parameters as a baseline. He is also interested in real-time dynamic variations in these parameters around their quasi-static values as the integration process goes on. The dynamic variations tell about the health of the hardware as the delicate integration process goes along. If we simply displayed the complete dynamic variations on the several hundred

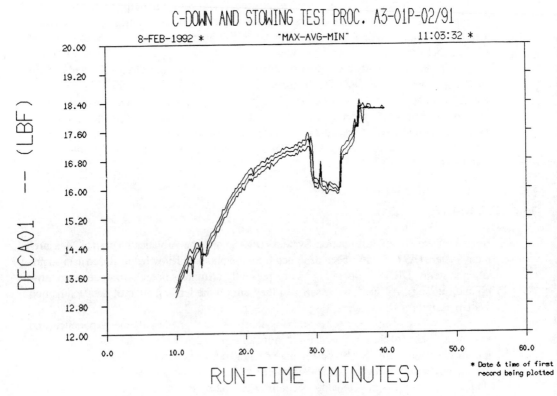

Figure 14.22   Example of utility enhancement of deployment data

channels as time histories, we would have been driven to very high sampling rates and the real-time graphics outputs would have become very busy and confusing to the customer.

In order to increase the real-time utility of this data to the customer, and to cut design complexity, we developed a digital filtering algorithm that notes the mean, maximum, and minimum values of all parameters over five second time slices and plots these values versus time. Figure 14.22 shows this for a load named DECA01. The three lines represent the mean value with the maximum and minimum values from the five second windows superimposed. This gives the test conductor the complete picture of the loading dynamics without obscuring the phenomena with too much meaningless data. This is a perfect example of *utility enhancement of experimental data,* here deployed in a non-Knowledge-Based data acquisition system.

Here is an update for the reader on the status of both the DRAS and ATMS systems since this paper was written. DRAS has been almost fully operational as described here for several years and has almost 50,000 analyses presently archived for instant recall. ATMS, renamed the Generic Thermal Data Acquisition System, has been used successfully on half a dozen systems level thermal tests. We think it is the best measurement system in the world for this class of testing. Our customers tend to agree.

NOTES

1. Charles P. Wright and Gerald A. Vellutini, *Knowledge-Based Systems for Test Data Acquisition and Reduction,* Proc. of the 11th Aerospace Testing Seminar, Institute of Environmental Sciences/The Aerospace Corporation/USAF Space Division, Manhattan Beach, CA, 1988.
2. Ronald. V. Trull, *Sweep Speed Effects in Resonant Systems,* Laboratory for Measurement Systems Engineering, Department of Mechanical Engineering, Arizona State University, Tempe, AZ, 1969.
3. David O. Smallwood, *The Shock Response Spectrum at Low Frequencies,* 56th Shock and Vibration Bulletin, Center for Shock and Vibration Information, Alexandria, VA.

Chapter 15

The Subject
of Software

Software. Software? Hmmmm? Software.

I think the purpose of software is to drive people responsible for deploying new digital measurement systems crazy.

Why am I even discussing software[1] in a book on effective measurement systems design? You can make the classical case for software having nothing to do with the engineering of measurement systems. I, too, can make that case but do not find it useful to do so. Software has become endemic in the measurements business. A measurements organization cannot survive without it. You simply cannot be effective in the test and measurements business without skillful design and deployment of measurements-related software. Once the valid data are in digital form, it can be easily cast in almost any form imaginable to help assimilation by the customer. Many data validation checks mentioned elsewhere in this book are best implemented and institutionalized in software. For these reasons, we always include computer assets and software within the overall measurement system design.

Having been responsible for developing a pot full of measurements-related software, and having written a little of it myself, I do not claim to be an expert on the subject. I know many people who develop it, but I don't know any who claim to be experts. Effective, robust software is a joy to operate, can make your eyes sparkle, and is absolutely necessary for Knowledge-Based measurement systems. My experience has been, however, that developing operational and complex measurements-related software is one of the most consistently difficult and least controllable portions of measurement system design and deployment.

No profound wisdom is offered on this subject. I do have many burnt and smoking tail feathers, numerous scars, and some *excellent* operational software! I'll share the causes for the burnt tail feathers and scars so you can avoid them in your work. These remarks pertain to three classes of measurements-related software: (1) standard, desktop commercial software; (2) custom software an outside vendor writes for you; and (3) more complex software developments requiring your in-house computer professionals.

15.1 STANDARD COMMERCIAL DESKTOP BASED SOFTWARE

This is the easiest class with which to operate. The market in this arena is exploding. I expect the rate of this explosion to continue as commercial vendors get smarter and computer hardware makers pack more for less into their offerings.

Our experience is that the best approach for these developments is to use a measurements engineer who enjoys computers and software for the development. In this approach, a professional computer person should not even be allowed in the room.

We've found two consistent shortcomings with commercial desktop data acquisition software. In the interest of selling their packages to as broad a market as possible, vendors have constructed their products to be *extremely* flexible. The price they invariably pay is efficiency. It takes inordinate amounts of labor to customize each job's setup. Here's an example. We recently deployed several single purpose measurement systems where 40 channels of strain gage bridges, measuring structural bending moments, are continually signal conditioned and monitored, limit checked and displayed in two formats. If a channel's amplitude exceeds one of its four (two high, two low) limit values, a logging session is initiated capturing the event on all channels. Simple, yes?

The best commercial desktop data acquisition software package (we think) forced us to limit check each parameter versus a five state limit model (four limits = five states)— *three times for each scan!* It was limit checked once for color-coded alphanumeric display, a second time for color-coded bar chart display, and a third time for triggering the logging session to disk storage. That is the definition of inefficiency and risk as the price of flexibility. We could assign all sorts of cute icons in 36,000 colors to the data, pop up cute flow charts and instrument simulators on the screen, but we still had to keep track of *three* separate limit checking data bases and functions! Being forced to keep track of three separate data bases and setups at least triples the risk of a data entry problem and resulting error. Simple, NO!

Vendors continually seem to focus on unneeded bric-a-brac and tinsel (particularly display tinsel), rather than on nuts and bolts functionality you can count on and that really makes a difference at test time. Why? I think it's because bric-a-brac sells to customers neither experienced nor perceptive enough to know the difference. I do not expect this condition to change any time soon. You have to know the difference.

There is an exception to this position in the area of creating computer-based virtual instruments for teaching purposes. Here, there is a potentially powerful opportunity to create instruments and measurement system components in software with the cute icons in 36,000 colors without committing to the purchase of the hardware. I do not yet know how this will work out. But it shows promise for underfunded educational programs and I hope it makes a contribution.

Even with the cute icons in 36,000 colors, if your application can stand the overhead labor associated with setting up such a system, they can make an elegant, low cost solution to your measurements software problem. If your application is of the high volume, quick response type, this solution may be too expensive for you in terms of recurring labor cost for continual customizing for your changing applications.

The second problem is that acquisition software vendors do not begin to understand the real-world applications existing in your laboratory. In particular, they do not understand the operational needs dictated by the existence of noise levels in your data. The covert software design assumption is that the sampled time series is perfect as it comes to your digital platform. As a result, they make few or no provisions for noise level documentation, nor for the diagnostic tools needed to perform this function. You should not even expect them to understand this vital aspect of measurement systems operations. That is our job.

15.2 CUSTOM COMMERCIAL SOFTWARE

This is software you've had a commercial vendor write to your specifications. This class of software operates at one level of effectiveness up from the previous class. When the development process is over, you should have what you want. But you won't. So, the job becomes one of creative and endless software maintenance.

There are two keys to procuring custom software. The first is finding the right vendor and establishing a powerful partnership with them. We have found that the right vendor will have three qualities: (1) they are very interested in your job for their self-serving reasons; (2) they have a product that you like and that they are trying to inject into the measurement marketplace and want to use your job to help do it; and (3) they are unbiased about the test measurement business.

We generally find that vendors of hardware/software with mature product lines in the measurements business are less interested in your job than you need them to be. They are less interested because your job probably does not support their primary business goal, usually increased market share. Why should they focus their expensive resources on your specific job if your requirements cannot be sold across their entire customer base? Your job probably won't maximize the bang for their development labor buck.

You need to find a vendor for whom your job is *really important*. This is a vendor who can tie your job to a marketing thrust for their product. Here, you stand a good chance at doing some cost sharing if you can convince the vendor your ideas support the general customer base they want to create. You offer them, in essence, your expertise in developing their product line. If they offer beta test site status, jump in. You could even suggest publishing the development jointly upon completion if warranted.

Last, you need a vendor who is unbiased about the test measurements business. You want a vendor who is willing to work your agenda as well as theirs. Mature vendors in the field tend to want to protect their existing product investments more than they want to solve your problem. Their tendency is to force your requirements into their predetermined solutions whether or not that is the right thing to do. To a vendor whose only solution is a hammer, all problems must look like nails. Here, their agenda clashes directly with yours. I'll go even farther than unbiased—what you really want is a vendor who is *ignorant* of the test measurements business and not afraid to admit it. Ignorance is merely a state of knowing that you don't know something. We *love* working with ignorant vendors. They want and appreciate your knowledge and experience. They are not sitting back constantly evaluating and second guessing everything you say or request.

In several cases we've created such powerful relationships with vendors. The results have been uniformly excellent with both sides experiencing a feeling of winning. Our systems for medium size (64 channels) static and dynamic measurements (up to 10 KHz) came from a relationship as I've described. I think the resulting software set is the best in the world for this class of quick reaction testing. It is fast, intuitive, bulletproof with very effective displays. It happened because we chose the right vendor and worked very hard with them.

The second key to procuring effective custom software is a complete *written* set of functional requirements. These specifications should define in detail what the item should do, not how it should do it. *What* is your business. *How* is their business. The specification should define exactly what functions you want to perform and how you want your data to look. You can't put too much detail into a successful functional specification.

Leave out *all* popular buzz phrase jargon. We recently wrote a functional specification saying we wanted some "brickwall" digital filters. We got filters looking nothing like we thought we were going to get, but they were "brickwall" filters alright! We should have, of course, specified the transfer function of the filters in detail. We wrote a lousy specification and the vendor fulfilled it with incorrect filters.

Include in the specification the details of the acceptance test you are going to run on the item and pass/fail criteria. Include the necessary and sufficient conditions in the specification because they are different. The *necessary* condition for bugging out a cable is that the right pins are connected to each other. The *sufficient* condition is that the right pins are also not connected to any other pins! The sufficient conditions take much more thorough thinking and craft than do the necessary conditions. You want your acceptance test conditions to be so clear that the following conversation happens when a performance problem occurs. You say, "The acceptance test says the right answer is X. Your unit says the answer is Y. Something is wrong. What are you going to do about it?" The vendor replies, "You're clearly right. Something is wrong. We'll fix it."

Writing a solid, technically competent functional specification that both you and the vendor can hang your hats on is hard work. It takes time and money. My experience has been that there is no substitute for a good specification. In the absence of this specification and its acceptance test, you're reduced to waving your arms at problems. We've found that every dollar spent on the development of a good functional specification saves about five dollars in problems downstream.

15.3 COMPLEX SOFTWARE DEVELOPMENT

Complex software developments are those requiring the talents of your organization's computer professionals. These cases will occur when you need significant software capabilities housed, usually, in fast minicomputers. These systems should operate at the highest level of effectiveness since the entire job is being done in-house.

A typical example of this class of measurements-related software would be the Knowledge-Based Generic Thermal Data Acquisition System (GTDAS) shown in Figure 15.1. This system has an architecture composed of a belchfire minicomputer networked with seven intelligent preprocessing workstations in three test site facilities, as well as several spacecraft test computer systems as shown in Figure 15.2. The system is required to service three 2000 channel thermal vacuum tests concurrently, each running 24 hours/day for up to 40 days, and each with a spacecraft telemetry link. About 80,000 lines of C code were developed for this system. This system required a team of six for development: a task leader, three measurements engineers defining requirements and verifying performance, and two full time software engineers for 24 months. This development class is the one to which the rest of my comments pertain.

The Generic Thermal Data Acquisition System software development occurred under a particularly difficult set of conditions over which we lacked control. These were driven by business and organizational conditions at the time, and accounting and programmatic constraints. It was a very stressful development process. The development personnel worked heroically and delivered a professional product with exceptional performance in spite of the process. They were, however, playing the game with a stacked deck.

Figure 15.1   The Generic Thermal Data Acquisition System (GTDAS)

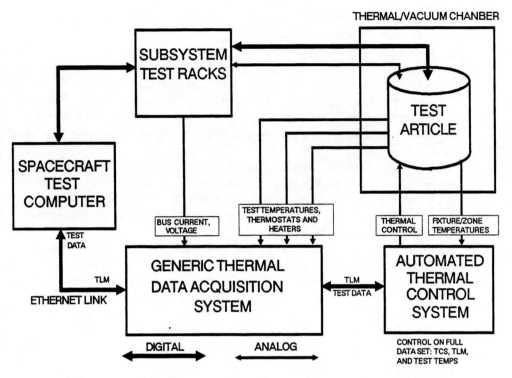

Figure 15.2   Generic thermal data acquisition system architecture in test configuration

My intention is to use this example as a model of how *not* to perform an effective, straightforward software development. I'm not trying to be cute here. It's just more fun, and I think more effective, to communicate what I need to communicate to you from the negative, tongue-planted-firmly-in-cheek, side of the net. Sin is more fun to write about than virtue. No matter how you run your software developments, if you avoid these mistakes you'll much better off in the long term.

15.3.1 Wright's Twelve-Point Plan for Stressful Software Development

1. **Be sure to use a computer architecture that is new to your software engineers**
   You always want to spend tight budget monies on a job where your software engineering staff is new to the chosen architecture. That way they can charge their necessary learning curve experiences to your budget. That's the budget you set up assuming full knowledge of the new architecture, thus no learning curve required.

2. **Use software engineers with no test experience**
   By not clouding the software engineers' minds with an appreciation of what is really important to the measurements engineering user and the customer, you assure that

arbitrary design decisions always get made in favor of programming ease and not eventual functional excellence.

3. **Do not put sufficient detail into the functional specification**

In particular, do not include the stuff you want to preclude the programmers from doing. This assures that users will ask hundreds of fun questions like, "Why in the world does it operate that way? No one in their right mind would do that!" This also assures lots of heated meetings with arm waving and tight throats for the design team.

4. **Assume that your test-inexperienced software engineers will have your set of "common sense"**

Nontest-experienced software engineers simply do not have the same "common sense" as measurements engineers. That which is simple common sense to a measurement engineer may seem absolutely bizarre to the software engineer—and vice versa. Assuming that these two sets of experiences are congruent guarantees many fun surprises during the development effort.

5. **Fund the project in a manner forcing an artificial end date**

Fund the project on overhead monies. In the aerospace industry at least, overhead budgets do not carry over year end. Use this funding method so the project's end date is New Year's Eve! This way, the team can work incredibly hard over the holidays knowing the end date for funding is driven by the accounting system and not a real need. This approach is excellent for team motivation.

6. **Have the project's sponsoring executive get promoted to a new job midway through the development**

Of course when the original sponsoring executive leaves, he'll take his carefully crafted network of commitments to your project with him. After the new executive stops all overhead projects, pending his review and assessment, you'll get the opportunity to sell him all over again as well as create an entirely *new* network of commitments. Since midway through your project is about the time nothing will be going well, you'll appreciate the break from holding the project together so you can focus on the reselling job. Be sure and share this with your development team who are working 60–70 hours per week so they don't get complacent about the job.

7. **Make sure your key user engineers are unavailable for the first half of the development**

Arrange things so your organization is so busy that your key user engineers are in and out of town for the first six months or so. In this way, the programmers can run open loop without needed feedback for this crucial period. They'll repay you by casting their early software, on which all later software is based, in concrete. An unavailable user group assures total programming freedom unfettered by user wishes, opinions, or reality.

8. **Provide an expensive set of software productivity tools, then give no direction for their use**

In the interest of minimizing development labor, provide your team with expensive productivity tools such as a relational data base manager with a surrounding spreadsheet-like environment, integrated workstation graphics, and best of all—a belchfire windows environment. Give them no direction about the use of, or judgment about, these tools. Make sure you choose software people completely unfamiliar with the productivity tools so they can get their learning curve experience on your job, thus eliminating all possibility of productivity improvements. Best of all, don't communicate to them your direction that the tools are to be used *only* if they will save labor. Put these conditions together and you get the tools used whether or not they're the right ones for your job. At the ultimate level, the job becomes *about the tools* and not about producing functional software. The programmers will, however, include familiarity with the tools on their resumes for their next job.

9. **Allow your software staff to make all arbitrary decisions and assumptions while your user staff is out of town**
    Allowing the programmers to make the necessary arbitrary decisions as they come up, without guidance or feedback from users, promotes the feeling of freedom among the software staff. Their set of "common sense," as mentioned earlier, will provide *unique* approaches to problems your experienced measurements engineering users would never even consider.

10. **Never educate your software staff on the pertinent issues of the systems architecture and software**
    By not sending your software staff to systems manager classes or, in our case, classes associated with our particular architecture before we began the development, we improved the opportunities for exciting on the job training experiences. Remember, learn by doing!

11. **Provide your software staff with myopic directions**
    In every software development, the person in the leadership position (guess who in this case) has certain axes to grind. "By golly, the next time I do this you can bet I'm going to solve *that* problem—no matter what! That problem has always driven me nuts."
    So, give very narrow directions to your staff to make sure your particular software itch is scratched *no matter what!* Do this even if your scratched itch causes extreme difficulties in down stream integration. You'll be able to say, "I sure fixed *that one* didn't I!" Unfortunately, you'll be saying this over the bleeding software bodies of those responsible for integrating that scratched itch into an effective working whole.

12. **Con your boss into thinking you are an expert at leading a complex software development effort**
    This assures an endless series of lively meetings during which he finds out the truth.

The major thing we've learned in the GTDAS development shows in Figures 15.3 and 15.4. Figure 15.3 shows who in the development team creates what knowledge in a measurements-related software project. Vertically on the left is a continuum of measurements engineering knowledge. Measurement engineers exist at the top of the continuum line. Software engineers exist at the bottom of this line. Software engineers tend to have little (at least initially) knowledge of measurement engineering principles. On the right is the continuum of software engineering knowledge. Your software professionals are at the

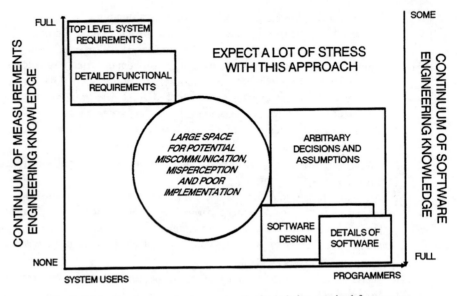

Figure 15.3 Continuum of who creates the knowledge required for measurement system implementation

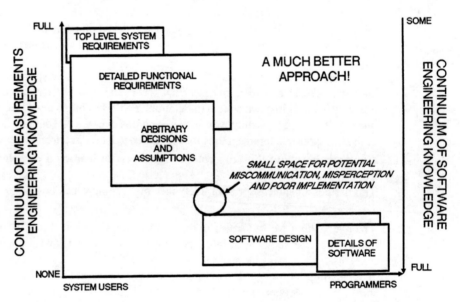

Figure 15.4 Continuum of who creates the knowledge required for measurement system implementation

bottom of this line with full knowledge. The measurement engineering user community lies at the top of the line with some software knowledge.

Across the bottom is the continuum line of who on the team creates what knowledge for a complex in-house software development. This knowledge includes, at least, the top level system requirements, detailed functional specification, the software design, the operational software, and a whole bunch of arbitrary decisions made along the way. Figure 15.3 shows a development space guaranteed to cause a lot of stress to the team. Top level requirements and functional specifications are created by the user community and thrown over the fence to the software developers. Arbitrary decisions, the software design and the software itself are created by the software staff. There isn't much communication on this chart. But there is a *large* space available for problems to occur. And problems will, in this model, fill that space!

A much better approach shows in Figure 15.4. Here, the detailed functional specifications are a shared responsibility, but weighted in favor of the users. The software design is also a shared responsibility, but weighted in favor of the software developers. Necessary arbitrary decisions and assumptions are a third shared responsibility, weighted now in favor of the users. The remaining space for problems is much smaller. This second model is what we are striving for in our process improvements in software developments. It reflects what we've learned by examining some of our burnt tail feathers.

15.4 CONCLUDING REMARKS

Software is funny stuff. I want to make two points before I leave the subject. Presume a case where you've got the most elegant system design you can imagine, exclusive of the software. Your transducer installations are perfect. They are creating errors so small you can't even see them over a frequency range wider than you need. Your overall experimental tolerance is verified at one-fifth your design goal. You've got complete analog visibility into your data stream. The suite of environmental noise levels is 0.1% of full scale and verified as such. In short, you've got an elegant design—so far. A poor software development will destroy this system's usefulness. Poor software can ruin an otherwise professional design job.

On the other hand, the best software development job in the world cannot save an otherwise poor system design. You can't hide your system design flaws with fancy icons on the screen in 36,000 colors. When people attempt to do this, they generally make 50% errors with 0.01% fancy displays and do it quickly. Computers and software are very good at creating junk quickly in the hands of amateurs. This is the ultimate in engineering self-delusion. In short, good software cannot save an unprofessional design job, and it should not be used to do so.

The second point regards how your software will affect others whose support you need. There will be a continuing cadre of people for whom you will demonstrate your new, spiffy digital measurement system. You may want to demonstrate to your customers, your sponsoring executive team, potential customers, new employees, or a tour of nuns. These people will not have the keen eye nor the time to appreciate the professionalism of your design. They will take away from the demonstration only *experiences of your software.* Your software will be the lens through which they see your system operate. Your software will be what they will remember.

For this additional reason, it is critical that your system's displays be crisp, clean, and impactful. A well-designed display will be understood by almost anyone. You want people to walk up to your system and have your displays knock their socks off. You want them to walk away with the experience that your design team really knows what they are doing, and that the issues your system is supposed to handle have been clearly *handled.* If this is public relations and not engineering, so be it. Smart design engineers with vision know it is vital to keep their customers' good will about their products. Your software is a crucial element in generating and maintaining that good will.

NOTES

1. These ideas were first published by the author in the following editorials in *Personal Engineering and Instrumentation News,* Rye, NH. They have been edited for inclusion here.

    "A Twelve Point Plan Guarantees Stressful Software Development," *Personal Engineering and Instrumentation News,* February 1993, p. 67.

    "The Key to Satisfactory Software Applications is Partnertship," *Personal Engineering and Instrumentation News,* December 1992, p. 65.

Chapter 16

Leadership
and Management Issues

16.1 KNOW YOUR MEASUREMENTS SOUL

There is much focus presently on improving the processes by which we all do our business. This is entirely appropriate and long overdue. Testing activities should be no exception to this trend. This is particularly true whenever the testing organization finds its work on the critical path for the development of whatever product is being tested. During these times the performance of the testing organization is driving the *entire project* and its staff. A day saved in test is a potential day savings to the project and may mean big dollars. I say *potential* because a test day saved gives the project manager a day he can use as he chooses. How he uses it is his business. The result of this is that testing organizations find themselves under intense yet appropriate pressure to reduce their cycle times. Cycle time is the serial time from the beginning of a work process (like run a test) to its completion. How is this to be done?

In my experience the task of understanding begins with the development of an "As Is" process flow chart. This flow chart details the entire process in intimate detail from beginning to end as it is now. It shows all tasks and defines the flow of information within, into, and out of the testing process. These flow charts are difficult to develop and take a lot of work. As an example, our As Is flow chart for the process that designs strain gage installations for flight hardware, gets the gages installed and the part calibrated had over 200 individual steps! Our discovery was that we usually had only a superficial idea of how the work actually got done, that it was never done the same way twice, and that there was no detailed guidance on how it should be done. Management and line engineering had top level ideas about how the work got done, but lacked real definition. When I discuss this point with my colleagues, I find that even in the best of testing organizations this condition is the general case.

Once the As Is process exists on paper, it can be attacked looking for tasks that add no value to the end product. Here, you're looking for the stuff you've always done because it is *traditional* in your organization. It had value in the past . . . maybe. Will it add value in the future? The As Is strain gage process noted above had more quality assurance inspection steps (21) than rigorous value-added steps (19)! We eventually improved the process to include only one QA inspection step!

These tasks are rightfully done in cross functional teams. One problem of cross functional teams is that not all team members understand your business as well as you do. That fact presents both a wonderful opportunity and a danger. The opportunity is to educate your customers about your business while they help you to tighten up your operations. Everybody wins. That opportunity should be welcomed and attacked with vigor. The danger is that in slimming the testing process down to its bare bones, measurements-related steps that add real and necessary value will be zealously deleted with the help of the team members not having the knowledge to judge the step's real contribution.

It is crucial for the serious, career measurements professional to understand thoroughly that subset of tasks are absolutely necessary to guaranteeing the validity of the data. The question you have to ask yourself is, "What are the things I *must* do to prove the data are valid over the valid data bandwidth and amplitude range agreed upon with the customer?" These things become your *measurements soul.* These are the things you cannot give up in the interest of improving the testing process. The identification of your measurement soul is a very interesting group process, and I recommend it. A staff that undergoes the struggle necessary, and it will be a struggle, to identify positively and collectively those tasks that comprise the organization's measurement soul is a better staff for the effort. They understand at a fundamental level what is *really* important and why. That is powerful information for a cohesive group to hold in common. It is particularly powerful information to hold in common in a world where most decision makers do not understand the nature of the measurements task or its importance, and think this work can be done by anyone. Their attitude is, "What's the big deal. This isn't *real* engineering anyway."

Once your measurements soul is identified, your job is not to give it away in the search for an optimized testing process. Other tasks may be negotiable, tweakable, optimizable, fixable, changeable, and deletable—*but not your measurements soul.* This set of tasks represents a line that you cannot ethically cross in good measurements engineering conscience.

What is the responsibility of a professional measurements engineer when presented with a team-developed process improvement that guts his measurements soul? If I were presented with this situation, I would thoroughly document my case in writing upward to my management, downward to my employees, and across to the process improvement team defining the reasons for my positions and the risks inherent in the "improved" process. I would plead that case in person and energetically to those who can affect the outcome.

If my management did not back me up on this fundamental professional and ethical issue, I would know that I had lost their confidence. Then, I would begin the search for another job. Working for and with people in decision-making positions who hold no confidence in your judgment is not satisfying.

16.2 CONTINUAL REEDUCATION OF SENIOR MANAGEMENT

For the last decade or so I've been a member of that pariah of management classes, the *middle* manager. In my job, I am the last level of management that understands the technical issues involved in our area of expertise. Above my level, things get aggregated very quickly and decisions are made based on other criteria. Above my level lies the arena of the professional manager whose career is about managing things, not about using, understanding, optimizing, and advancing technology. Decisions at this level should be driven by what is best for the customer and the enterprise—not necessarily what is best for an individual organization. I've had some great ideas in my career that, unfortunately, made no business sense whatsoever! They died stillborn as they should have. One or two levels above the middle manager level lies the stratospheric level of the *vice president*.

It is in your interest to provide for the continual education of these professional managers for several reasons. Continual education means creating an increasing awareness of what your organization does and what it contributes to the enterprise. It is an exceptional vice president who will stick his head into your office and ask, "What are you doing and how can I help you do it better?" I'm still waiting for that VP to get the job. They have other things on their minds on a day-to-day basis. Further, most of them are passing through your organization on their way to the next *real* management job. Test and measurements work is never considered a *real* management position in any organization that I know. The job is one where a potential very senior manager gets "rounded out" on the way somewhere more important. So, it is highly unlikely that your very senior management, the general manager or CEO for example, came from the testing organization. Professional test and measurements folks generally do not rise into senior management. They tend to like what they're doing and remain, or they move on to other fields.

16.2.1 Getting Management's Attention

Since your senior management probably has little idea what your organization contributes to the enterprise, you have to tell them. I've found the best way to do this is to invite them around to meet your troops and see what it is you do. Enroll them in what you are doing as important for his job performance, for his bonus. Get him on your team. Share your successes and problems with him. Ask for his help. In short, *get visible*.

Most of your less aggressive colleagues will never do this. Most would rather hide out. Most would rather be *in*visible. Invisibility is very limiting. You cannot turn your significant ideas into reality from invisibility.

As an example of what getting visible looks like, I'll relate the history of the Generic Thermal Data Acquisition System. After a very complex spacecraft level thermal vacuum test several years ago, it became abundantly clear the 2000 channel measurement system used on the test was obsolete. Its 10 year old software had been fixed, tweaked, bubble gummed, cross strapped, fiddled with, and jury rigged to such an extent that it was extremely brittle. In test, it would take hours to make a minor change to a data page display. The system required intensive, expensive software maintenance. In spite of it all, the test was a smashing success as was the spacecraft mission after launch.

We planned to replace the system with a Knowledge-Based system that would handle three 2000 channel thermal vacuum tests concurrently. I did not want to make this painful change more than once! To sell the idea that a replacement system was needed (read: educate senior management) in the face of extremely tight capital and overhead budgets, I delivered, over a 12-month period, *47* separate briefings at which I was the lowest ranked person. Each was an opportunity to give up. I even had to brief some vice presidents twice! Upon returning to one vice president's office to ask for his support a second time, I was greeted with, "What part of NO didn't you understand?" The willingness to take only yes for an answer when you know you're right, and to give the required 47 briefings, is what visibility looks like.

16.2.2 The Problem of Management Mobility

One of the reasons that American industry is perceived as lacking ability to react quickly to the marketplace is the short-term moves of senior management. Senior management moves are a severe problem source for the measurements engineer trying to turn a significant idea into reality. In the aerospace industry at least, it seems that it takes *two to four years* to turn a significant idea with complex issues into a functioning reality.

You start by enrolling necessary people in your idea. If the idea requires capital funds, you must enroll the senior manager whose signature is required on the purchase documents. This takes a while. He must own your idea with you. Unfortunately, the life of an aerospace vice president is about half the gestation period of a good idea! Once you've got him educated and in your corner, he gets promoted or moved elsewhere. The new vice president comes in and naturally puts all on-going improvement activities on hold. He wants to put his own English on what's happening. He wants to revisit everything with his own set of priorities and marching orders from management Olympus. As a result, another year of development time is wasted. This is extremely frustrating for someone wanting to make a difference.

I see no way out of this predicament. This appears to just be the way it is. Your only hope is to reeducate continually the present senior management knowing they will leave while you still have irons in the fire. That is your responsibility to your organization's good ideas. Are their ideas worth a lot of work on your part? If so, get on your vice president's calendar and begin the education process.

While you're at it, get clear about what the VP's goals are and what is important. Then pitch your improvement idea directly into that goal statement and the statement of what is important. Use that information as a big bull's eye. If your idea cannot be pitched clearly into the center of this target, you probably have the wrong idea.

16.3 GET YOUR CUSTOMER TO SELL YOUR IDEAS FOR YOU

Eventually, you have to sell your improvement ideas to the executives who can fund it for you. This is inevitable. But first, you have to sell it to the direct customer for the improvement. In the absence of customer acceptance of your idea, you have an idea that should and will go nowhere. An improvement idea must rise or fall on its merits. That is really all you can ask. The problem is who gets to judge the merits! The first judge is your customer.

When the time comes to present the improvement idea to the executives who can fund it, take your customer along as your partner. I never pitch a major idea to my management without the benefiting customers in the room. This pitch is the public ceremony where commitment begins.

Better yet—have your customer pitch the idea to your management for you. Only customers who own your improvement as their own will be willing to do this. Your customer's pitch for your idea is much more powerful than your own. Some executives will experience your pitching the idea as self-serving. Having your customer make this pitch cuts that perception off at the pass.

In the best of all worlds, your customer says something like, "Look, Mr. Executive, we want you to help these people make this improvement. We like it. We think it's a good idea. We want it. It will benefit our project in the following manner." Most executives will experience this as your having done some very effective homework.

16.4 THE BUREAUCRACY

Many organizations are whacking away at the levels of bureaucracy they've installed over the years. But too many still exist to work effectively. Since you'll probably never be given the alluring opportunity to eliminate it all, the next best alternative is to figure out how to play the bureaucracy like a violin. I'll give you an example.

The largest purchase requisition I've ever originated was for around $1,000,000 for the entire subcontracted data acquisition system for a test facility for acoustics and shock testing. The purchase requisition required *fourteen* signatures! Mine was the first and represented the *last* person on the list understanding what was being purchased and why. The last signature was that of the Chief Executive Officer, the CEO. In between were twelve layers of people who, as far as I could tell, added no value to the process. My boss, who was *responsible* for bringing the test facility on-line on time, was not required to approve.

How do you operate within this mass of non-value-added bureaucracy? I have only developed one solution to this problem over the years. That solution, labor intensive as it is, is to take total ownership of the entire process. When one of the fourteen people had signed, someone on our staff was immediately at their shoulder to answer questions and be ready to pick up the package and move it to the next person. We had to commit to doing everybody else's legwork.

I guess it depends on how much you want to have the process turn out the way you desire. If you really want it to turn out, you must own it—warts and all.

16.5 THE MEASUREMENTS CONTRACT—WHO OWES WHAT TO WHOM?

I recently had the privilege of teaching our internal Measurements Engineering Educational Course to about 30 customer engineers from a sister division. This group of customers used our services as well as those at off-site testing facilities and subcontractors. Our relationship with them was and remains excellent. They were experiencing, however, a continuing series of woes with the experimental data they were getting from other organizations and their subcontractors. In working with us the subject of our educational course came up and they asked us to deliver it to them. The intent of the course was to educate them on the basic principles of measurements engineering so they could be better customers for us and other experimental groups, and to increase their skills as customers in working with all experimental groups. This was an experience in educating your customers about your business. During the course, I made a consistent effort to identify those things that the customer and experimenter owe each other that tend to take the mystery and stress out of the job.

As part of the educational course, we developed a *Measurements Contract* defining who owes what to whom. We developed the contract by looking closely at those jobs that went smoothly and resulted in a *delighted customer*. Delight is one qualitative level above satisfaction. I believe in having delighted customers. A delighted customer is one who says, "That was fun. Let's do it again!"

The Measurements Contract defines those things that must be delivered and shared between the measurements folks and the customer if the job is to run smoothly.

The Customer Owes the Experimenter:

1. **A well thought out problem—What exactly do I need and why do I need it?**
   When the customer comes into the interaction with a problem thought through, both parties can immediately focus on the best measurements solution. The experimenter is assured that there will be answers to the important questions when they are needed. (Note: This is very important. The list of questions asked when a strain gage installation on flight hardware is to be designed has 207 questions on it! No answers—no installation.) Knowing why the customer needs the data helps the measurement engineer formulate a solution feeding directly and effectively into the next step in the design process.

2. **A statement of how the data will be used**
   This information helps in the formulation of the optimum response. It helps keep the measurements criteria sane. Assume we received a request to measure strain on a structure to experimental tolerances of $\pm 1\%$. This would be difficult since the gage factor alone has a $\pm 0.5\%$ tolerance (two sigma) from the manufacturer. It would take some real doing (read as money and time)—but it could be done. If we then found out the customer was going to use the strain data to calculate stresses using $\pm 5\%$ (!) handbook modulus and Poisson's ratio values, we would have wasted a lot of time and money for nothing. Knowing how the data would be used in the first place would result in a lower cost test program for the customer with the same level of satisfaction.

3. **An open mind regarding the best solution to the measurements problem**

   Nothing irritates professional measurement engineers more than the customer who comes into the office with a preconceived, and possibly uninformed, notion of how his problem should be solved. This is typified by the fellow who came into my office wanting 36 channels of strain gages on his hardware in predetermined locations. He even had the installation drawings done! He left with a plan to get the data he really needed from four channels of strain gages and one of deflection for one-fifth the cost.

   What really works is for the customer to come with a clear understanding of the problem to be solved, and then work in partnership with the measurements engineer to develop an optimum solution. This lets both parties add their value to the process.

4. **A specific measurements request for every parameter**

   This request must have several pieces of information for each measured parameter or group of similar parameters: (1) the explicit frequency and amplitude range over which valid data is required; (2) whether the waveshape or the frequency content is to be reproduced; (3) an allowable experimental tolerance that makes sense; and (4) a statement of the conditions under which the measurement is to be made.

5. **A statement of the customer's wishes regarding the proof of validity**

   A professional measurement engineer is prepared to prove the validity of the data. How much of this proof is to be supplied to the customer? Is the customer prepared to take the experimentalist's word for it?

6. **An open mind regarding the planning of the experiment so that measurements needs are reflected**

   This results in a viable test plan that supports acquiring the necessary data as a *primary* objective.

7. **A viable test plan reflecting measurements needs**

   Test plans are usually developed and optimized based on some parameter associated with the test itself: minimum cycle time, minimum cost, an available skill set, etc. Unless the test is associated solely with seeing if the test article survives the exposure, the test's purpose is to get the data. The data are necessary to see if the test objectives have been met and *should be*, therefore, *a primary test objective.* The test plan should reflect this. If the attitude is, "We'll get whatever data we can along the way," then the test should not be run.

8. **Acknowledgment that a cost estimate is a cost *estimate***

   Self-explanatory.

The Experimenter Owes the Customer:

1. **A specific plan for verifying the measurements requirements: How am I going to fulfill the customer's requirements?**

   The customer has given the experimenter a specific set of measurement requirements for each parameter. The experimenter owes the customer an explanation of how those requirements are to be met. This defines what proof of validity will be available and/or delivered to the customer. This product says to the customer, "This is how I am going to measure your parameters and here is the proof of validity."

2. **A statement and assessment of necessary technical assumptions**

 What must the experimenter assume to make the measurements to the agreed upon tolerances? Overt assumptions are usually easy. Covert assumptions must be discovered and shared.

 Here is an example of a covert assumption. We occasionally use the difference between two measured accelerations to calculate rotational acceleration. The covert assumption is that rigid body motion occurs over the entire valid frequency range between the two acceleration transducers. Only under this condition is the differencing process valid. If flexible body motion occurs, all bets are off.

3. **A statement and assessment of risks to the data product**

 Are there any aspects of the test that are a risk to the data product? These sound like, "If this occurs during the test, we will probably lose the data over there."

4. **Data that are demonstrably valid and of high utility**

 This is what this entire book is about. You owe this to the customer.

5. **Data that meet the customer's schedule needs**

 The data must be timely in terms of the overall testing program. The customer is the sole arbiter of timeliness.

6. **Data clearly labeled as "not validated" if that is so**

 You owe the customer the clear identification of data that cannot be validated. The customer uses this data at his own risk.

7. **Flexibility regarding evolving measurements requirements as the test effort progresses**

 Effective testing requires flexibility on the part of the measurement engineer. This is so because you and the customer will always learn things as the test progresses. This new knowledge allows you both to look at the process in new ways.

8. **A realistic estimate of costs updated as makes sense and a commitment to meet that cost goal**

 Self-explanatory

9. **An overall commitment to delighting the customer all the time, every time**

 This is the most important commitment of all. Such delight should be delivered always, every time. Delight is a higher level response than satisfaction. In this book, satisfaction is not good enough. Delight keeps them coming back for more!

We have found that it is most useful to discuss the terms and conditions of the Measurements Contract early in your relationship with your customer on almost any measurements job. They are universally impressed that you've taken the time and care to lay a contract like this out on paper and are willing to use it in the test process. Early agreement on the terms and conditions, modified to meet your needs, will go a long way toward creating the delighted customers you want for your laboratory. The complete Measurements Contract shows in Figure 16.1.

THE MEASUREMENTS CONTRACT

The customer owes the measurements engineer:

- A well thought out problem
- A statement of how the data will be used
- An open mind regarding the best solution to the measurements problem
- A specific measurements request for every parameter
- A statement of the customer's wishes regarding proof of validity
- An open mind regarding the planning of the experiment so that measurements requirements are reflected
- A viable test plan reflecting measurements requirements
- Acknowledgment that a cost estimate is a cost <u>estimate</u>

The measurements engineer owes the customer:

- A specific plan for verifying the measurement requirements
- A statement and assessment of necessary technical assumptions
- A statement and assessment of the risks to the data product
- Data that are demonstrably valid and of high utility
- Data clearly labeled as "not validated" if that is so
- Flexibility regarding evolving measurements requirements
- A realistic estimate of costs updated as makes sense and a commitment to meet those costs
- An over arching commitment to delighting the customer all the time, every time

Figure 16.1   The measurements contract

Chapter 17

The Words
They Never Talked
about in College

17.1 SOME PHILOSOPHY

George Carlin, a comedian, used to have a routine where he discussed the seven forbidden words you never hear on radio or television. I'm going to discuss six forbidden words you never hear in an engineering education.

The undergraduate engineering curriculi in American universities is full to the breaking point. Some universities are even considering going to a five-year program for BS degrees in engineering. This is, it seems, in recognition of at least three very real pressures on those institutions. The first is the simple fact that a significant percentage of students do not get BS degrees in engineering in the standard four years, eight semesters or twelve quarters. It seems to take at least one summer session or a fifth year for a number of otherwise smart, dedicated students. Engineering is, after all, a difficult knowledge set to master. Second, many universities cannot, due to budget pressures, offer all the courses a student needs to graduate in four years. Where there used to be three sections every year of necessary course X, now there might be one section every other year. Third, with technology exploding, educators try to cram more and more subjects into the four-year program. And if you think your company or organization is political, you should see how university faculties work when jockeying the curriculum around!

As a result, certain subjects or ideas that are fundamental, in my opinion, to a successful career are never addressed in any engineering curriculum known to me. But I do talk to a lot of people in the field, and I notice that these foundational subjects are never addressed in engineering schools. The teaching professors are professional educators. They may never have been a practicing engineer. They *teach* engineering, but they may never have *done* engineering. There is a difference. There is nothing new about universities missing this subject matter. They missed it when I was in school too. When I bring this up to educators now, they tend to wince in discomfort and tell me these subjects have no part in an engineering curriculum.

Nonsense.

The purpose of engineering education is to prepare someone to practice an old and noble profession. Crucial knowledge should not be left out in the interest of teaching the sexy technical subjects. In retrospect, I wish someone had given me a course in this subject matter. It would have made my later professional life much more simple. It would have helped me to answer the fundamental questions every engineer must ask every now and then: What is expected of me and how should I act while providing it? In other words, what is expected out here in the cruel world where there are no answers in the back of the book, and what is *really* important?

If you assume that a graduating engineer is skilled in the theory, if not the practice, of the subject matter taught in his or her university, my observation over 25 years is that a successful, contributive, satisfying, respected career is not guaranteed. How come? What's missing?

I'm talking about the following six subjects: skill, craft, excellence, responsibility, vision, and professionalism. You can take these subjects for granted if you wish. If you do, you can stop reading this book right here. But I notice that the measurement engineers I respect, *the really good ones*, live their professional lives from these qualities on a daily basis. They don't talk about it and probably don't even think about it. They just *do* it. These are the qualities I've distilled from 25 years of watching measurements engineers from a number of companies and national laboratories do extremely creative work in a pressure packed, tough, edge-of-the-envelope environment. These qualities are, of course, not limited to the practice of measurement engineering or the design of measurement systems. This is just where I noticed them. How far you expand their relevance into other engineering areas is up to you. My sense is that they are general in nature. I'm sharing them to take some of the mystery out of it for you. I wish someone had taken some of the mystery out of it for me!

The best way I know to communicate these qualities of the respected engineers I know is to begin, simply, with those parts of the dictionary definitions that apply to the practice of engineering. These definitions are right out of *Webster's New Collegiate Dictionary*. English is a wonderfully powerful and discerning language when used well. That power and discernment begins in the precise definitions of its words.

17.2 SKILL

skill, n.: ability to use one's knowledge effectively and readily in execution or performance

skill, vt: (archaic, middle English) to make a difference

The key word is in this definition is *effectively*. Here, effectively means coming up with a robust solution to a problem in a relatively short amount of time. Engineers whose abilities allow them to come up with robust solutions in large amounts of time will simply find themselves unemployed. The worldwide marketplace drives this situation.

I have always been enchanted by the secondary definition of skill from the middle English. In those times skill was a verb—to skill. It meant to *make a difference*. The respected engineers I know all are driven to make a difference by their personal commitment and contribution.

17.3 CRAFT

craft, n: skill in planning, making or executing

We think of craftsmen as silversmiths, jewelers, potters or, maybe, cabinet makers. But engineers? Our colloquial use of the word craft has a sense of the creative arts about it. Inspection of the word's definition shows that it is absolutely applicable to the practice of engineering. Craft is something that is, in the engineering context however, somewhat hard to connect with. The approach I use when lecturing is to say that, although you may not be able to define craft, you'll absolutely know it when you see it. I use a series of 10–15 photographs of measurement engineering related objects such as strain gage installations, data cable harnessing jobs, rack wiring jobs, data pages from an XY plotter, graphical data, pages from a personal engineering log book, etc. No one ever fails to see the craft in the photos meant to show it, nor the lack of craft in the photos meant to show that. Everyone invariably scores 100%. Figures 17.1 through 17.4 are examples of these photos. Is there any doubt in your mind which is which?

Craft is important for several reasons in the practice of measurements engineering. Strain gage installations designed or implemented without a high level of craft simply will not perform up to their capabilities. The quality of the data will invariably show this. Installations that look beautiful to a practiced eye will perform well 99% of the time. Installations that look messy will never perform. That's why some people say the practice of installing strain gages is more art than science. Precision load cells installed without craft will simply not perform up to the manufacturer's specifications. Complex measurement activities not planned with craft will be ineffective, disjointed, risky, and will take too long to execute.

The second reason craft is important is that it supports effective trouble shooting. If a measurement system has been designed and built with a high level of craft, troubles will occur less often and it will be much easier to fix when troubles do occur. And please note that troubles in measurement systems only occur at the worst possible time—when you need them operating flawlessly.

A sports writer, watching Mario Andretti practice on a Formula One racetrack the year he won the world driving championship, noticed that two crew members would immediately start polishing the car whenever Andretti pulled into the pits after a couple of practice laps. He asked Andretti why they feverishly polished his car after every couple of laps when his was the only team even at the track, there was no crowd watching and no photographs being taken. Who cares about some dust or a little road grime from a practice run? Andretti's answer was very revealing. He wanted his crew's hands and eyes on the car as

Figure 17.1   Test site consoles for Generic Thermal Data Acquisition System (customer on the left)

much as possible because that was how they noticed *small* problems. You can see a wheel that has fallen off from 50 yards away. But you have to have your hands and eyes intimately on the car to notice a missing 25 cent cotter pin. Those were the small problems that, undiscovered, could kill him and those were the problems Andretti wanted found.

It's exactly the same with the measurement systems you'll design. If you were to open the racks of one of our systems and peek in you would see a beautiful, carefully routed harnessing job. Why? Because to perform a carefully designed harnessing job, you have to have your hands and eyes on the cabling all the time during integration. By having your hands and eyes on the cabling you notice the things that will kill your data at test time—the almost broken shields, the connectors that touch when they shouldn't, the cable that will fray in use, the control on a instrument case with its internal lock loose. These are the annoying things that will take hours to trouble shoot at test time when you don't have hours. Check the photographs. Which system would you rather trouble shoot?

Last, craft is important to your customer—especially *visible* craft. You want to create an impression in your customer's mind the moment they step into your laboratory space. You want to create this conversation in your customer's head: "Gee, this setup for my test

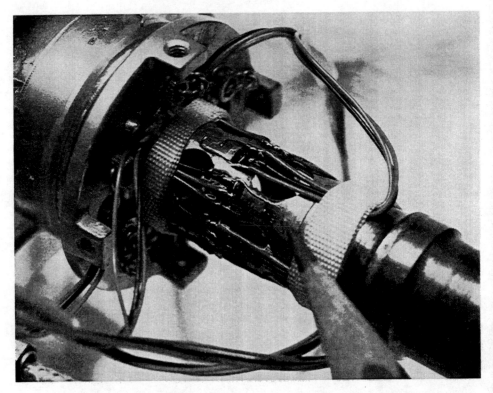

Figure 17.2   Strain gage installation on spacecraft thruster (.015" gage is shown at pencil point)

looks really nice. These people must really understand what they're doing and the problem I'm trying to solve. Looks like they've got this under control. I feel better about bringing my expensive hardware over here for test." When you create that impression in your customers very early, you've gone a long way toward having a very positive relationship with them. You create that impression with craft.

17.4 EXCELLENCE

excel, vt: to be superior to; to surpass in accomplishment or achievement

excellent, adj: (archaic) superior; very good of its kind; eminently good; first in class

excellence, n: the quality of being excellent

I am privileged to work with some excellent people in my department. These people tackle measurements problem others would not know how to approach. They stack up favorably with any other measurements engineering organization in the world in terms of vision, cost

Figure 17.3 Back of thermal control racks during thermal/vacuum test (not TRW!)

and quality of service, and customer satisfaction. If there were an Olympics of experimental mechanics measurements, these people would win medals. But there isn't. So how do you know whether your organization is excellent or not?

Worse that that, there really isn't an objective way to judge your own organization's excellence versus anyone else's. The best way I know of is to publish your work and note how your peers in professional societies react to it. Are they positive? Cheering? Silent? Laughing derisively? (Editorial: only publish work that is a contribution to the literature knowing that 75% of what is published is not a contribution and much of it is embarrassingly trivial.) Another way is to ask your external customers, who have cognizance over your products and those of other companies or laboratories, to tell you how you stack up.

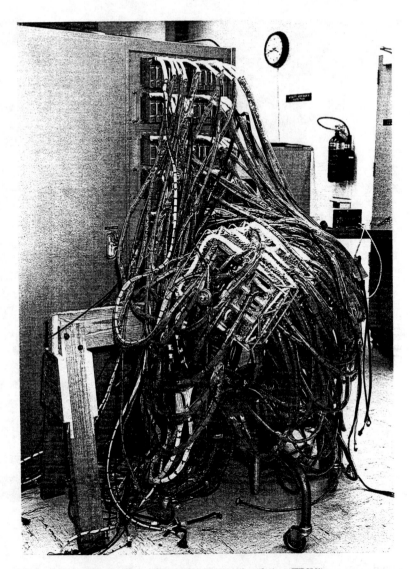

Figure 17.4   Back of test control rack (not TRW!)

They'll tell you because it's in their interests to have you improve if you need it, or remain the best if you are. You win either way.

Excellence is a matter of survival. It should be the *expected* level of performance in any professional measurements engineering organization. It is in ours. With American industry under external pressure from the worldwide marketplace, if your products are judged to be less than excellent your career will be short. This is reflected internally with continual and appropriate pressures to improve processes and service. If your measurements services are not considered to be excellent, someone else in your company will point that out to your leadership and try to deliver those services in your place. The internal competition usually thinks this stuff is easy to do. They, unfortunately, only find out how hard it is after you've lost the job. You must continually improve your processes and service to stay in front of your external and internal competition. You rest on your laurels in the unemployment line.

My experience is that professionals enjoy working in an organization where excellence is expected, high standards of performance are the norm, and average work is deplored. I like to hear our first line supervision saying to the troops every now and then, "That's just not good enough around here. Here's what is expected. Here's how to do it better."

17.5 RESPONSIBILITY

responsible, adj: able to answer for one's conduct and obligations; able to choose for oneself between right and wrong; trustworthy

responsibility, n: the quality or state of being responsible

A colleague related this story to me recently. His measurements organization had to extend a digitally networked measurement system for acquisition and reduction of vibration and shock data into another division's building. In order to get their system up and running, they were forced to send digital signals through five separate network architectures in two cities through utilities owned by two different phone companies. What a mess!

Through discussions with the communications organization in the other division about using their building's internal data networks to support the task, the measurements folks learned that it would take about six months to get it planned and configured for their use. Thank you very much. They needed it in *two weeks* to meet major test commitments. In order to make that facility operable, the measurements folks formed the Midnight Communications Company and installed their own networks in the building. They told the other division that they had found a network the division did not know about and that when it was no longer needed it would, good naturedly, be returned to the other division. What a deal! The facility was operable in two weeks as planned.

That is what responsibility looks like in the test measurement business. It looks like a willingness to do whatever it takes to make it happen within the bounds of ethics and legality. It looks like total ownership of a service area. It looks like creative, off-the-wall solutions to arbitrary barriers set up by entrenched, uninterested, and high inertia bureaucracies. It looks like the tenacity to never give up in the face of the odds when you think you're right. This is difficult because the system you're working in will give you unlimited opportunties to give up (see chapter on Leadership and Management Issues: note the 47 briefings done to get GTDAS funded). If you're right, convince them you're right. That is your responsibility to a good idea and to the people who will come after you. Your superiors and those who follow you will, in the end, respect you for your conviction.

The other and necessarily coexisting side of responsibility is the willingness to take the heat publicly when it all goes south. If you are doing difficult work you *will* get into trouble once in a while. In this case, responsibility looks like standing up and being willing to say, "Yes, this one got screwed up. We dropped the ball and I am responsible." When you stand up and say that to your leadership (or more likely your leadership's leadership!) you had best add three other pieces of information. These are a statement about why it got screwed up, what you're going to do to fix it, and what action you're going to take to prevent that one from ever happening again.

The last part of responsibility that's important is the word trustworthy. A professional measurements engineer *never* lies to the customer. When it's turned out perfectly, tell them that and celebrate your joint success. When it has gone south, tell them that. When you can't trust your data, or it can't be validated, tell them that and tell them why. We have institutionalized that message with rubber stamps stating "These data cannot be validated." They get used now and then too. Trust is a priceless yet fragile commodity in the test business. If you lose your integrity with a customer because they catch you shading the truth, you've probably lost it forever with that customer. It's the old joke about being a little bit pregnant. Being pregnant is digital—you are or you aren't. Trustworthiness is the same. You are either trusted or you're not. The test business is very difficult if your customer does not trust you and your data.

17.6 VISION

vision, n: unusual discernment or foresight

discernment, n: the quality of being able to grasp or comprehend what is obscure

foresight, n: an act or the power of foreseeing; a view forward

foresee, vt: to see (as a development) beforehand

One of the things your employer is going to pay you for is vision. Vision is crucial to a successful career in test measurements. Testing requirements and cost pressures are changing so fast that without vision you are doomed to solve next year's problems with the last decade's solutions. The problem with that is the other guy is solving next year's problems with the next decade's solutions. He will be doing your job for you if you don't display some vision.

Vision has to do with having some feeling about what will be needed in five or ten years and having a sense about what you should be doing about that now. The measurement system designers and users I respect all have this ability to comprehend what is obscure beforehand. When they are working in this difficult area, certainty is not part of the process. How can you expect to be certain when there might not even be a solution to your problem yet? They never seem to let that uncertainty stop them even though it is uncomfortable for engineers. It is our nature to want to *know*—not be uncertain.

You've got be be leading the target by 3–5 years to be successful over the long term in this business. One of the problems with American industry is that it takes much too long to get anything meaningful done. It takes us one to four years to turn a significant idea into a measurements reality operating out on the test floor producing value for a customer. And we are very good at doing this. If it takes one to four years to turn an idea into a reality, you had better have your thinking and questioning out there five years. That takes vision.

Another part of vision is the ability to see disparate processes or things or capabilities, and identify new ways they can go together synergistically. You have to be willing to fail to be good at this part of vision. In the cases of most of the people I know who have this ability, it expresses itself almost subconsciously. The new and outrageous idea pops into someone's head while driving to work and thinking about their taxes.

I wish I knew how to educate myself and our staff about improving our abilities in this facet of vision. It may very well be that you either have it or you don't. You do not seem to hear most engineers say, "Look at Joyce over there. She must have had some transformational experience. She's become so visionary over the last month!" What you hear is, "Well, Joyce seems to have another great idea. Why didn't I think of it. I never would have put those two concepts together."

17.7 PROFESSIONAL

profession, n: a calling requiring specialized knowledge and often long and intensive academic preparation

professional, adj: characterized by or conforming to the technical and ethical standards of a profession; having a particular profession as a permanent career; participation for gain or livelihood in a field of endeavor often engaged in by amateurs

professional, n: one that engages in a pursuit or activity professionally

amateur, adj: one lacking in experience or competence in an art or science

When I look at the professionals in this business who are respected, they all have the qualities noted above. I don't know of any engineer who attained "professional" status in the engineering design and use of measurements systems that did it just passing through the field on the way somewhere else. A lot of engineers pass through and leave for other areas. But they don't become professionals. It seems to take a love for the field and a long-term commitment to get there.

There's another difference between professionals and amateurs in this field. The professionals in measurement system design all know that they are amateurs in other collateral areas of engineering. The expert measurement engineers that I know are absolutely clear that they are amateurs at finite element modeling and analysis, dynamics analysis, the design of composite structures, the internal design of high speed A/D converters, or the ones and zeros of computer science. They can talk effectively to all of those people and understand the phenomena they worry about—but they do not consider themselves professionals in those areas.

The amateur working in the field of measurement system design and use may not even realize that there are professionals in this engineering area. Some amateurs don't even consider themselves amateurs because they don't appreciate what a professional's work looks like. If you've never seen the work of a master silversmith you cannot know what professional silversmithing looks like. The contrast is not apparent to the amateur. The contrast is apparent to the professionals. After all, the popular position in industry is that anybody can do this stuff. Just buy a PC, a couple of boards, some software, and some transducer things and you're off and running.

I stated in the preface that my intention for this book was to shorten the apprenticeship of those thousands of engineers who entered this field through the back door and know they are still amateurs. Another way of saying this is that I want to help move engineers, from amateur to professional status as measurement systems designers and users, faster. If I've helped you do that, it has been my privilege.

Index